T0320390

Power System Protection and Relaying

This textbook provides an excellent focus on the advanced topics of the power system protection philosophy and gives exciting analysis methods and a cover of the important applications in the power systems relaying. Each chapter opens with a historical profile or career talk, followed by an introduction that states the chapter objectives and links the chapter to the previous ones, and then the introduction for each chapter. All principles are presented in a lucid, logical, step-by-step approach. As much as possible, the authors avoid wordiness and detail overload that could hide concepts and impede understanding. In each chapter, the authors present some of the solved examples and applications using a computer program.

Toward the end of each chapter, the authors discuss some application aspects of the concepts covered in the chapter using a computer program.

In recognition of requirements by the Accreditation Board for Engineering and Technology (ABET) on integrating computer tools, the use of SCADA technology is encouraged in a student-friendly manner. SCADA technology using the Lucas-Nülle GmbH system is introduced and applied gradually throughout the book.

Practice problems immediately follow each illustrative example. Students can follow the example step by step to solve the practice problems without flipping pages or looking at the book's end for answers. These practice problems test students' comprehension and reinforce key concepts before moving on to the next section.

Power System Protection and Relaying: Computer-Aided Design Using SCADA Technology is intended for senior-level undergraduate students in electrical and computer engineering departments and is appropriate for graduate students, industry professionals, researchers, and academics.

This book has more than ten categories and millions of power readers. It can be used in more than 400 electrical engineering departments at top universities worldwide.

Based on this information, targeted lists of the engineers from specific disciplines include the following:

- Electrical, computer, power control, technical power system, protection, design, and distribution engineers

Designed for a three-hour semester course on "power system protection and relaying," the prerequisite for a course based on this book is knowledge of standard mathematics, including calculus and complex numbers.

Power System Protection and Relaying

Computer-Aided Design Using SCADA Technology

Samir I. Abood and John Fuller

CRC Press is an imprint of the
Taylor & Francis Group, an informa business

First edition published 2024
by CRC Press
2385 Executive Center Drive, Suite 320, Boca Raton, FL 33431

and by CRC Press
4 Park Square, Milton Park, Abingdon, Oxon, OX14 4RN

CRC Press is an imprint of Taylor & Francis Group, LLC

ISBN: 9781032495507 (hbk)
ISBN: 9781032495521 (pbk)
ISBN: 9781003394389 (ebk)

DOI: 10.1201/9781003394389

Typeset in Times
by codeMantra

Dedicated to my great parents who never stop giving of themselves in countless ways; my beloved brothers and sisters; my dearest wife, who leads me through the valley of darkness with the light of hope and support; and my beloved kids Daniah and Mustafa, whom I can't force myself to stop loving.

Samir I. Abood

Dedicated to my family and especially to my wife Sherylle, who has been beside me in reaching this point in life, and highest regards to my mother, Bernice Fuller, who sacrificed so that this period in time is a reality.

John Fuller

Contents

Preface

The protection is the electric power engineering branch that deals with equipment design and operation (called "relays" or "protective relays") that detects abnormal conditions of the power system and initiates corrective action as quickly as possible to bring the power system back to its normal state. A key aspect of protective relaying systems is the rapid response, often requiring a few milliseconds response times in order. Consequently, human interference in the operation of the defensive mechanism is not feasible. The response must be automated and swift and trigger minimal power system disruption. These general criteria rule the whole subject: accurate diagnosis of trouble, speed of response, and minimal disruption to the power system. We need to analyze all potential faults or irregular conditions in the power system to achieve these objectives.

This book aims to provide a good understanding of the power system's protection and its applications and optimization. This book begins with the study of the concept of protection and relays. It then presents their applications in the different types of configurations shown in lucid detail. It optimizes the protective scheme's location in power and uses the power electronics devices installed to protect the power system. This book is intended for college students, both in community colleges and universities. This book is also intended for researchers, technicians, technology, and skills specialists in power and control systems. This book presents the relationship between the power system's quantities and their protection and management. This book's major goal is to briefly introduce protecting the power system covered in two semesters. This book is appropriate for juniors, senior undergraduate students, graduate students, industry professionals, researchers, and academics.

This book is organized into 12 chapters. Chapter 1 introduces the philosophy of power system protection and the effect of faults on the protection system's power systems and performance requirements. The relay connection to the power system and protective zones is also introduced in this chapter.

Chapter 2 concerns some aspects of data required for the relay setting and relay types; it also discusses the digital relay operation, signal path for microprocessor relays, and digital relay construction.

In recognition of requirements by the Accreditation Board for Engineering and Technology (ABET) on integrating computer tools, using Lucas-Nülle SCADA software is encouraged in a student-friendly manner. The reader does not need to have previous knowledge of SCADA. The material of this text can be learned without SCADA. However, the authors highly recommend that the reader studies this material in conjunction with the SCADA system. Chapter 3 provides a practical introduction to SCADA technology.

Chapter 4 deals with fault analysis, first, introduction to the faults in power systems, transient phenomena, and three-phase short circuit – an unloaded synchronous machine. Second, the short circuit theory consists of balanced and unbalanced fault calculation in general and conventional methods for small systems. These fault types involve single line-to-ground faults, line-to-line faults, and double line-to-ground faults. The last three unsymmetrical fault studies will require the knowledge and use of tools of symmetrical components. This chapter also deals with network models, shunt elements, fault analysis, and algorithms for short-circuit studies.

The description of fuses and circuit breakers, their types, and their specifications are discussed in Chapter 5. This chapter includes an introduction to the construction and working of a fuse, its characteristics, and its applications. It also discusses high-voltage circuit breakers. A directional overcurrent time protection design using Lucas-Nülle GmbH power system/SCADA network devices is explained.

Chapter 6 presents the overcurrent relay, PSM, time grading, and relay coordination method; this chapter also discusses requirements for proper relay coordination and hardware and software for overcurrent relays. Overvoltage and undervoltage protection using Lucas-Nülle GmbH power system/SCADA network devices is explained. Also, the directional power protection system using SCADA technology is discussed in this chapter.

Chapter 7 describes the preceding transmission lines protection, distance relay as impedance, reactance, and MHO relay. The fundamentals of differential protection systems used to protect transmission lines are also discussed. Protection of parallel lines (parallel operation) and parametrizing non-directional relays using the SCADA system are discussed. Directional time overcurrent relays and high-speed distance protection using SCADA technology are explained in this chapter.

Chapter 8 presents transformer protection, types, connection, and mathematical models for each type of device. Overcurrent relays, differential relays, and pressure relays may secure the transformer and be controlled with winding temperature measurements, and chemical analysis of the gas above the insulating oil for incipient trouble is discussed in this chapter.

Chapter 9 deals with generator protection and generator fault types. It also presents motor and busbar protection. This chapter briefly discusses the types of internal faults and various abnormal operating and system conditions. Additional protective schemes, such as overvoltage, out of step, and synchronization, should also be considered depending on the generator's cost and relative importance.

Chapter 10 presents the concept of feeder configuration, HIF modeling, nonlinear load modeling, and capacitor modeling. The three test case systems' designs and data are presented in this chapter. Also, the technique validation and algorithm verification are presented in this chapter.

Finally, descriptions about earthing in the power system, types, and specifications are discussed in Chapter 11. Also, it includes analysis procedures of electric power system grounding. Besides, it identifies techniques that can be applied to evaluate substation grounding systems (as well as indication line towers grounding). It also presents soil resistivity measurement methods; two measurement methods are defined: the three-limit method and the four-limit method.

Acknowledgments

The authors acknowledge Prairie View A&M University for providing a platform where faculty, administrators, PhD students, and private and governmental entities can contribute to a book that will be used in the education of future power engineers and for providing a resource to contribute to advancing the knowledge of power systems and continuing the technical foundation building for the production of future engineers.

It is our pleasure to acknowledge the outstanding help and support of the team at CRC Press in preparing this book, especially from Nora Konopka and Prachi Mishra.

The authors appreciate the suggestions and comments from several reviewers, including Professor Zainab Ibrahim, Electrical Engineering Department, University of Baghdad, and Dr. Muna Fayyadh, American InterContinental University. Their frank and positive criticisms improved this work considerably.

Finally, we express our profound gratitude to our families, without whose cooperation this project would have been difficult, if not impossible. We appreciate feedback from professors and other users of this book. We can be reached at siabood@pvamu.edu and jhfuller@pvamu.edu.

Authors

Samir I. Abood received his BS and MS from the University of Technology, Baghdad, Iraq, in 1996 and 2001, respectively. He earned his PhD in the Electrical and Computer Engineering from Prairie View A&M University. From 1997 to 2001, he worked as an engineer at the University of Technology. From 2001 to 2003, he was a professor at the University of Baghdad and Al-Nahrain University. From 2003 to 2016, he was a Middle Technical University/Baghdad-Iraq professor. He is an electrical and computer engineering professor at Prairie View A&M University in Prairie View. He is the author of 30 papers and 12 books. His main research interests are sustainable power and energy systems, microgrids, power electronics and motor drives, digital PID controllers, digital methods for electrical measurements, digital signal processing, and control systems.

John Fuller is an electrical and computer engineering professor at Prairie View A&M University in Prairie View, Texas. He received a BSEE degree from Prairie View A&M University and a master's degree and a PhD degree from the University of Missouri, Columbia. He has researched some funded projects over a 48-year teaching career in higher education. Some of the major projects of his research efforts are hybrid energy systems, stepper motor control, the design and building of a solar-powered car, nuclear survivability and characterization on non-volatile memory devices, nuclear detector/sensor evaluation, and some other electrical and computer-related projects. Dr. Fuller is presently the coordinator of Title III funding to the Department of Electrical and Computer Engineering in developing a solar-powered home. He is also the Center for Big Data Management associate director in the Department of Electrical and Computer Engineering. In addition to teaching and research duties with college-level students, he is also active in the PVAMU summer programs for middle and high school students. Dr. Fuller has also held administrative positions as head of the Department of Electrical Engineering and as interim dean of the College of Engineering at Prairie View A&M University. In 2018, he was recognized as the Texas A&M System Regents Professor.

1 Introduction to Power Protection Systems

1.1 INTRODUCTION

Power system protection is a philosophy of system reliability with maximum safety protection and other aspects related to protection coordination. It is a science of monitoring power systems, detecting faults, initiating an operation to isolate faulted parts, and ultimately tripping the circuit breaker.

The aims of power system protection are to:

i. Minimize dangerous effects on the workers and establish techniques and procedures due to the abnormal current in the power system.
ii. Avoid damage to power system components involved in failure and human injury prevention.
iii. Limit the service duration interruption whenever equipment fails, adverse natural events occur, or human error occurs on any portion of the system.

Protection engineering and technicians are involved in designing and implementing "protection schemes." Protection schemes are specialized power systems control that monitors the system, detects faults or abnormal conditions, and initiates corrective action. In power system networks, the configuration should be given all the equipment necessary to generate, transmit, distribute, and utilize electric power.

This chapter focuses on the philosophy of power system protection and the effect of faults on the power system and performance requirements. The relay effect of connection to the power system and protective zones is also introduced in this chapter.

1.2 PHILOSOPHY OF POWER SYSTEM PROTECTION

A fault is an abnormal power system state that generally consists of short circuits and open circuits. An open-circuit condition minimally occurs and is normally less severe than a short-circuit condition. If short circuits are allowed to persist on a power system for an extended period, many or all of the following undesirable effects are likely to occur:

i. Reduced stability of the power system.
ii. Damage to the equipment that is in the vicinity of the fault due to heavy current, unbalanced current, or low voltage.
iii. Explosions that may occur in equipment with hazards.
iv. Disruptions in the entire power system service area by the action of cascaded protective systems in cascading.

DOI: 10.1201/9781003394389-1

TABLE 1.1
Typical Short-Circuit Type Distribution

Type	Rate of Occurrence (%)
Single-phase ground	70–80
Phase-phase ground	17–10
Phase-phase	10–8
Three-phase	3–2

Causes of short-circuit faults are as follows:

i. Insulation failure due to lightning.
ii. Birds and animals bridging insulators.
iii. Dig-ups for underground cables.
iv. Collapsing poles.
v. Breaking of conductors.
vi. Effect of a vehicle collision.
vii. Wind-borne debris.
viii. Incorrect operation by personnel.

The frequency of the faults incidence on different items in a power system is given in Table 1.1.

1.3 EFFECTS OF FAULTS

The effect of the fault on the power system includes the following:

i. Huge currents can flow through parts of the power network.
ii. These huge currents can only flow for a very short time (within 10 ms to 3 s); otherwise, equipment and generators would be damaged.
iii. Arcs, sparking, and the heating effect of short-circuit currents.
iv. Significant mechanical forces can be caused by short-circuit currents, which can potentially damage equipment.
v. Fault currents can escape from the network conductors and flow through paths that could create a hazard to people or livestock.

Fault occurrence can be minimized and controlled by:

i. Adequate insulation and coordination with lightning arresters.
ii. Overdesigning for mechanical strength.
iii. Provision of overhead ground wires.
iv. Blocking or interlocking of undesirable switching operations.
v. Regular maintenance practices.

Protective gear detects a fault using the:

i. Current magnitude utilizing overcurrent protection.
ii. Current in the abnormal path.
iii. Current balance.
iv. Voltage balance using overvoltage or undervoltage protection.
v. Power direction.
vi. Change of parameters.
vii. Damage to equipment.
viii. None electrical parameters.
ix. The magnitude of current in the earth and neutral – Earth fault protection.
x. Magnitude and angle of impedance (ratio *V/I*) protection.
xi. Difference between two currents – Differential protection.
xii. Difference between phase angles of two currents – Phase comparison protection.
xiii. Temperature – Thermal protection.
xiv. Specials, i.e., transformer gas protection.
xv. The magnitude of frequency – Overvoltage or under-frequency protection.
xvi. The magnitude of the negative sequence current.

1.4 PERFORMANCE REQUIREMENTS OF PROTECTION SYSTEM

Speed, selectivity, sensitivity, security, and reliability are the keys required for reliable operation and the safety of a power system. Selectivity requires that the protection framework accurately detects faults in its protection zones. Sensitivity is the relay's ability to pick up even the smallest faults possible. Safety is a property that describes a false trigger or the defense system's ability to refrain from working when it is not meant to do. So, dependability is the degree of confidence that the relay will function properly. Reliability requires the protection system's operability to ensure that the overall design provides sufficient protective measures, even though some of the protective apparatus might have failed.

The area of power engineering dealing with the design, implementation, and operation of safety devices, called "relays" or "protective relays," is power system protection. The purpose of these devices is to detect irregular conditions in the power system and take appropriate steps as quickly as possible to restore the power system to its usual operating mode. The relays have to meet the following criteria to achieve the desired performance:

i. *Sensitivity*: This term is sometimes used when referring to the minimum operating standard of relays or full safety schemes (current, voltage, power, etc.). When their primary operational parameters are tiny, relays or security schemes are said to be vulnerable.
ii. *Speed*: It is the relay's capacity to isolate faults as quickly as possible, mitigate harm to power system equipment, safeguard supply continuity, and prevent the loss of synchronism and consequent power system failure.

iii. *Stability*: It is the relay's ability to remain unaffected by incidents outside its security region, including external faults or heavy load situations.
iv. *Selectivity*: It is the ability to isolate only the faulted zone.
v. *Safety*: It is the ability to secure against improper activity.

$$\%\text{Security} = \frac{\text{No. of correct trippings}}{\text{Total no. of trippings}} \times 100 \tag{1.1}$$

It should be affordable and should not restrict the rating of primary plants and equipment. It should not have any "blind spots," i.e., unprotected zones.
vi. *Discrimination*: It is between load (normal) and fault (abnormal) conditions. It should not be confused with non-damaging transient conditions. Discrimination is a relay system's ability to discriminate between internal and external faults to its intended protective zones.
vii. *Dependability*: A relay is dependable if it trips only when expected. Dependability is the degree of certainty that the relay will operate correctly. It can be improved by increasing the sensitivity of the relaying scheme.

$$\%\text{Dependability} = \frac{\text{No. of correct trippings}}{\text{Total no. of desired trippings}} \times 100 \tag{1.2}$$

viii. *Reliability*: It is the ability not to "fail" in its function. It can be achieved by redundancy. Redundancy in protection depends on the criticality of the apparatus. Reliability can be improved by providing backup protection.

$$\%\ \text{Reliability} = \frac{\text{No. of correct trips}}{\text{No. of desired trips} + \text{No. of incorrect trips}} \times 100 \tag{1.3}$$

The number of the desired tripping can be greater than or equal to the correct tripping.

The optimal implementation and coordination of protective relays are obtained considering the objectives, the system's topology to be protected, the typical operation scenarios, and the probable fault occurrences.

Example 1.1

An overcurrent relay was monitored and had an observed performance over one year. It was found that the relay operated 15 times, out of which 13 were correct trips. If the relay failed to issue a trip decision on four occasions, compute the relay's dependability, security, and reliability.

Solution

$$\text{Number of correct trips} = 13$$

$$\text{Number of desired trips} = 13 + 4 = 17$$

$$\text{Dependability} = \text{Number of correct trips/Number of desired trips}$$

$$= 13/17 = 76.47\%$$

$$\text{Security} = \text{Number of correct trips/Total number of trips} = 13/15 = 86\%$$

$$\text{Reliability} = \text{Number of correct trips/}\left(\text{Number of desired trips} + \text{number of incorrect trips}\right)$$

$$= 13/(17+2) = 68.42\%$$

1.5 BASIC PROTECTION SCHEME COMPONENTS

Protective gear is a collective term that covers all the equipment used for detecting, locating, and initiating the removal of a fault from the power system. The term includes all types of relays, direct a.c. trips, fuses, accessories such as CTs, VTs, d.c. and a.c. wiring, and any other device related to the protective relays. The protective gear does not include circuit breakers, the main switchgear, or other devices used to open the contacts.

The isolation of the faults and abnormalities requires protective equipment that senses an abnormal current flow and removes the system's affected portion. The primary protective equipment components are shown in Figure 1.1.

i. PR – Protection relay.
ii. CB – Circuit breaker.
iii. Equip protected item.
iv. CT – Current transformer.
v. VT – Voltage transformer.
vi. DC Aux – DC auxiliary supply.
vii. HMI – Human–machine interface.
viii. PCL communications link.
ix. Tr CB – Trip coil.

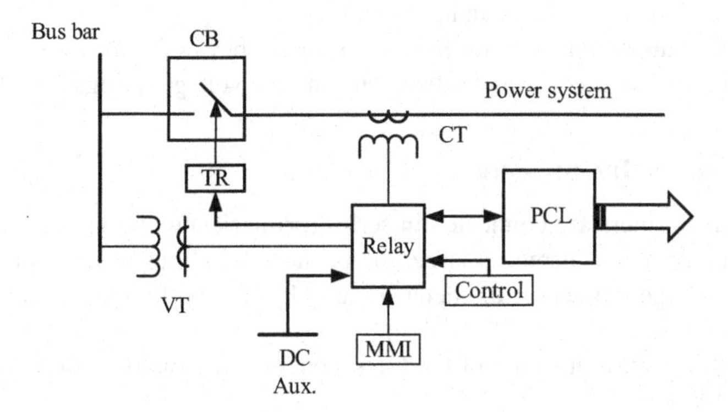

FIGURE 1.1 The basic arrangement of a protection scheme.

1.6 PROTECTIVE RELAY

It is a device that receives a signal from the power system and determines whether a "normal" or "abnormal" (measuring function) condition exists and initiates relay signal circuit breakers to disconnect the equipment that may be affected (switching or signaling function) if an irregular condition is present. Signal "relays" from the system activate the circuit breaker.

The aim of the protective relaying systems is only to isolate the defective power system portion.

Relaying devices are divided into two groups:

1. Equipment for primary relaying.
2. Backup relaying units.

The first line of defense for protecting the devices is the primary relay.

Backup safety relaying only works when (they are slow in action) the primary relaying system fails.

1.7 TRANSDUCERS

Apart from nonelectrical quantities (temperature, pressure, etc.), the principal item in this category are transformers.

There are:

i. Current transformer (CT).
ii. Voltage transformer (VT) or potential transformer (PT).
iii. Linear coupler.

These equipment required to reduce the sampled quantity in their secondary as faithfully as possible.

CTs and VTs are used.

i. To reduce the power system currents and voltages to a safe, adequate low value for measurements and protection use.
ii. To insulate the relay circuit from the primary power circuit.
iii. To permit the use of standardized current and voltage ratings for relays.

1.7.1 CURRENT TRANSFORMER

Current transformers are connected in series (primaries) with the protective circuit. Because the primary current is large, the primary windings usually have very few turns and a large conductor diameter. Figure 1.2 shows the CT connection to the power system.

The nominal current rating of the CT secondaries is usually 1 or 5 A (e.g., 50:5, 250:1, 1,200:5).

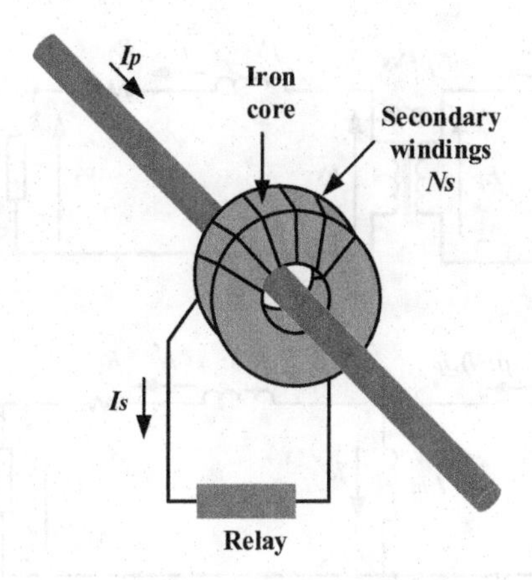

FIGURE 1.2 CT connection in the system.

The initial problem is how to connect a low-voltage device to the high-voltage system and have the ability to handle large fault currents (kilo-Amps). How can we make the relay measure the currents flowing in the high-voltage system to detect these faults?

The solution uses a special type of transformer called a current transformer.

The main parts of a current transformer are:

- Iron core.
- Secondary winding.
- Primary conductor.
- External insulation.

Some current transformers do not have a primary conductor. In those cases, the primary is the line or bus itself. The core and secondary winding are sometimes directly installed in the circuit breakers or transformers' bushing. These CTs are called "bushing CTs."

Some current transformers may have a primary that consists of several turns. Typically, the primary number of turns is 1.

The total load connected to the CT terminal (*g* and *h* in this case) is called "burden."

Ideally, the secondary current of a CT is perfectly proportional to the primary current. It will be shown later that, in reality, this is sometimes not true.

Figure 1.3 shows an equivalent circuit of CT, an exact circuit, and an approximate circuit.

1.7.1.1 IEC Standard Accuracy Classes

The IEC (International Electrotechnical Commission) specifications for the current transformer is 15 VA Class 10 P 20, where 15 VA is the continuous VA; 10 represents

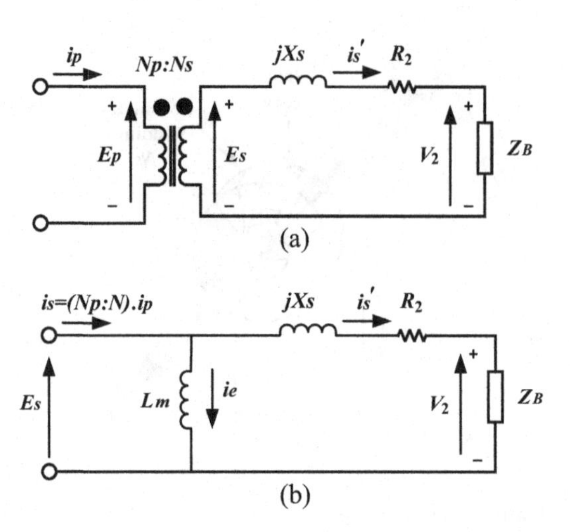

FIGURE 1.3 Equivalent circuit of CT. (a) Exact circuit. (b) Approximate circuit.

the accuracy class; P represents protection; and 20 represents the accuracy limit current factor.

Table 1.2 shows standard current transformer ratios, and Table 1.3 shows CT classes and accuracies (accuracy class).

TABLE 1.2
Standard Current Transformer Ratios

50:5	100:5	150:5	200:5	250:5	300:5
400:5	450:5	500:5	600:5	800:5	900:5
1,000:5	1,200:5	1,500:5	1,600:5	2,000:5	240:5
2,500:5	3,000:5	3,200:5	4,000:5	5,000:5	6,000:5

TABLE 1.3
CT Classes and Accuracies (Accuracy Class)

Class	%Error	Application
0.1	±0.1	Metering
0.2	±0.2	
0.5	±0.5	
1.0	±1.0	
5P	±1	Protection
10P	±3	

TABLE 1.4
Standard VT Ratios

1:1	2:1	2.5:1	4:1	20:1	25:1	40:1	60:1	200:1
300:1	400:1	500:1	600:1	800:1	1,000:1	2,000:1	3,000:1	4,500:1

1.7.2 Voltage Transformer

These provide a voltage that is much lower than the system voltage, the nominal secondary voltage being 115 V (line–line), or 66.4 V (line to neutral) in one standard and 120 V (line–line) or 69.4 V (line to neutral) in another.

There are two VT types, conventional electromagnetic and capacitor, for high-voltage levels (132 kV and above). Table 1.4 shows the standard VT ratios.

There are three main types of voltage transformers:

i. Magnetic voltage transformers (ordinary two-winding types – used for LV and MV).
ii. Capacitive voltage transformers (CVT) are used for high and extra-high voltages.
iii. Magneto-optic voltage transformers (new).

1.7.3 Magnetic Voltage Transformer (VT)

Figure 1.4 shows a VT connection to the power system, and Figure 1.5 shows a VT equivalent circuit referred to as the secondary side.

1.7.4 Capacitive Voltage Transformers (CVT)

Figure 1.6 shows capacitive voltage transformers, which are classified into two types:

i. Coupling-capacitor voltage transformer.
ii. Capacitor – bushing voltage transformer.

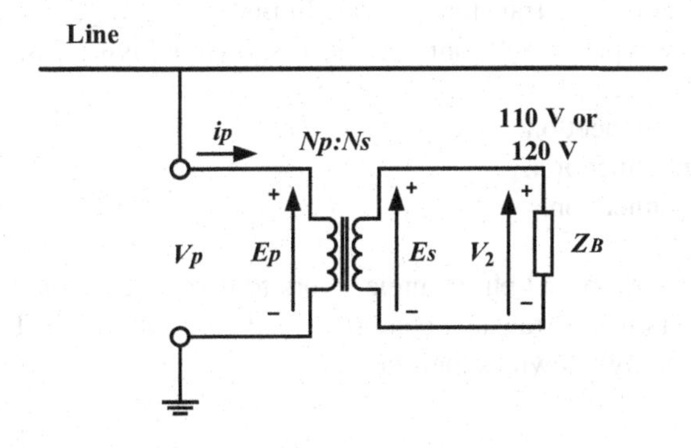

FIGURE 1.4 VT connection to the system.

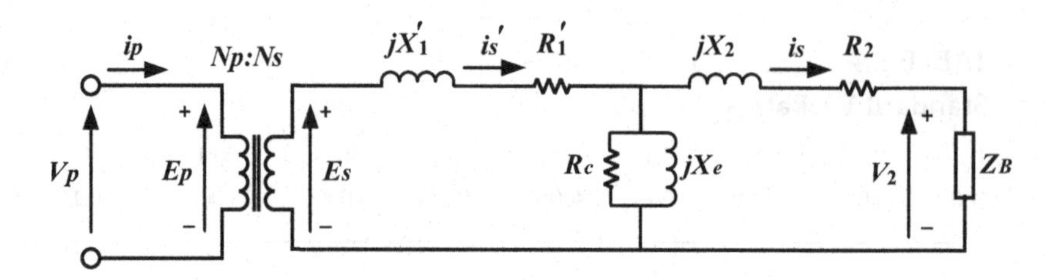

FIGURE 1.5 VT equivalent circuit referred to the secondary side.

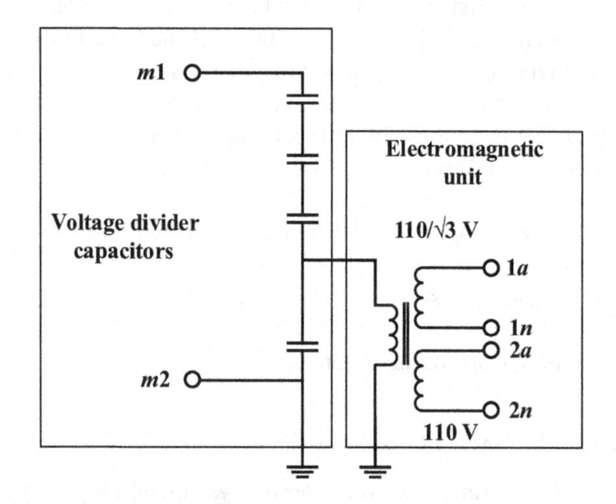

FIGURE 1.6 Capacitive voltage transformers.

These types are shown in Figure 1.7.

VT_{error}: Magnitude error can be calculated from:

$$VT_{error} = \frac{nV_s - V_p}{V_p} \times 100\% \tag{1.4}$$

Table 1.5 gives the voltage transformer error limits.

There are three types of VT connections, as shown in Figure 1.8.

i. Open delta connection.
ii. Delta–star connection.
iii. Star–star connection.

CTs and VTs have ratio and phase angle errors to certain degrees. Errors are more pronounced in CTs under transient conditions and core saturation. The load on CTs and VTs is commonly known as their burden.

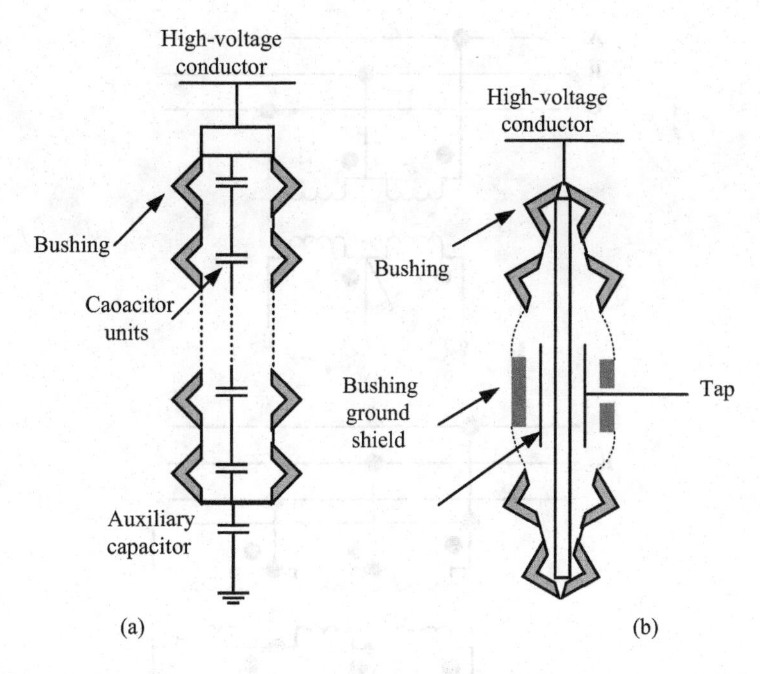

FIGURE 1.7 Capacitor voltage transformers: (a) coupling-capacitor voltage divider. (b) Capacitance-bushing voltage divider.

TABLE 1.5
Voltage Transformers' Error Limits

Class	Primary Voltage	Voltage Error (%)	Phase Error (±minutes)
0.1	0.8 V_n, V_n, and 1.2 V_n	0.1	0.5
0.2		0.2	10
0.5		0.5	20
1.0		1.0	40
0.1	0.5 V_n	1.0	40
0.2		1.0	40
0.5		1.0	40
1.0		2.0	80
0.1	V_n	0.2	80
0.2		2.0	80
0.5		2.0	80
1.0		3.0	120

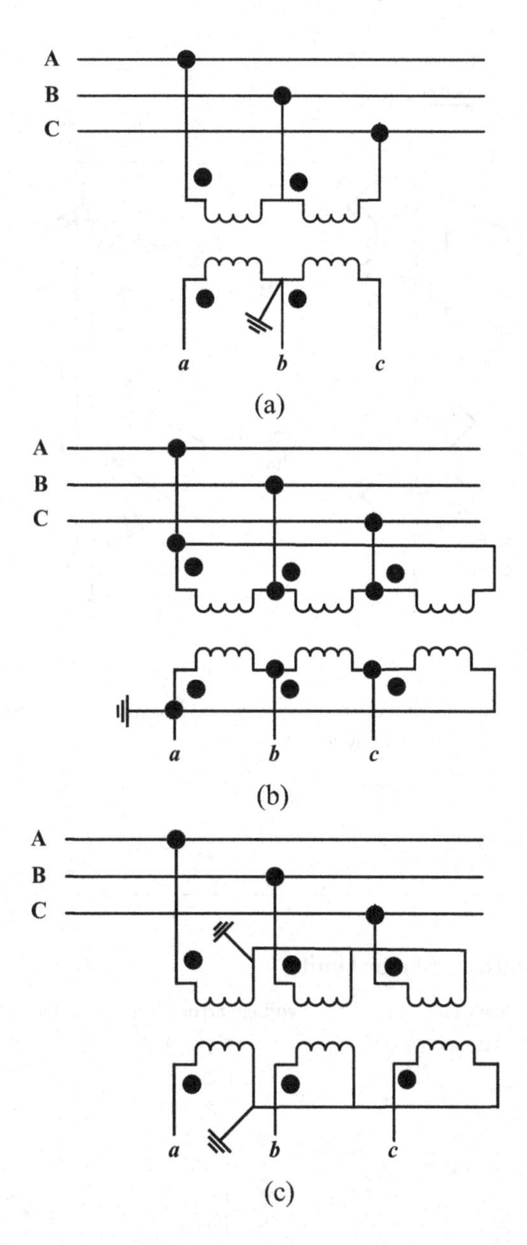

FIGURE 1.8 Different types of VT connections. (a) Open delta connection, (b) delta–star connection, and (c) star–star connection.

Example 1.2

The delta VT connection is shown in Figure 1.9, suppose $V_{AB} = 230\angle 0°$ kV, $V_{BC} = 230\angle -120°$ kV, $V_{CA} = 230\angle 120°$ kV, the VT ratio is 110 kV/120 V, calculate v_{ab}, v_{bc}, and v_{ca}. If the dot mark is moved to *b*, recalculate the above voltage.

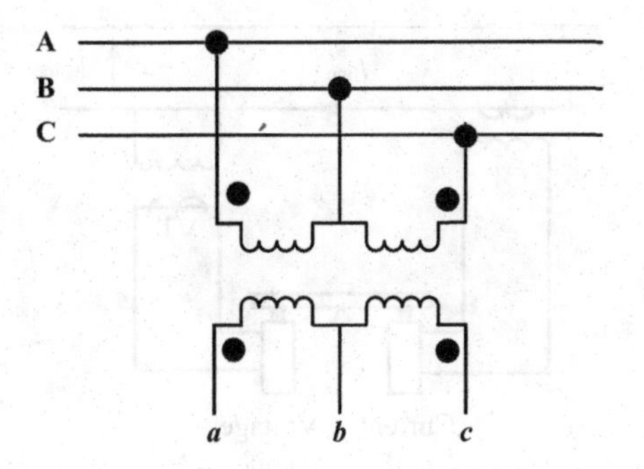

FIGURE 1.9 Configuration of Example 1.2.

Solution

$$\text{VT ratio} = \frac{110,000}{120} = 916.6$$

$$\vec{v}_{ab} = \frac{1}{916.6}(230\angle 0°\ \text{kV}) = 250.92\angle 0°\ \text{V}$$

$$\vec{v}_{bc} = \frac{1}{916.6}(230\angle -120°\ \text{kV}) = 250.92\angle -120°\ \text{V}$$

$$\vec{v}_{ca} = -(\vec{v}_{ab} + \vec{v}_{bc}) = 250.92\angle 120°\ \text{V}$$

If the dot mark moved to *b* (Figure 1.10),

$$\vec{v}_{ab} = \frac{1}{916.6}(230\angle 0°\ \text{kV}) = 250.92\angle 0°\ \text{V}$$

$$\vec{v}_{bc} = -\frac{1}{916.6}(230\angle -120°\ \text{kV}) = 250.92\angle 60°\ \text{V}$$

$$\vec{v}_{ca} = -(\vec{v}_{ab} + \vec{v}_{bc}) = 434.6\angle -150°\ \text{V}$$

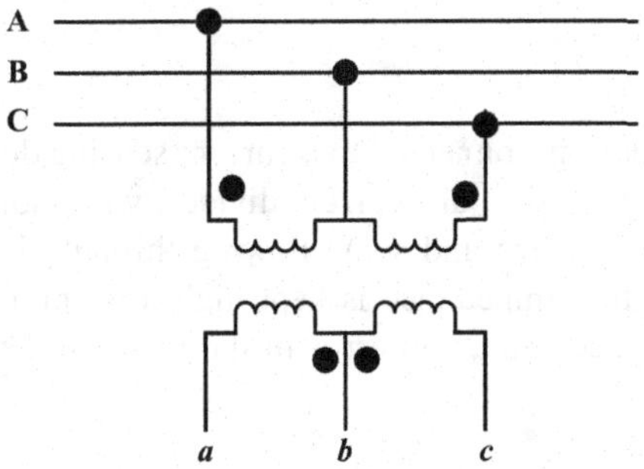

FIGURE 1.10 Another sequence connection diagram of Example 1.2.

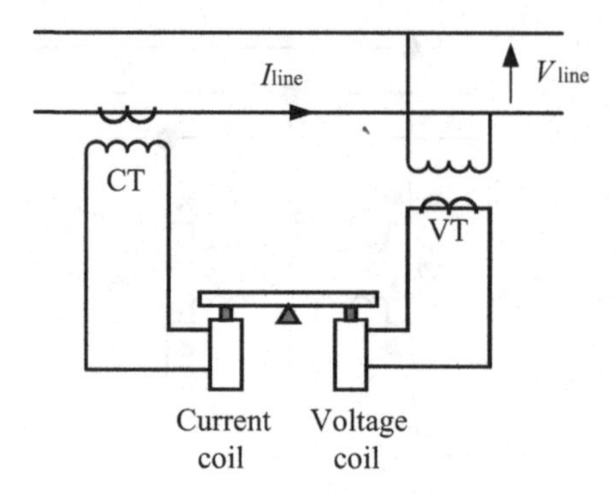

FIGURE 1.11 Relay connection to the power system.

1.8 RELAY CONNECTION TO THE PRIMARY SYSTEM

Suppose the relay is connected to the power system via a current and voltage transformer, as shown in Figure 1.11.

$$V_{\text{line}} = V_{\text{relay}} \times K_{\text{VT}} \tag{1.5}$$

$$I_{\text{line}} = I_{\text{relay}} \times K_{\text{CT}} \tag{1.6}$$

$$K_{\text{VT}} = \text{Voltage transfomer ratio.}$$

$$K_{\text{CT}} = \text{Currnt transformer ratio.}$$

The operating impedance of the line is

$$Z_{\text{line}} = \frac{V_{\text{line}}}{I_{\text{line}}} = \frac{V_{\text{relay}}}{I_{\text{relay}}} \times \frac{K_{\text{VT}}}{K_{\text{CT}}} = Z_{\text{relay}} \times \frac{K_{\text{VT}}}{K_{\text{CT}}} \tag{1.7}$$

1.9 CT ERROR

Ideally, the CT secondary is connected to a current-sensing device with zero impedance, but in practice, the secondary current divides, with most flowing through the low-impedance-sensing device and some flowing through the CT shunt excitation impedance. CT excitation impedance is kept high to minimize excitation current. The excitation impedance causes an error in the reading of the secondary current (Tables 1.6 and 1.7).

TABLE 1.6
Current Transformer Specification IEC 185–IEC 44-1

Accuracy Class	Current Error at Rated Primary Current (%)	Phase Displacement at Rated Primary Current		Composite Error (%) at Rated Accuracy Line Primary Current
		Minutes	Centiradians	
5P	±1	±60	±1.8	5
10P	±3			10

TABLE 1.7
Current Transformer Specification IEC 44-6

Class	Ratio Error (%) at Primary Rated Current	Phase Displacement at Rated Primary Current		Maximum Instantaneous Value Error (%) at Rated Accuracy
		Minutes	Centiradians	
TPX	±0.5	±30	±0.9	$\varepsilon=10$
	±1.0	±60	±1.8	$\varepsilon=10$
	±1.0	180±18	5.3±0.6	$\varepsilon=10$

Example 1.3

A CT has a rated ratio of 500:5 A. A secondary side impedance is $Z_2 = 0.1 + j0.5\ \Omega$, and the magnetizing curve is given in Figure 1.12. Compute I_2 and the CT error for the following cases:

i. $Z_L = 4.9 + j0.5\ \Omega$, $I_1 = 400$ A (load current).
ii. $Z_L = 4.9 + j0.5\ \Omega$, $I_1 = 1{,}200$ A (fault current).
iii. $Z_L = 14.9 + j1.5\ \Omega$, $I_1 = 400$ A (load current).
iv. $Z_L = 14.9 + j1.5\ \Omega$, $I_1 = 1{,}200$ A (fault current).

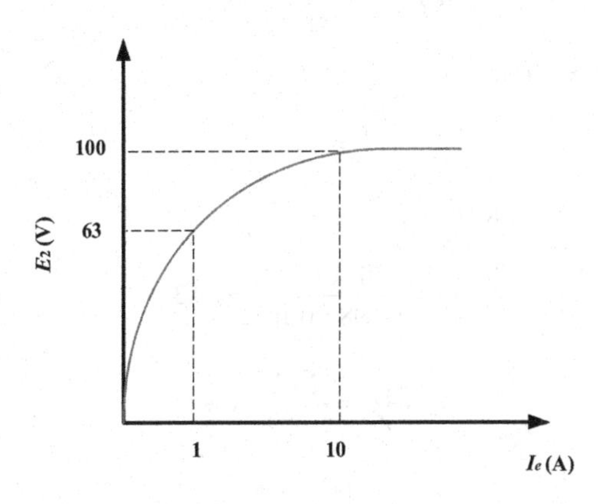

FIGURE 1.12 Magnetization curve of Example 1.3.

Solution

From the magnetizing curve, substitute the points (1, 36) and (10, 100) in Frohlich's equation:

$$E_2' = \frac{A \cdot I_e}{B + I_e}$$

$$63 = \frac{A \cdot 1}{B + 1}$$

$$100 = \frac{A \cdot 10}{B + 10}$$

Solving to get the constants A and B, where $A = 107$ and $B = 0.698$, Frohlich's equation will be

$$E_2' = \frac{107\ I_e}{0.698 + I_e}$$

i. $Z_T = Z_2 + Z_L$

$$= 4.9 + j0.5 + 0.1 + j0.5$$

$$= 5.0 + j1.0\ \Omega = 5.009\angle 11.3°\ \Omega$$

$$E_2 = I_2 \cdot Z_2$$

$$= \left(\frac{5}{500} \times 400\right) \times 5.009 = 20.4\ \text{V}$$

$$I_e = \frac{20.4}{\sqrt{5^2 + \left(1 + \dfrac{107}{0.698 + I_e}\right)^2}}$$

Using trial and error to find

$$I_e = 0.163\ \text{A}$$

From Frohlich's equation

$$E_2' = \frac{107 \times 0.163}{0.698 + 0.163} = 20.3\ \text{V}$$

$$I_2 = \frac{E_2'}{Z_T} = \frac{20.3}{5.009} = 3.97\ \text{A}$$

$$\text{CT error\%} = \frac{|4 - 3.97|}{4} \cdot 100 = 0.7\%$$

ii. For the same steps in part (i) $E_2 = 61.2$ V

$$I_e = 0.894 \text{ A}$$

$$E_2 = 60.1 \text{ V}$$

$$I_2 = 11.78 \text{ A}$$

$$\text{CT error } \% = 1.8\%$$

iii. CT error % = 3.5%
iv. CT error % = 45.1%

Example 1.4

Assume that a CT has a rated current ratio of 500/5 A. The impedance of the secondary winding $Z_2 = 0.242\ \Omega$, and the burden impedance $Z_B = 0.351\ \Omega$. The core area $A = 0.00193\ m^2$. The CT must operate at a maximum primary current of 10 kA. If the frequency is 60 Hz and the core is built from silicon steel:

i. Determine whether or not the CT will saturate.
ii. Determine the CT error.

Solution

i. The secondary current (relay side) is given by:

$$I_s = 10{,}000 \times (5/500) = 100 \text{ A}$$

If we neglect the excitation current, I_e.

$$I_s' \approx I_s = 100 \text{ A}.$$

The secondary side voltage is:

$$E_s = I_s (Z_B + Z_2)$$
$$= 100 \times (0.351 + 0.242) = 59.29 \text{ V}$$

$$E_s = 4.44 \times f \times N_2 \times A \times Bm$$

The lower limit for the silicon steel saturation is

$$Bm = 1.2 \text{ T}$$

Hence, the CT will not saturate.

From the saturation curve, the core corresponding I_e for

$$E_s = 59.29 \text{ V} \quad \text{is} \quad I_e = 0.1 \text{ A}.$$

ii. $CT_{error}\% = \frac{I_e}{I_s} \times 100\% = \frac{0.1}{100} \times 100\% = 0.1\%$

1.10 PROTECTIVE ZONES

A complete power system is divided into "zones," associated, for example, with an alternator, transformer, busbar section, or a feeder end; each zone has one or more coordinated protective systems.

The zones of the protective relay are the distance that the relay can cover the protection.

All network elements must be covered by at least one zone or more in the power systems. The more important elements must be included in at least two zones, where the zones must overlap to prevent any element from being unprotected. The overlap must be finite but small to minimize the likelihood of a fault inside this region. A relay location usually defines the zone boundary.

1.10.1 Backup Protection

If the primary protection fails to operate, the backup protection will operate to remove the faulty part from the system. The primary and the backup are independent (relay, breaker, CT, PT). The backup relay is slower than the primary relay, but sometimes backup protection opens more circuit breakers than necessary to clear the fault. The backup protection provides primary protection when the usual primary apparatus is out of service.

There are two types of backup protection depending on the method of installation:

i. *Local backup*: Clears fault in the same station where the failure has occurred (see Figure 1.13).
ii. *Remote backup*: Clears fault on station away from where the failure has occurred (see Figure 1.14).

1.10.2 Selectivity and Zones of Protection Selectivity

It is defined in terms of regions of a power system (zones of protection) for which a given relay is responsible. The relay will be considered **secure** if it responds only to faults within its zone of protection

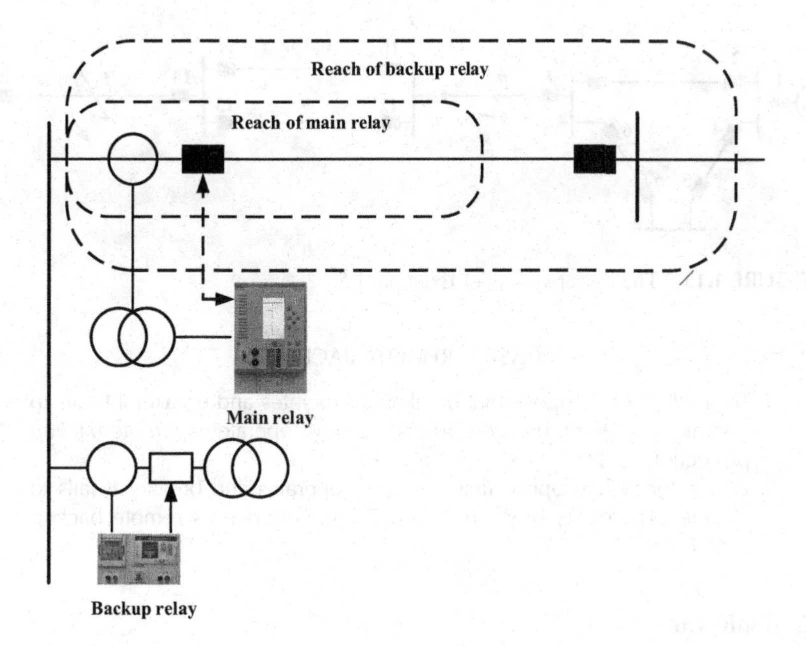

FIGURE 1.13 Local backup protection at different locations.

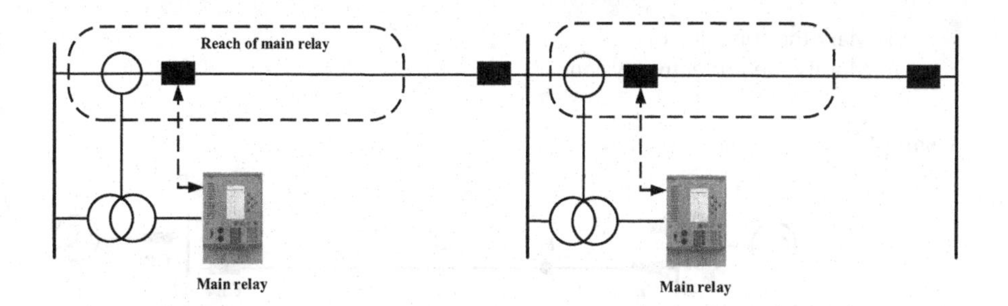

FIGURE 1.14 Remote backup protection.

Example 1.5

Consider the following simple power system, and discuss the local and remote backup protection for two fault locations in Figure 1.15.

CASE 1: LOCAL BACKUP

i. *For fault at F1*: Suppose that breaker 15 operates and breaker 14 fails to work. Therefore, breakers 11 and 13 must work as *local backup* protection.
ii. *For fault at F2*: Suppose that breaker 9 operates and breaker 8 fails to operate. Therefore, breakers 3 and 6 must operate as *local backup* protection.

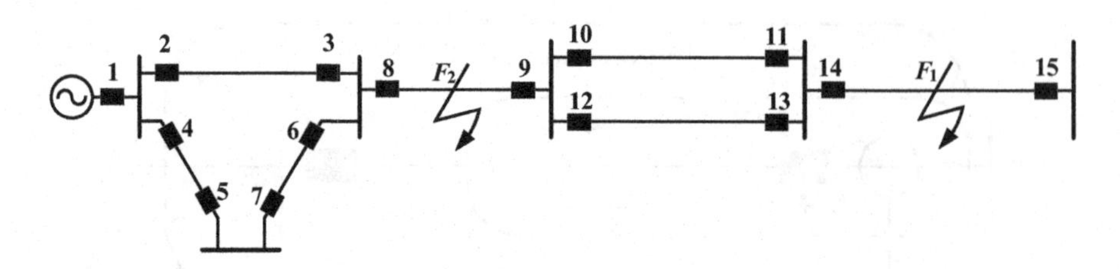

FIGURE 1.15 The power system of Example 1.5.

CASE 2: REMOTE BACKUP

i. *For fault at F1*: Suppose that breaker 15 operates and breaker 14 fails to operate. Therefore, breakers 10 and 12 must operate as *remote backup* protection.
ii. *For fault at F2*: Suppose that breaker 9 operates and breaker 8 fails to operate. Therefore, breakers 2 and 7 must operate as remote backup protection.

Example 1.6

Consider the power system shown in Figure 1.16.

i. Mark the suitable zones.
ii. Modify the protective system.

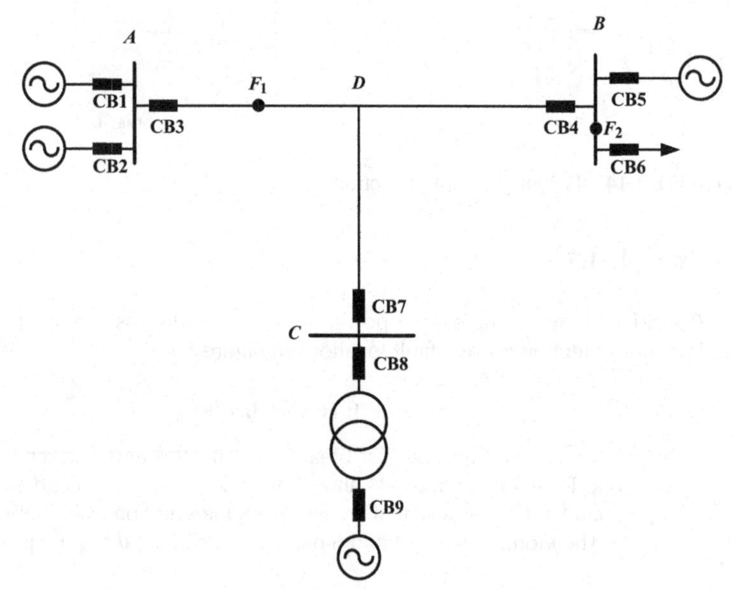

FIGURE 1.16 The power system of Example 1.6.

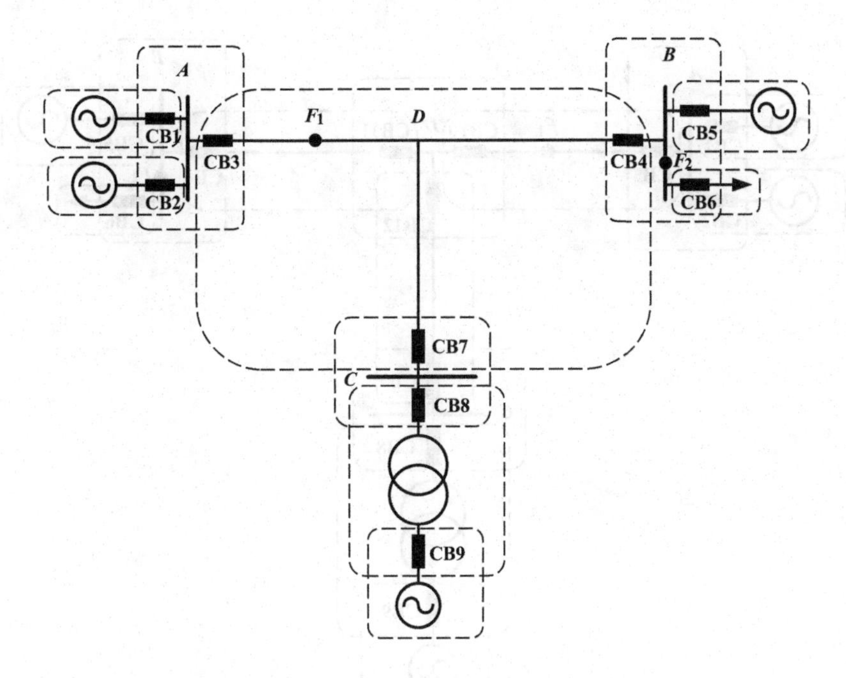

FIGURE 1.17 Protection zones for the power system of Example 1.6.

iii. Mark that all circuit breakers will operate when faults P_1 and P_2 occur, as shown in parts (i) and (ii).

Solution

i. Figure 1.17 illustrates the distribution of the protective zone on the power system.
ii. To modify the system in Figure 1.17, add three circuit breakers at node D (CBs 10, 11, and 12), as shown in Figure 1.18.
iii. For part (i), if the fault occurs at point F1, circuit breakers 3, 4, and 7 will operate, and for a fault at F2, circuit breakers 4, 5, and 6 will operate.

For part (ii), if the fault occurs at point F1, circuit breakers 3 and 10 will operate, and for the fault at F2, circuit breakers 4, 5, and 6 will operate.

1.11 *R–X* DIAGRAM

A relay and a system's characteristics can be graphically represented in only two variables (R and X or $|Z|$ and θ) rather than three (V, I, and θ). The R–X diagram or Z-plane, or simply the complex plane. The complex variable Z is determined by dividing the RMS voltage by the RMS current. The resulting Z can be expressed in rectangular, polar form as

$$Z = R + jX = |Z|e^{j\theta} \tag{1.8}$$

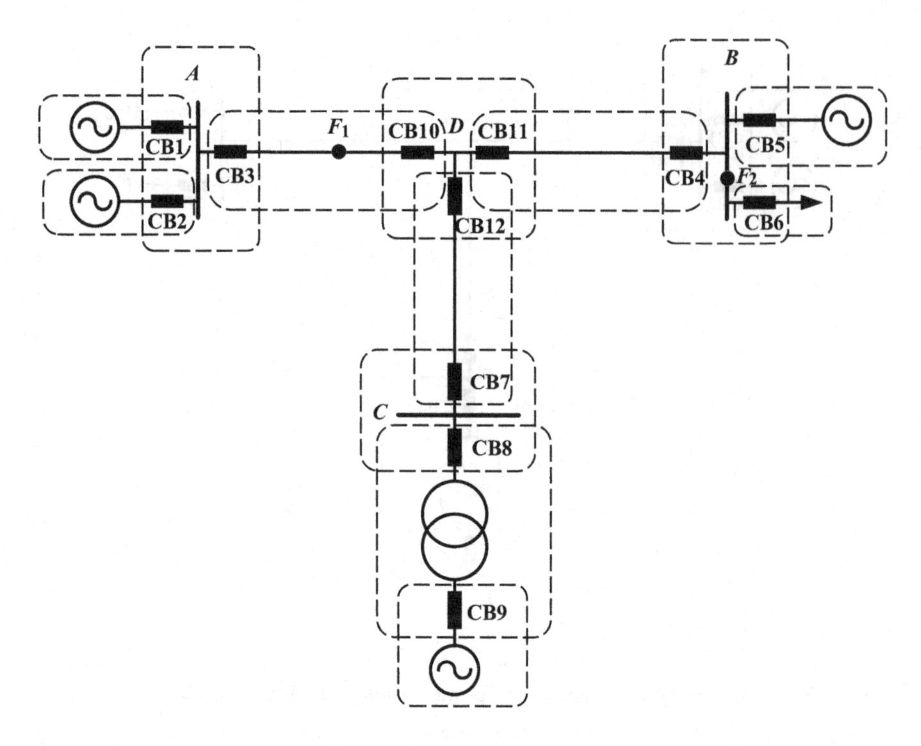

FIGURE 1.18 Modify the power system of Example 1.6.

In addition to the plot of the operating characteristics of a given relay, the system condition affecting the operation of this relay can be superimposed on the same *R–X* diagram so that the response of the relay can be determined. The system characteristics must be within the relay characteristics' operation region to achieve this relay operation. Note that the superimposed system and relay characteristics have to be in terms of the plane quantities and the same scale. These can be both in p.u., where Ohm is used. Both have to be on either a primary or a secondary basis, using

$$\text{Secondary } \Omega_s' = \text{Primary } \Omega_s' \times \left(\frac{\text{CT ratio}}{\text{VT ratio}} \right) \tag{1.9}$$

For example, in the short transmission line shown in Figure 1.19.

$$V_s = I \cdot Z_L + V_R \tag{1.10}$$

FIGURE 1.19 Short transmission line.

and

$$Z_s = \frac{V_s}{I} \tag{1.11}$$

$$Z_R = \frac{V_R}{I} \tag{1.12}$$

also

$$Z_s = Z_L + Z_R \tag{1.13}$$

The receive end load impedance can also be expressed as

$$Z_R = R_R + jX_R \tag{1.14}$$

where

$$R_R = \frac{|V|^2 P}{P^2 + Q^2} \tag{1.15}$$

and

$$X_R = \frac{|V|^2 Q}{P^2 + Q^2} \tag{1.16}$$

Example 1.7

Assume that a short transmission line has the receiving end load apparent power S_R and voltage $|V_R|$ of $2.5 + j0.9$ and 1.0 p.u., respectively. If the line $Z_L = 0.1 + j0.25$ p.u., determine

i. Receive end impedance Z_R.
ii. Send-end impedance Z_s.
iii. Draw the R–X diagram.
iv. The power angle δ.

Solution

i. The real part of load impedance is:

$$R_R = \frac{|V|^2 P}{P^2 + Q^2}$$

$$= \frac{|1.0|^2 (2.5)}{2.5^2 + 0.9^2} = 0.3541 \text{ p.u.}$$

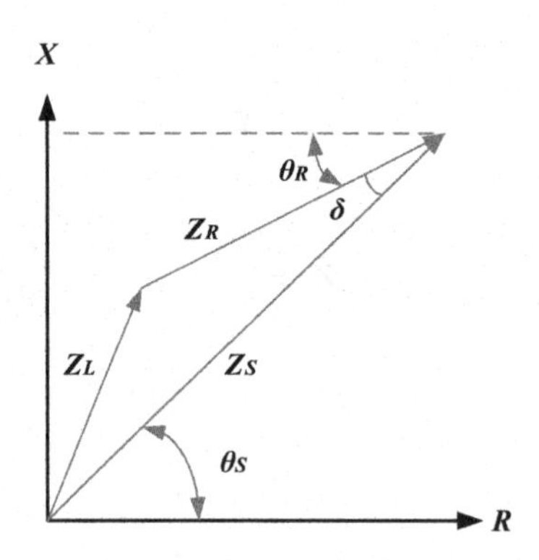

FIGURE 1.20 The R–X diagram of Example 1.7.

The imaginary part of load impedance is:

$$X_R = \frac{|V|^2 Q}{P^2 + Q^2} = \frac{|1.0|^2 (0.9)}{2.5^2 + 0.9^2} = 0.1275 \text{ p.u.}$$

The load impedance is:

$$Z_R = 0.3541 + j0.1275 \text{ p.u.} = 0.3764\angle 19.8° \text{ p.u.}$$

ii. The send-end impedance is:

$$Z_s = Z_L + Z_R = 0.1 + j0.25 + 0.3541 + j0.1275$$
$$= 0.4541 + j0.3775 \text{ p.u.} = 0.5905\angle 39.7° \text{ p.u.}$$

iii. The R–X diagram (Figure 1.20)

iv. The power angle is:

$$\delta = \theta_s - \theta_R = 39.7° - 19.8° = 19.9°$$

PROBLEMS

1.1. A CT has a rated current ratio of 500/5 A, $Z_2 = 0.1 + j0.5\ \Omega$, and a magnetization curve equation as

$$E_2 = \frac{105 I_e}{0.65 + I_e}$$

Compute I_2 and the CT error when used to drive an overcurrent relay of PS = 150% and its $Z_L = 14.9 + j1.5\ \Omega$, with a fault current I_f = 1,200 A. Will the relay detect the fault?

1.2. A CT has a rated ratio of 500:5 A, $Z_2 = 0.1 + j0.5\ \Omega$, and the magnetization curve is shown in Figure 1.21, compute the primary current I and the CT error for the following cases. Will the relay detect the primary current?

a. $Z_B = 4.9 + j0.5\ \Omega, I' = 4\ \text{A}(\text{Load current})$.

b. $Z_B = 4.9 + j0.5\ \Omega, I' = 12\ \text{A}(\text{Fault current})$.

c. $Z_B = 14.9 + j1.5\ \Omega, I' = 12\ \text{A}(\text{Fault current})$.

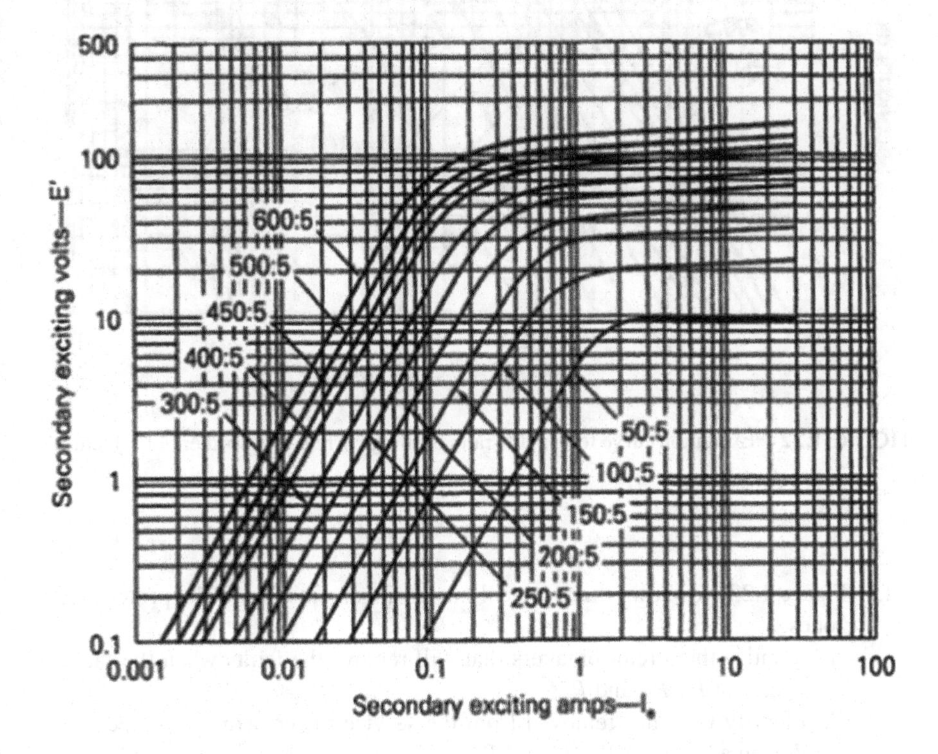

FIGURE 1.21 CT exciting voltage–current characteristics of Problem 1.2.

1.3. An overcurrent relay set to operate at 6 A is connected to a multi-ratio current transformer (whose excitation curve is given in Figure 1.22 with CT ratio = 150:5, $Z_s = 0.082\ \Omega$, and $Z_B = 0.85\ \Omega$). The core area is $2.8 \times 10^{-3} \text{m}^2$ and $f = 60\,\text{Hz}$.

i. Will the relay detect a 200 A fault current on the primary side?

ii. Determine whether the CT will saturate.

iii. Compute the percentage CT error.

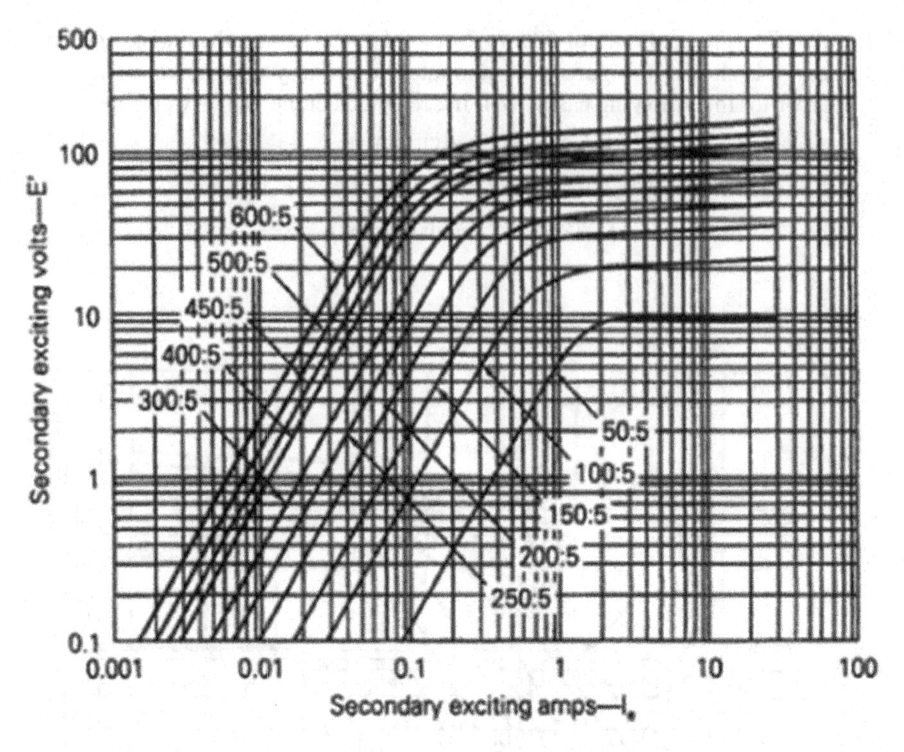

FIGURE 1.22 Excitation curve for a multi-ratio bushing current transformer of Problem 1.3.

1.4. Figure 1.23a–c shows three typical bus arrangements. Draw the protective zones.

i. Identify the circuit breakers that will remove the faults when the fault occurs at P_1, P_2, and P_3.

ii. Identify the lines removed from the service under primary protection during a bus fault P_1, P_2, and P_3.

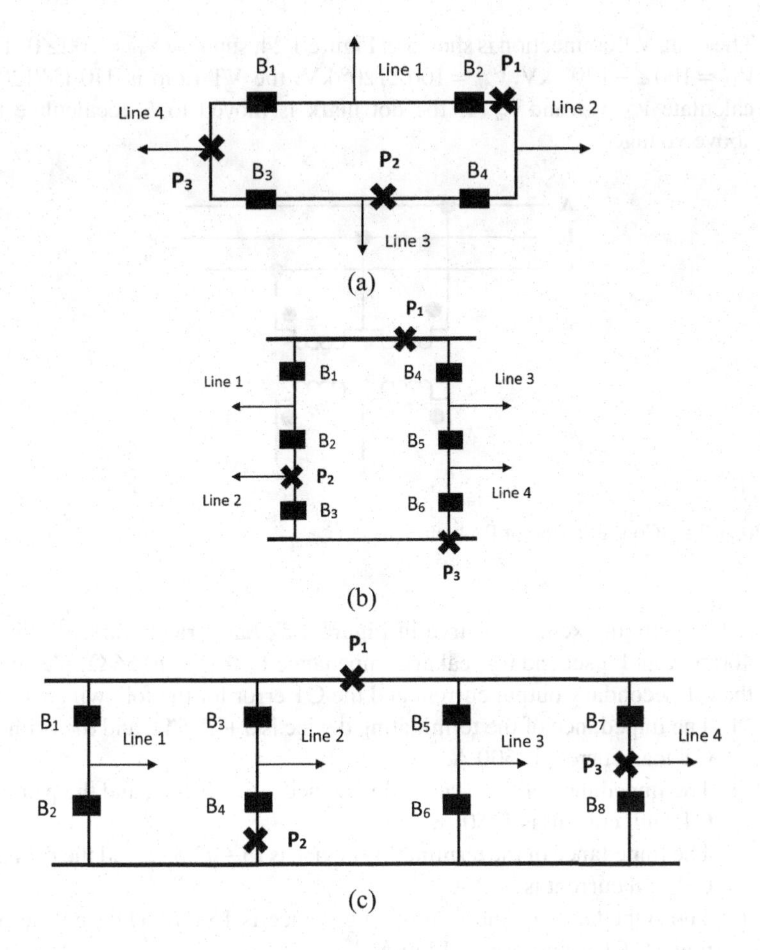

FIGURE 1.23 Power system configurations for Problem 1.4. (a) Ring bus. (b) Breaker and a half double bus. (c) Double breaker double bus.

1.5. Assume that a short transmission line has the receiving end load SR and |VR| of $2.8+j1.0$ and 1.0 p.u., respectively. If the line $Z_L = 0.15 + j0.3$p.u., determine
   i. Receive end Z_R.
   ii. Send-end impedance Z_s.
   iii. Draw the R–X diagram.
   iv. The power angle δ.
1.6. What is the effect of the fault on the power system?
1.7. An overcurrent relay was monitored and performed over one year. It was found that the relay operated 16 times, out of which 13 were correct trips. If the relay failed to issue a trip decision on four occasions, compute the relay's dependability, security, and reliability.

1.8. The delta VT connection is shown in Figure 1.24, suppose $V_{AB} = 100\angle 0°$ kV, $V_{AB} = 100\angle -120°$ kV, $V_{AB} = 100\angle 120°$ kV, the VT ratio is 110 kV/120 V, calculate v_{ab}, v_{bc}, and v_{ca}. If the dot mark is moved to b, recalculate the above voltage.

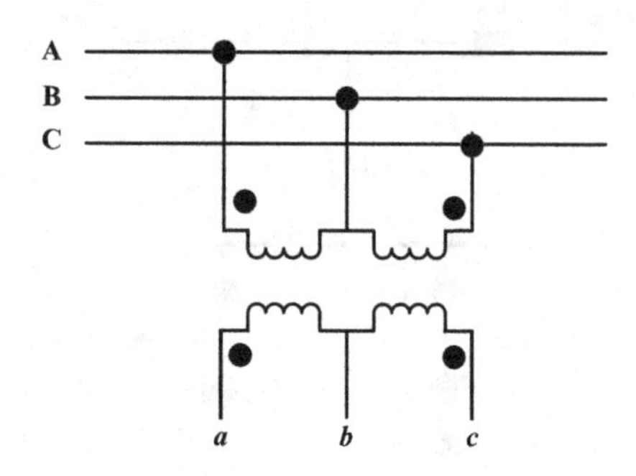

FIGURE 1.24 Configuration of Problem 1.7.

1.9. A CT with an excitation curve in Figure 1.22 has a rated current ratio of 400:5 A and a secondary leakage impedance of $0.15 + j0.56\,\Omega$. Calculate the CT secondary output current and the CT error for the following cases:

i. The impedance of the terminating device is $5 + j1.5\,\Omega$, and the primary CT load current is 500 A.
ii. The impedance of the terminating device is $5 + j1.5\,\Omega$, and the primary CT fault current is 1250 A.
iii. The impedance of the terminating device is $15 + j2.5\,\Omega$, and the primary CT load current is 400 A.
iv. The impedance of the terminating device is $15 + j2.5\,\Omega$, and the primary CT fault current is 1250 A.

2 Protective Relays

2.1 INTRODUCTION

A protective relay is an electrical system designed to respond to an abnormal input condition, remove it, and trigger a contact operation after a specified condition is met. Relays are used in all power transmission systems, in-home communications systems, and industries.

In an irregular or unsafe device condition and an appropriate control circuit initiation, a protective relay is identified as a relay whose function is detecting a faulty line, apparatus, or other power condition. A fuse is often used as security and is characterized as an overcurrent protective device with a fuse component in a circuit opening due to heating and severed by an overcurrent condition.

This chapter presents aspects of data required for the relay setting and types of relays; it also discusses the digital relay operation, signal path for microprocessor relays, and digital relay construction.

2.2 DATA REQUIRED FOR THE RELAY SETTING

Data required for relay settings are

i. Short-circuit level of the power system involved, showing the type and rating of the protection devices and their CTs.
ii. The maximum and minimum values of short-circuit currents.
iii. The impedances of the system components.
iv. Starting current and settling time of the induction motor.
v. The maximum load current.
vi. Decrement curve showing the rate of decay of the fault current supplied by the generator.
vii. The performance curve of the CTs.

The relay setting is first determined to give the shortest operating time at maximum fault levels and then checked to see if the operation will also be satisfactory at the minimum fault current expected.

All power systems aim to maintain a very high standard of service status and reduce downtime when intolerable conditions arise. Power loss, voltage drop, and overvoltage will occur due to natural disasters, physical injuries, equipment malfunction, and human error disturbance.

Safety is the science, skill, and art of applying, setting, and/or fusing to provide full sensitivity to a fault and unwanted situation.

DOI: 10.1201/9781003394389-2

Protective relays are typically linked to the power system via CT and/or VT. In normal operation and when the circuit breaker CB is closed, the contact closes to energize the CB trip coil, which operates to open the breaker's main contact and energize the attached circuit.

The fault sequence of the event and disturbance recording indicate

- What happened?
- What did the current and voltage signals look like (CT saturation)?
- When did the protection device issue a trip signal?
- How long did the circuit breaker need to operate?
- What was the magnitude of the interrupted current?
- How did the system behave after the circuit breaker tripped?

2.3 CLASS OF MEASURING RELAYS

1. *Current relay:* Operate at a predetermined threshold value of current.
2. *Voltage relay:* Operate at a predetermined threshold value of voltage.
3. *Power relay:* Operate at a predetermined threshold value of power.
4. *Directional relay:*
   - *Alternating current:* Operate according to the phase relationship between alternating quantities.
   - *Direct current:* Operate according to the direction of the current and are usually of the permanent-magnetic, moving-coil pattern.
5. *Frequency relays:* Operate at a predetermined frequency. These include over-frequency and under-frequency relays.
6. *Temperature relays:* Operate at a predetermined temperature in the protected component.
7. *Differential relays:* Operate according to the scalar or vectorial difference between two quantities, such as current, voltage, and so forth.
8. *Distance relays:* Operate according to the "distance" between the relay's current transformer and the fault. The "distance" is measured in terms of resistance, reactance, or impedance.

2.4 BASIC DEFINITIONS AND STANDARD DEVICE NUMBERS

2.4.1 Definitions of Terms

Defined subsequently are some of the important terms used in protective relaying:

i. *Normally open contact (N/O):* is open when the relay is not energized.
ii. *Normally closed contact (N/C):* is closed when the relay is not energized.
iii. *Operating force or torque:* that which tends to close the relay contacts.
iv. *Restrain force or torque:* opposes the operating force or torque and prevents the relay contacts' closure.
v. *Pick-up:* A relay is said to pick up when it moves from OFF to ON. The value of the characteristic quantity above this change occurs, known as the pick-up value.

vi. *Pick-up level:* The value of the actuating quantity (current or voltage) on the border above which the relay operates.
vii. *Drop-out or reset level:* The current or voltage value below which a relay opens its contacts and returns to its original position.
viii. *Operating time:* The time that elapses between the instant when the actuating quantity exceeds the pick-up value and when the relay contacts close.
ix. *Reset time is* when the actuating quantity falls below the reset value to when the touch of the relay returns to its normal location.
x. *Primary relays are the relays* connected directly in the circuit to be protected.
xi. *Secondary relays* are connected in the circuit to be protected through CTs and VTs.
xii. *Auxiliary relays* work to assist another relay in performing its role in response to its operating circuit's opening or closing. This relay may be instantaneous or may have a time delay.
xiii. *Reach:* A distance relay operates whenever the relay impedance is less than a prescribed value; this impedance or reactance corresponding distance is known as the relay's reach.
xiv. *Instantaneous relay*: One with no intentional time delay operates in <0.1 seconds.
xv. *Blocking prevents the protective relay from tripping due to its characteristics or* an additional relay.
xvi. *Time delay relay:* One which is designed with a delaying means.
xvii. *Protective relay:* An electrical device designed to initiate isolation of a part of an electrical installation or to operate an alarm signal in the event of an abnormal condition or a fault.
xviii. *Energizing quantity:* The electrical quantity (current, voltage, phase shift, or frequency), alone or in combination, is required for the relay's functioning.
xix. *Characteristics quantity:* The quantity to which the relay is designed to respond, e.g., current in overcurrent, phase angle in a directional relay, and so forth.
xx. *Setting:* The actual value of the energizing or characteristics quantity at which the relay is designed to operate under given conditions.
xxi. *Flag or target:* A device used for indicating the operational relay.
xxii. *Overreach or underreach:* Errors in relay measurements resulting in operation or failure, respectively.
xxiii. *It reinforces relay*, which is energized by the main relay's contacts and its contacts in parallel with those of the main relay.
xxiv. *Pilot channel:* A means of interconnection between relaying points for protection.
xxv. *Protective scheme:* The coordinated arrangements for the protection of a power system. It may include several protective systems.

2.4.2 Devices Numbers

The list of devices used in the protection scheme is given in Table 2.1.

TABLE 2.1
List of Device Numbers

Code	Types	Code	Types
1	Master element	52	AC circuit breaker
2	Closing relay/time delay starting	53	DC generator/exciter relay
3	Interlocking relay or checking	54	High-speed DC circuit breaker
4	Master contactor	55	Power factor relay
5	Stopping device	56	Field application relay
6	Starting circuit breaker	59	Overvoltage relay
7	Anode circuit breaker	60	Current/voltage balance relay
8	Control power disconnecting device	61	Machine split phase current balance
9	Reversing device	62	Time delay stopping/opening relay
10	Unit sequence switch	63	Pressure switch
12	Overspeed device	64	Ground detector relay
13	Synchronous-speed device	65	Governor
14	Under-speed device	66	Starts per hour
15	Frequency and speed matching device	67	AC directional overcurrent relay
20	Elect. operated valve (solenoid valve)	68	Blocking relay
21	The distance relay	69	Permissive control device
23	Temperature control device	71	Level switch
25	Synchronism check device/ synchronizing	72	DC circuit breaker
26	Apparatus thermal device	74	Alarm relay
27	Under-voltage relay	75	Position changing mechanism
29	Isolating contactor	76	DC overcurrent relay
30	Annunciator relay	78	Phase-angle measuring/out-of-step protective relay
32	Power directional relay	79	AC-reclosing relay
36	Polarizing voltage/polarity device	81	Frequency relay
37	Under-current or under-power relay	83	Automatic selective control/transfer relay
38	Bearing protective device	84	Operating mechanism
39	Mechanical conduction monitor	85	Carrier/pilot-wire receiver relay
40	Field relay	86	Lockout relay
41	Field circuit breaker	87	Differential protective relay
42	Running circuit breaker	89	Line switch
43	Manual transfer/selector device	90	Regulating device
46	Phase-balance/reverse-phase relay	91	Voltage directional relay
47	Phase-sequence voltage relay	92	Power and voltage directional relay
48	Incomplete-sequence relay	94	Trip-free relay/tripping
49	Transformer thermal relay	95	Reluctance torque synchro check
50	Instantaneous overcurrent	96	Autoloading relay
51	AC time overcurrent relay		

2.5 CLASSIFICATION OF RELAYS

Protection relays can be classified based on the *function* they carry out, their *construction*, the *incoming signal*, and the type of *protection*.

1. *General function:* Auxiliary. Protection. Monitoring. Control.
2. *Construction*:
   i. Electromagnetic.
   ii. Solid-state.
   iii. Microprocessor.
   iv. Computerized.
   v. Nonelectric (thermal, pressure, etc.).
3. *Incoming signal*:
   i. Current
   ii. Voltage
   iii. Frequency
   iv. Temperature
   v. Pressure
   vi. Velocity
   vii. Others.
4. *Type of protection*:
   i. Overcurrent
   ii. Directional overcurrent
   iii. Distance
   iv. Overvoltage
   v. Differential
   vi. Reverse power
   vii. Other.

2.6 TYPES OF RELAYS

The fundamental objective of system protection is to isolate a fault area quickly so that the rest of the power system is protected as much as possible.

The main types of protective relays are summarized in Figure 2.1.

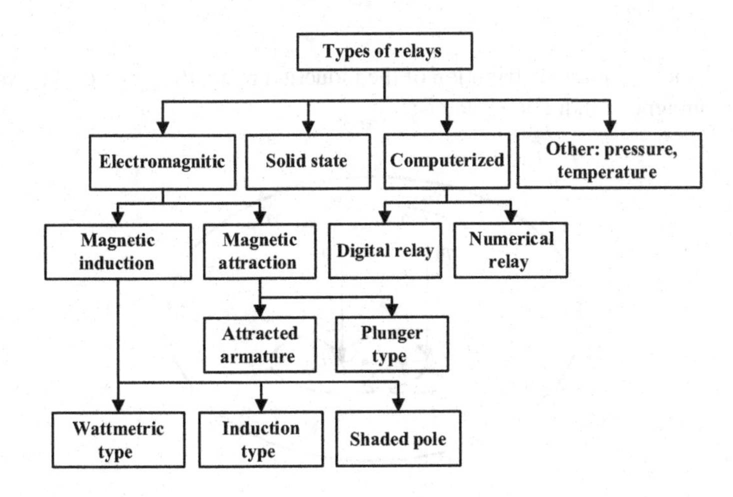

FIGURE 2.1 Types of relays.

2.6.1 Electromagnetic Relays

2.6.1.1 Electromechanical Relays

Research started at the end of the 19th century. In the 1930s, the relay family was completed, and they are still in use. These relays were the oldest types used for power systems safety, dating back almost 110 years. They work on a mechanical force theory that causes a relay contact to function in response to a stimulus. Hence, in the electromechanical relay, the mechanical force is produced by current flow in one or more windings on a magnetic core or cores. The main advantage of such relays is that they provide galvanic isolation in a simple, cheap, and reliable way between the inputs and outputs, so they are still used for simple on/off switching functions where the output contacts have to bear significant currents.

It is possible to classify electromechanical relays into many different types:

i. Magnetically attracted armature relays.
ii. Magnetic induction relays.
iii. Moving coil.
iv. Thermal.

However, only attracted armature and induction forms have an important application; more modern equivalents have replaced all other types. Electromagnetic relays are constructed of electrical, magnetic, and mechanical parts, durable, inexpensive, and effective, and have an operating coil and different contacts.

2.6.1.2 Magnetic Induction Relays

An induction relay only operates with alternating current. It is possible to divide magnetic induction relays into three groups:

i. Wattmeter type.
ii. Induction type.
iii. Shaded pole.

Figure 2.2 shows a force distribution of the induction relay; the principle operation of the electromagnetic induction relay is:

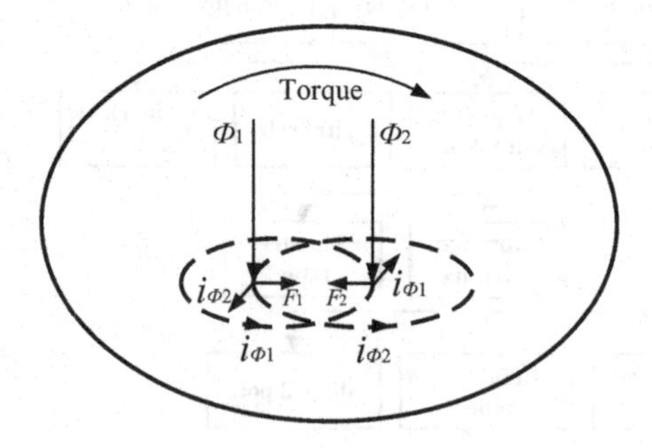

FIGURE 2.2 Force distribution of the induction relay.

The two magnets of the induction-type relay produce two alternating magnetic fields φ_1 and φ_2

$$\varphi_1 = \varphi_{1m} \sin \omega t \tag{2.1}$$

$$\varphi_2 = \varphi_{2m} \sin(\omega t + \theta) \tag{2.2}$$

where φ_2 leads φ_1 by an angle θ.

φ_1 and φ_2 produce eddy currents in the rotating disk, which are i_{φ_1} and i_{φ_2}

$$i_{\varphi_1} \propto \frac{d\varphi_1}{dt} \tag{2.3}$$

$$i_{\varphi_2} \propto \frac{d\varphi_2}{dt} \tag{2.4}$$

or

$$i_{\varphi_1} \propto \varphi_{1m} \cos \omega t \tag{2.5}$$

$$i_{\varphi_2} \propto \varphi_{2m} \cos(\omega t + \theta) \tag{2.6}$$

F_1 is the force produced by the intersection of F_1 and i_{f2}.

F_2 is the force produced by the intersection of F_2 and i_{f1}.

The net force is

$$F = F_2 - F_1 \propto \left[\varphi_2 \cdot i_{\varphi_1} - \varphi_1 \cdot i_{\varphi_2}\right] \tag{2.7}$$

Thus,

$$F \propto \varphi_{1m}\varphi_{2m}\left[\sin(\omega t + \theta) \cdot \cos \omega t - \sin \omega t \cos(\omega t + \theta)\right] \tag{2.8}$$

Solving to get

$$F \propto \varphi_{1m}\varphi_{2m} \sin\theta \tag{2.9}$$

So, the net torque will be

$$T = F_r \propto F \tag{2.10}$$

And can be represented in Figure 2.3.

or

$$T = K_1\varphi_{1m}\varphi_{2m} \sin\theta \tag{2.11}$$

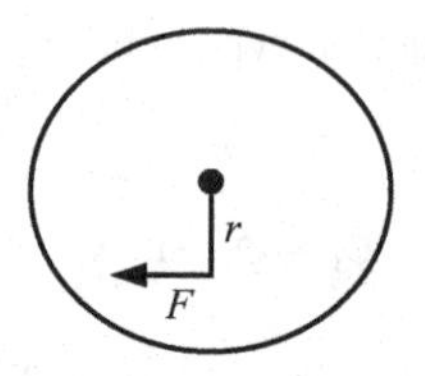

FIGURE 2.3 Force direction in the rotation disk.

Let Φ_1 be the RMS value of φ_1 and Φ_2 the RMS value of φ_2
or

$$T \propto \Phi_1\Phi_2 \sin\theta \tag{2.12}$$

$$T \propto I_1 I_2 \sin\theta \tag{2.13}$$

$$T = K I_1 I_2 \sin\theta \tag{2.14}$$

The characteristics of the induction principle are suitable for AC systems. The torque does not differ with time: no vibration, DC offset inherent rejection, and low overreach.

2.6.1.2.1 Wattmetric-Type Relay

It consists of an electromagnetic system consisting of two electromagnets installed on a disk, as shown in Figure 2.4, acting on a moving conductor, and Figure 2.5 shows an induction-type relay with plug settings.

2.6.1.2.2 Induction-Cup Relay

The process is like an induction disk; here, in a bell-shaped cup that rotates and carries the moving contacts, two fluxes at right angles cause eddy currents. Figure 2.6 displays a four-pole relay.

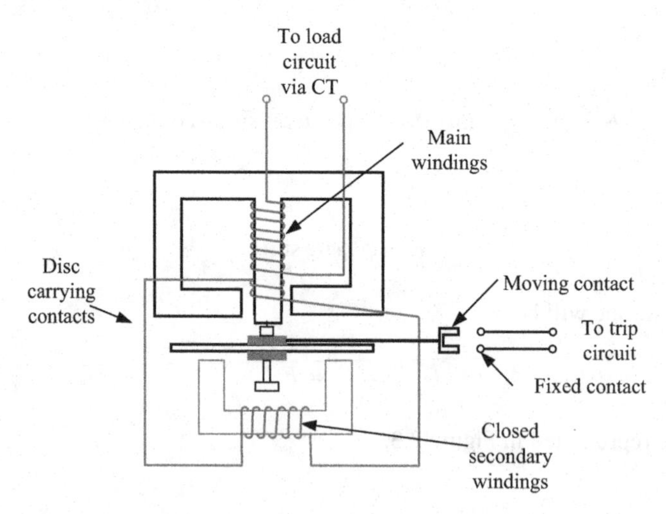

FIGURE 2.4 Induction-type overload relay.

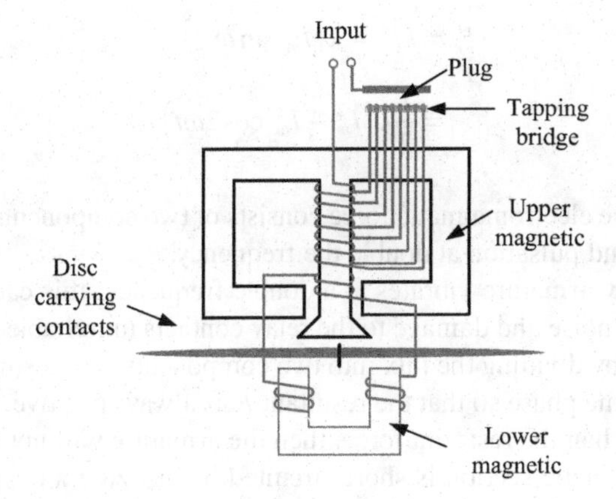

FIGURE 2.5 Induction-type relay with plug settings.

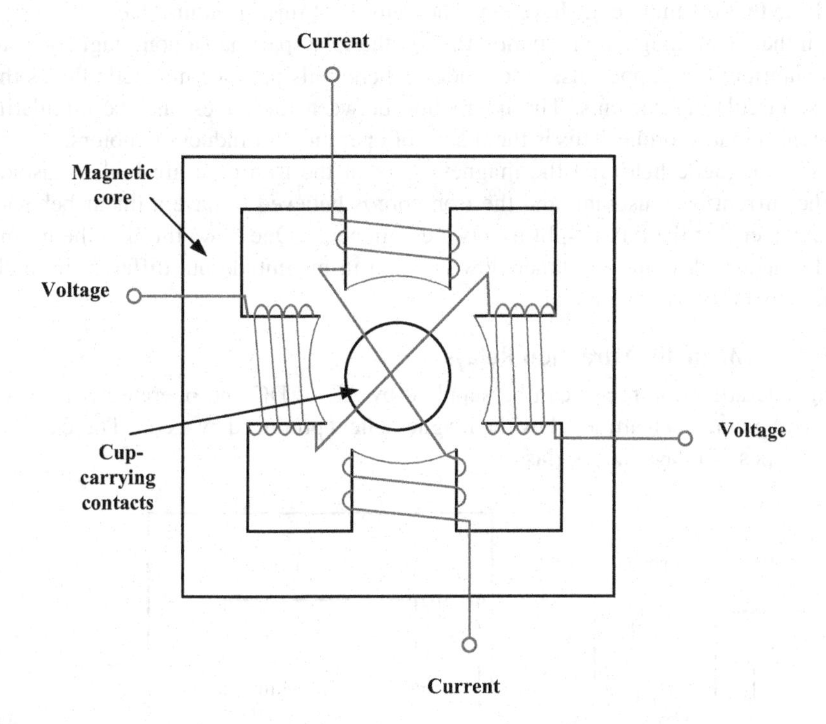

FIGURE 2.6 Four-pole induction-cup relay.

2.6.1.2.3 Shaded Pole Relay

The force is directly related to time to a constant time the square of flux or the current square. In DC, this force is constant; if the force exceeds the restraining force, the relay operates. In AC, the force is given by

$$F_e = kI^2 = k(I_m \sin\omega t)^2$$

$$= \frac{1}{2}k\left(I_m^2 - I_m^2 \cos 2\omega t\right)$$

This shows that the electromagnetic force consists of two components: constant independent of time and pulsation at double the frequency.

Hence the relay armature vibrates at a double frequency; this causes the relay to hum and produce noise and damage to the relay contacts (unreliable operation). This can be overcome by dividing the flux into two components acting simultaneously but differing in the time phase so that the resultant F_e is always positive, and if this force is always greater than the restraint force, then the armature will not vibrate.

The electromagnetic section is short-circuited by utilizing a copper ring or coil. This creates a flux in the area influenced by the short-circuited section (the so-called shaded section), which lags the flux in the non-shaded section, see Figure 2.7.

Note that the main coils have taps; the number of turns is adjustable.

In the electromagnetic induction theory, the relay part has a non-magnetic rotor (an aluminum or copper disk or cylinder) where coils produce magnetic fluxes that cause circulating currents. The interaction between the fluxes and the circulating currents creates torque. This is the theory of operation for induction motors.

The magnetic field and the magnetic flux in the iron core are both sinusoidal if the current is sinusoidal, and the iron core is believed to have a linear behavior. Remember that the flux is split into two components: One flows through the normal and shaded poles, and the other flow is equal in magnitude but different in angle. These two fluxes are identical.

2.6.1.3 Magnetic Attraction Relays

Magnetic attraction relays can be supplied by AC or DC and operate by moving a piece of metal when attracted by the magnetic field produced by a coil. There are two main types of relays in this class.

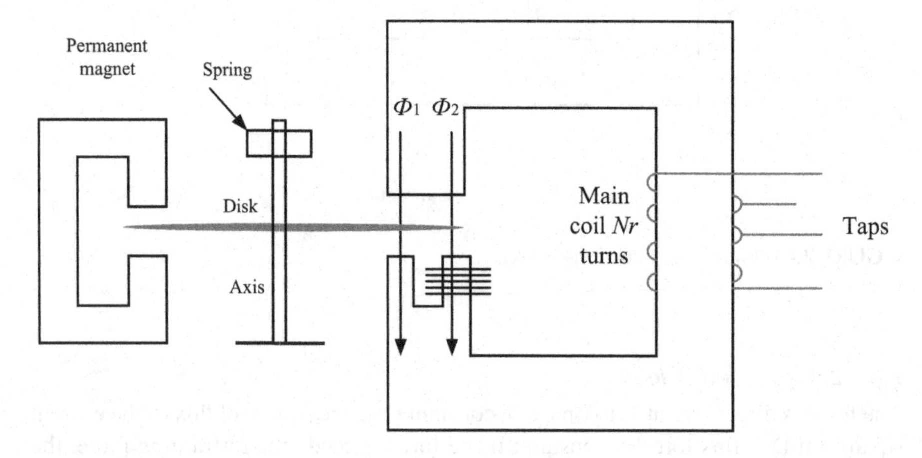

FIGURE 2.7 Shaded pole relay.

i. Attracted armature type (clapper type).
ii. Plunger type.

2.6.1.3.1 The Attracted Armature Relays

As shown in Figure 2.8, it consists of a bar or plate of metal that pivots when it is attracted to the coil.

As the current approaches a certain predetermined value, the armature is drawn to the electromagnet (i_{op} – operating current). The armature force will cause the circuit breaker's connection mechanism to work as a relay and close the contacts of a separate tripping circuit. The armature is drawn against gravity or a spring. The trip's current can be varied to fit the circuit conditions by changing the armature's distance from the electromagnet or the spring's voltage.

The armature carries the moving part of the contacts, which, when the armature is attracted to the coil, is closed or opened according to the design.

2.6.1.3.2 Plunger-Type Relay

The other type is the piston or solenoid relay, illustrated in Figure 2.9, in which the α bar or piston is attracted axially within the solenoid field. In this case, the piston also carries the operating contacts. This is called a plunger-type relay.

The force of attraction can be equal to $K_1I^2 = K_2$, where $\emptyset_1$ depends, among other things, on the number of turns on the working solenoid, the air distance, the effective area, and the reluctance of the magnetic circuit. The restraining force is K_2, typically created by a spring. The resulting force is zero when the relay is balanced, and thus,

$$K_1I^2 = K_2$$

so

$$I = \sqrt{\frac{K_2}{K_1}} \quad \text{constant} \tag{2.15}$$

This equation can be proved as follows:

Attracted armature relay analysis

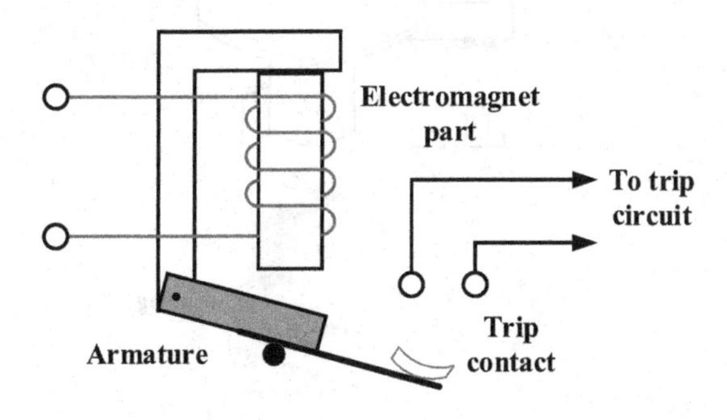

FIGURE 2.8 Attracted armature-type relay.

In general, the mechanical force produced by an electric magnet is proportional to φ^2, i.e.,

$$F(t) \propto \varphi^2 \tag{2.16}$$

$$\varphi = \frac{\text{mmf}}{R} = \frac{Ni}{R} \tag{2.17}$$

where

$$R = \frac{l_g}{\mu_o \cdot A} \propto l_g = \text{reluctance} \tag{2.18}$$

so

$$\varphi \propto \frac{Ni}{l_g} \tag{2.19}$$

or

$$\varphi^2 \propto \frac{N^2 i^2}{l_g^2} \tag{2.20}$$

Hence

$$F(t) = k_\varphi \cdot \varphi^2 \tag{2.21}$$

where

k_φ is the constant

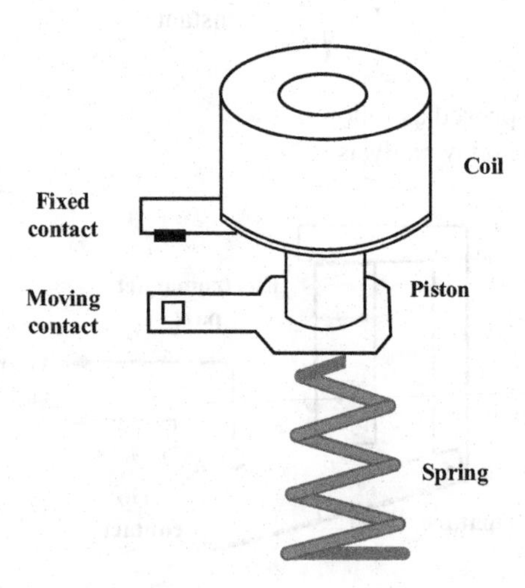

FIGURE 2.9 Solenoid-type (plunger) relay.

$$F(t) = k_{\varphi} \cdot \frac{N^2 i^2}{l_g^2} = k_1 i^2, \quad k_1 = k_{\varphi} \cdot \frac{N^2}{l_g^2} \tag{2.22}$$

The net force is

$$F_n(t) = F(t) - k_2 = k_1 i^2 - k_2 \tag{2.23}$$

where K_2 is the restraining force produced by the spring

When the relay is balanced, $F_n(t) = 0$.

$$0 = k_1 i^2 - k_2$$

$$K_1 I^2 = K_2,$$

So that,

$$I = \sqrt{\frac{K_2}{K_1}} \text{ constant} \tag{2.24}$$

I = RMS value of i

The restraining tension of the spring or the solenoid circuit's resistance may be varied to control the value at which the relay begins to work, thus changing the restricting power. Effectively, attraction relays have no time delay and are commonly used when instantaneous operations are needed.

Example 2.1

An electromagnetic relay of attracted armature type has constants $k_1 = 0.6$ and $k_2 = 10$. Find whether the relay will operate or not when:

i. A current of 3 A flows through the relay winding.
ii. A current of 5.5 A flows through the relay winding.
iii. Find the minimum current required to operate the relay.

Solution

i. For 3 A current:

$$F_n(t) = k_1 i^2 - k_2 = 0.6 \times 3^2 - 10 = 4.6 \text{ N}$$

The restraining force is greater than the operating force.
So, the relay will not operate.

ii. For 5.5 A current:

$$F_n(t) = k_1 i^2 - k_2 = 0.6 \times 5.5^2 - 10 = 8.15 \text{N}$$

The restraining force is less than the operating force.
So, the relay will operate.

iii. The minimum current required to operate the relay is when the relay becomes a balanced condition
or

$$I = \sqrt{\frac{10}{0.6}} = 4.08\,\text{A}$$

2.7 COMPARATOR RELAYS

Comparator relays measure functions of applied quantities and compare two or more inputs.

Type of measurements: magnitude, product measurements, ratio measurement.

$$F_A = k_1 I_A \tag{2.25}$$

$$F_B = k_2 I_B \tag{2.26}$$

Under balanced state (Figure 2.10)

$$\frac{I_A}{I_B} = \frac{k_2}{k_1} = \text{constant} \tag{2.27}$$

Force has been overcome, which requires a current representing the calibration level of the relay. Since the fault current level changes with generating conditions, obtaining selectivity based on current magnitude is seldom possible. Most applications require adding a timer function so that the relay nearest the fault location, having the most current, will trip first.

There are difficulties in obtaining selectivity in the relay of a single quantity measuring element. Most high-speed relays measure a derived quantity, a combination of several simple quantities, such as impedance, admittance, the current

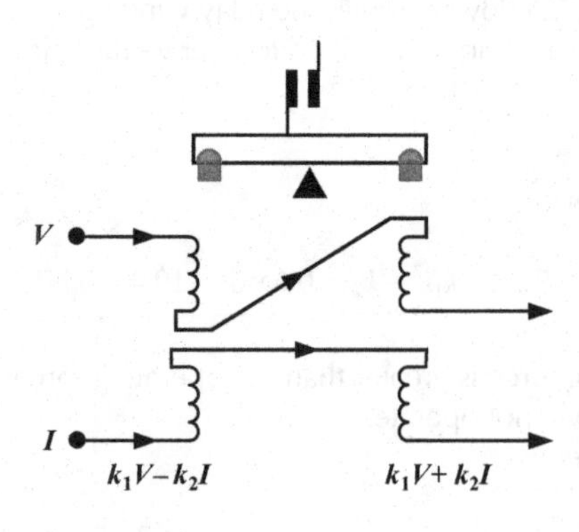

FIGURE 2.10 Comparator relay.

ratio, and the like, in which two simple quantities are compared in magnitude and phase relation.

The relay usually makes the comparison by turning the electrical quantities into force, with MMF or EMF being proportional to the two quantities compared.

The phase relation and amplitude are a function of the system conditions; the relay compares these two fundamental aspects. The principal part of the relay is the comparator, which can be a phase or amplitude comparator.

2.7.1 Generalized Amplitude Comparator

The two sides of a comparator are fed with mixed signals originating from current and voltage sources that will match when the magnitudes of these signals are equal

$$|k_1 I + k_2 V| = |k_3 I + k_4 V| \tag{2.28}$$

Divide by I and substitute $Z = \dfrac{V}{I} = R + jX$

$$|k_1 + k_2(R + jX)| = |k_3 + k_4(R + jX)|$$

Simplifying and calculating the absolute value

$$\left(k_2^2 - k_4^2\right)R^2 + \left(k_2^2 - k_4^2\right)X^2 + 2(k_1k_2 - k_3k_4)R + \left(k_1^2 - k_3^2\right) = 0 \tag{2.29}$$

Comparing this with the general equation

$$x^2 + y^2 + 2gx + 2hy + c = 0$$

where

$$x = R, \quad y = X$$

$$g = \frac{k_1k_2 - k_3k_4}{k_2^2 - k_4^2}$$

$$h = 0$$

$$c = \frac{k_1^2 - k_3^2}{k_2^2 - k_4^2}$$

So, the characteristic is a circle on the R–X plane.

2.8 ADVANTAGES OF ELECTROMECHANICAL RELAYS

The advantage and limitations of static relays are:

- *Advantage:*
  i. The low burden on CT and VT.
  ii. Low mechanical inertia and bouncing.
  iii. High-speed operation and long life.
  iv. Low maintenance.
  v. Quick reset action.
  vi. Unconventional characteristics are possible.
  vii. Low energy is required.
  viii. Easy amplification.
  ix. High sensitivity.
- *Limitation:*
  i. Temperature-sensitive.
  ii. Low short-term overload capacity.
  iii. Aging effect on the relay characteristics.
  iv. Vulnerability to voltage spike (use of filters and shielding to overcome this limitation).

2.9 SOLID-STATE RELAYS

The study of solid-state relays started in the late 1950s, with the first commercial products starting in the 1940s. In the 1960s, complete production took place. The term solid-state relay refers to an electronic device composed of electronic components that provides an electromechanical relay with a similar purpose but does not have any moving parts, improving long-term reliability. In early 1960, the invention of static relays began. It is generally referred to as a relay incorporating solid-state components such as IC transistors, diodes, resistors, capacitors, and so forth; in this type of relay, the function of comparison and measurement is performed as static circuits (no moving parts).

The architecture uses analog electronic equipment instead of coils and magnets to build the relay. Early models used isolated devices such as transistors and diodes combined with resistors, capacitors, inductors, and the like. However, advancements in electronics have made it possible for later versions to use linear and digital ICs for signal processing and logic circuitry implementation.

2.9.1 Solid-State Relay Principle of Operation

Solid-state relays (static relays) are extremely fast in their operation. They have no moving parts and have a very quick response time, and they are very reliable.

Figure 2.11 shows the elements used in a single-phase time lag overcurrent relay.

A protective relay is an analog–binary signal converter with measuring functions. The variables such as current, voltage, phase angle, frequency, and derived values obtained by differentiation, integration, or other mathematical operations always appear as analog signals at the input of the measuring unit. The output will be attained as a binary signal.

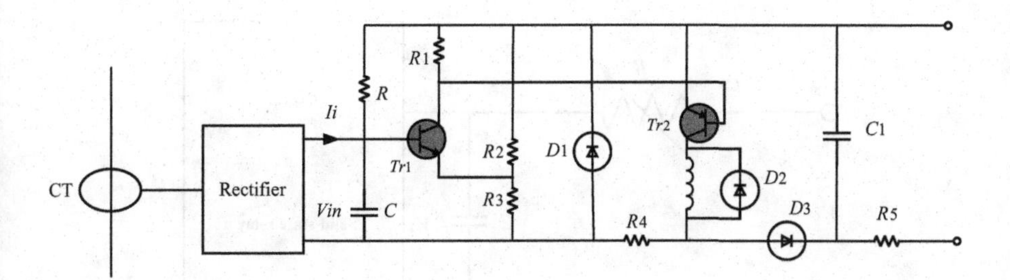

FIGURE 2.11 Solid-state relays (static relays) circuit diagram.

The AC input from the current transformer CT is rectified and converted to DC voltage V_{in} through shunt resistance.

A delay time circuit (RC) produces the required time delay.

If $V_{in} < V_R$, the base emitter of transistor TR1 is reversed, forcing the transistor to be in the cut-off state.

When $V_{in} > V_R$, transistor TR1 will be in the ON state, turn on TR2, and activate the output relay. R_1 and R_2 set V_R.

Example 2.2

An RC circuit used to produce time delay for a solid-state relay is shown in Figure 2.12. For a step input voltage $V_i(t) = 6\ U(t)$, the output voltage $V_o(t) = 3$ V and $C = 10$ µF. Determine the time delay (t_{delay}) for the following cases (Figures 2.13 and 2.14):

i. $R = 10$ kΩ;
ii. $R = 1$ MΩ.
iii. Sketch the output $V_o(t)$ versus time for cases (i) and (ii).

Solution

i. When $R = 100$ kΩ

$$\tau = R \cdot C = 100 \times 10^3 \times 10 \times 10^{-6} = 1 \text{ seconds}.$$

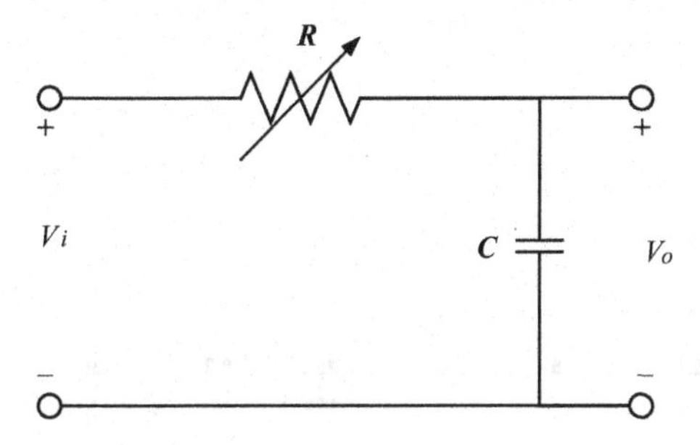

FIGURE 2.12 Circuit diagram of Example 2.2.

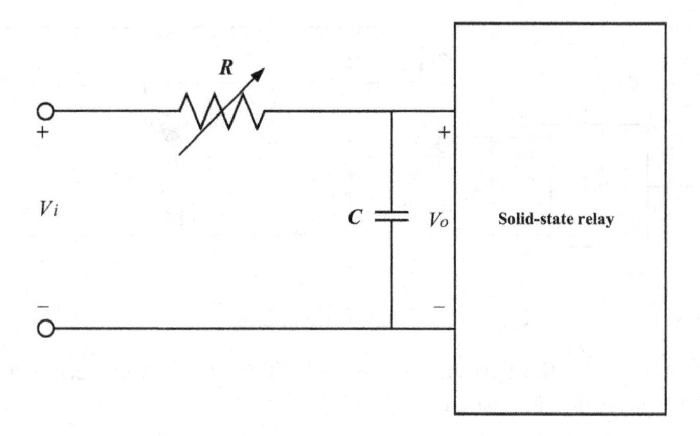

FIGURE 2.13 The RC circuit connects to the relay.

$$V_o = V_c\left(1 - e^{-t/\tau}\right)$$

$$3 = 6 \times \left(1 - e^{-\frac{t_{\text{delay}}}{1}}\right)$$

$$t_{\text{delay}} = 0.693\,\text{seconds}.$$

ii. When $R = 1\ \text{M}\Omega$

$$\tau = R \cdot C = 1 \times 10^6 \times 10 \times 10^{-6} = 10\ \text{seconds}.$$

$$V_o = V_c\left(1 - e^{-t/\tau}\right)$$

$$3 = 6 \times \left(1 - e^{-\frac{t_{\text{delay}}}{10}}\right)$$

$$t_{\text{delay}} = 6.93\ \text{seconds}.$$

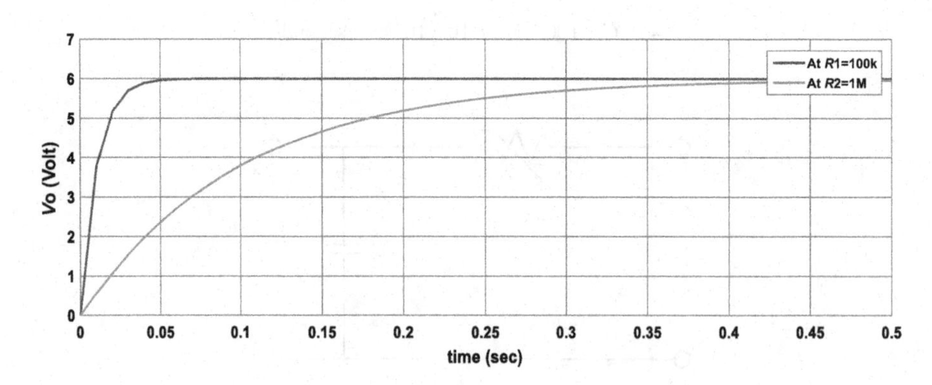

FIGURE 2.14 The output voltage across the RC circuit.

2.10 COMPUTERIZED RELAY

2.10.1 Digital Relays

A microcomputer-operated relay is a digital protective relay. Research began in 1960. Basic developments: Early 1970. Figure 2.15 shows a block diagram of a digital relay. The data acquisition device gathers information from the transducers and transforms it into the microcomputer's correct form. CT and PT information and other devices are amplified and sampled at many kHz levels. With the A/D converter, the sampled signals are digitized and fed to registers in the microprocessor framework. The microprocessor may use a counting technique or the discrete Fourier transform (DFT) to compare the data with overcurrent, over/under voltage, and the like, preset limits, and then send commands to the circuit breakers through the D/A converter to alarm or trip signals.

2.10.2 Digital Relays Operation

The relay applies A/D (analog/digital) conversion procedures to voltages and currents.

The relay analyzes the A/D converter output using the Fourier transform principle to extract the magnitude of the incoming quantity (RMS value). The Fourier transform is frequently used to extract the signal's relative phase angle to some reference.

Based on current and/or voltage magnitude (and angle in some applications), the digital relay may analyze whether the relay can activate or hold back from triggering. Figure 2.16 provides examples of digital relays.

2.10.3 Signal Path for Microprocessor Relays

The signal path for voltage and current input signals is shown in Figure 2.17.

The signals are filtered with an analog filter after the currents and voltages are reduced to acceptable levels by the instrument transformers. The signal is then

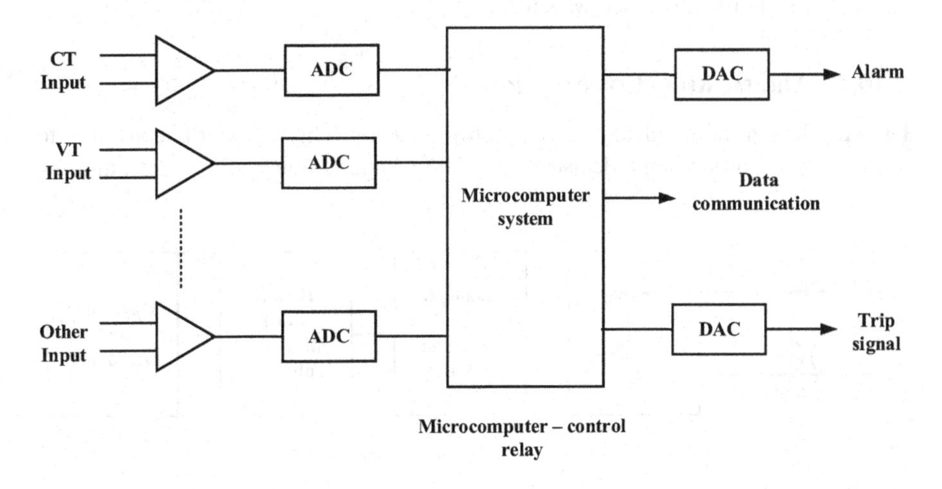

FIGURE 2.15 Block diagram of a digital relay.

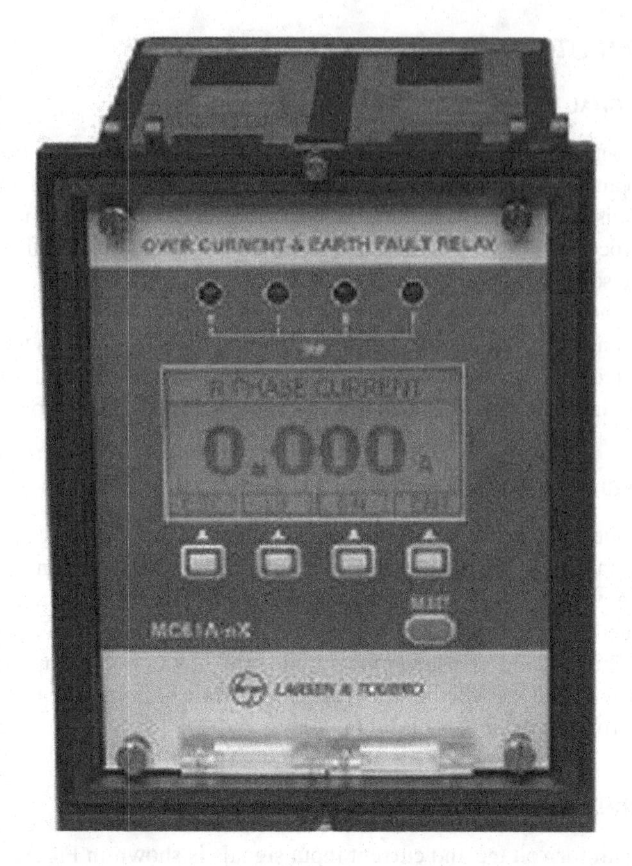

FIGURE 2.16 Digital relays (numerical relays for protection, monitoring, and control).

digitized and re-filtered with a digital filter. Numerical operating quantities are then calculated from the processed waveforms.

2.10.4 Digital Relay Construction

Digital relays have an advantage in recording events. This allows the consumer to see the timing of key logic decisions, relay I/O (input/output) adjustments, and see

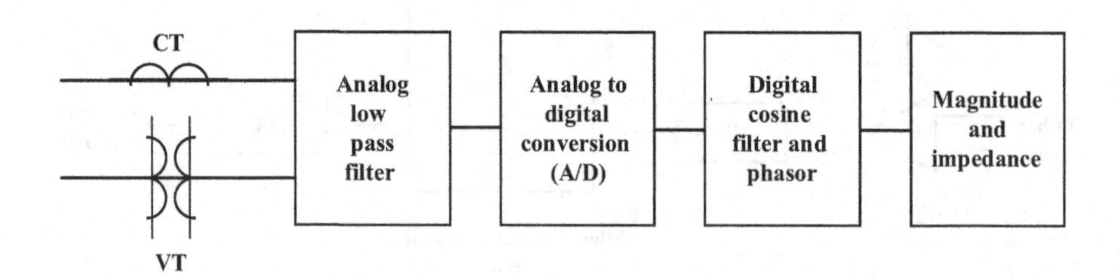

FIGURE 2.17 The signal path for voltage and current input signals to the microprocessor relays.

at least the incoming AC waveform's basic frequency portion in an oscillographic fashion that would be included in the event recording.

The relay has a comprehensive array of settings beyond what can be entered via front panel knobs and dials. These settings are passed to the relay via a PC (personal computer) interface. This same PC interface is used to obtain event reports from the relay.

The digital filter smoothens the signal by eliminating DC and frequency components different from the fundamental (when required).

The more recent digital relay models would have advanced metering and communication protocol ports, enabling the relay in a SCADA system to become a focal point.

Figure 2.18 shows a block diagram for digital relay architecture. Figure 2.19 shows a digital relay algorithm.

2.10.5 Advantages of Digital Relays

i. High reliability.
ii. Use for protection and control purposes.
iii. Measurement and fault recording.
iv. Capability for communication.
v. Compatibility with integrated digital systems.
vi. Integration and self-testing.
vii. The relays supervise the protection system.
viii. High sensitivity and selectivity.
ix. Modern protection principles.
x. Modern relay operating characteristics.
xi. Low maintenance cost.
xii. Reduced burden on CTs and VTs.
xiii. Adaptive protection.
xiv. Low cost.

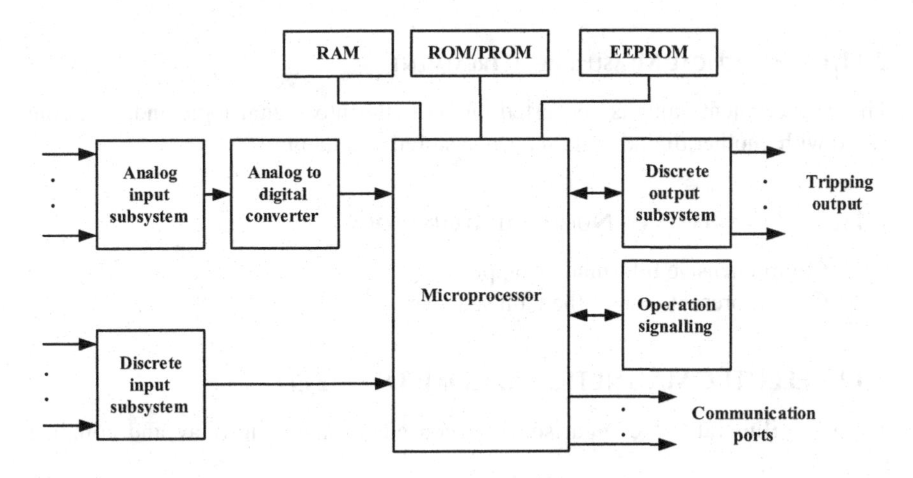

FIGURE 2.18 Digital relay architecture.

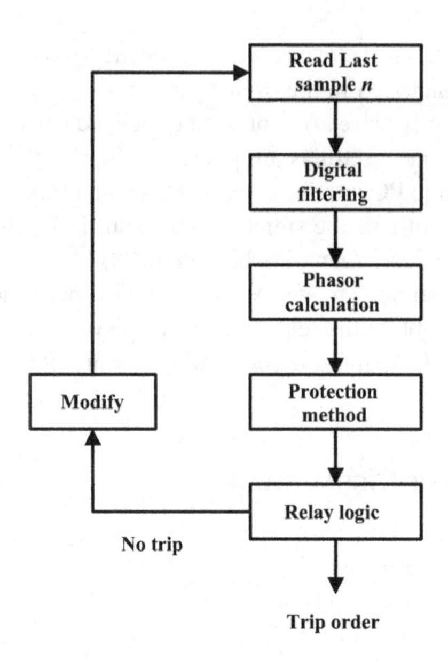

FIGURE 2.19 Digital relay algorithm.

2.11 NUMERICAL RELAYS

The difference between digital and numerical relays is based on points of sufficient technical detail and is seldom seen in areas other than defense. Because of advances in technology, they can be called normal developments of optical relays. They usually use a specialized digital signal processor (DSP) as the computational hardware and related software equipment.

2.11.1 Numerical Measurement Treatment

The measurement value is converted numerically into digital logic and then compared with another digital logic sequence stored in a memory.

2.11.2 Advantages of Numerical Technology

i. Comprehensive information supply.
ii. Clear representation of the fault sequence.

2.12 ELECTROMAGNETIC VS. COMPUTERIZED

Table 2.2 illustrates a comparison between electromagnetic relay and computer relay.

TABLE 2.2
Comparison between Electromagnetic Relay and Computer Relay

Characteristics	Electromagnetic Relay	Computer Relay	
		Digital Relay	Numerical Relay
Function	Mechanical operation	Static	Static
Mechanical vibration	High vibration	No vibration	No vibration
Burden effect	High	Low	Low
Time type	Clock	Count	Count
Reset time	High	Low	Low
Accuracy	Depend on temperature	Stable	Stable
Programming	No	Yes	Yes
Auxiliary supply	Necessary	Necessary	Necessary
SCADA connection	No	Yes	Yes
Response time	Slow	Fast	Very fast
Size	Large	Small	Compact
Operation reliability	High	High	High
Indication of the fault	By flag	By light	By light
Monitoring	No	Self	Self
Setting range	Limited	Wide	Wide
Internal resistance	Small	High	High
Maintenance	Frequent	Low	Very low
Events recording	No	Yes	Yes

PROBLEMS

2.1. What are the advantage and limitations of static relays? Sketch the different scheme connections for stator protection of a delta-connected generator.

2.2. What does the fault sequence of event and disturbance recording indicate?

2.3. Draw the signal path for voltage and current input signals to the microprocessor relays.

2.4. An RC circuit used to produce time delay for a solid-state relay is shown in Figure 2.20. For a step input voltage $V_i(t) = 10\,U(t)$, the output voltage $V_o(t) = 6$ V and $C = 12$ μF. Determine the time delay (t_{delay}) for the following cases:

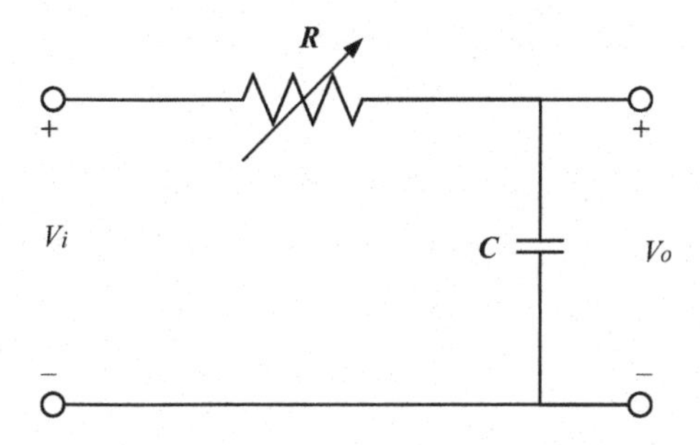

FIGURE 2.20 Circuit diagram of Problem 2.4.

i. $R = 100\ \text{k}\Omega$;
ii. $R = 1\ \text{M}\Omega$.
iii. Sketch the output $Vo(t)$ versus time for cases (i) and (ii).

2.5. An RC circuit used to produce time delay for a solid-state relay is shown in Figure 2.21. For a step input voltage $V_i(t) = 8\ U(t)$, the output voltage $V_o(t) = 4\,\text{V}$ and $C = 6\ \mu\text{F}$. Determine the time delay (t_{delay}) for the following cases:
i. $R = 1\ \text{k}\Omega$;
ii. $R = 10\ \text{k}\Omega$.
iii. Sketch the output $V_o(t)$ versus time for cases (i) and (ii).

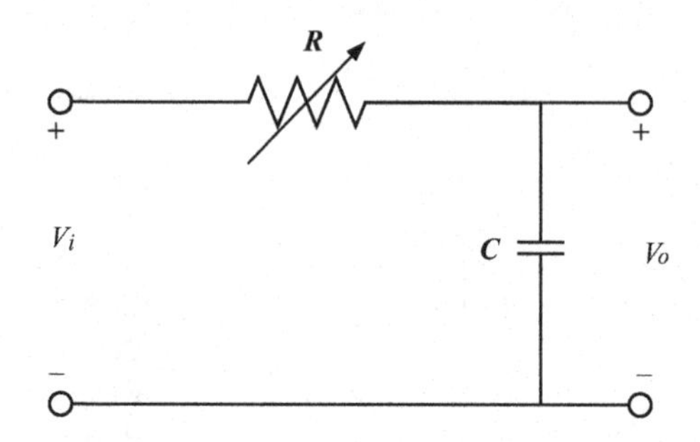

FIGURE 2.21 Circuit diagram of Problem 2.5.

2.6. Compare wattmeter type, induction type, and shaded pole relay.
2.7. What is the class of measuring relays?
2.8. What are the data required for the relay setting?

3 Protection Systems with SCADA Technology

3.1 INTRODUCTION

The term "SCADA" means supervisory control and data acquisition, which is a software that allows the supervision and control of industrial processes, including manufacturing, production, power generation, fabrication, and refining.

Through SCADA programs, a simple personal computer (PC) can become a workstation to watch and control processes to communicate with the field. Programmable logic controllers (PLCs) and remote telemetry units (RTUs), also known as intelligent devices, are essential components utilized in diverse industries for process automation, machinery control, and system monitoring. Allowing for the graphic representation of a plant, the supervision of its processes, modification of certain parameters, elaboration of reports, emission of alarms, and other functions.

The processes can be controlled remotely or locally to make changes to the parameters of the process on-site (local way) or adjustments in a control room (remote way).

3.2 BACKGROUND

SCADA is a control system architecture that uses computers, networked data transmission, and graphical user interfaces for high-level process supervisory management.

It utilizes additional auxiliary devices such as PLCs and discrete PID controllers to connect to equipment or processing plants.

Each of these levels will be covered in depth in this chapter, along with how they work, how SCADA has developed over the last 30 years, and how security needs and regulatory compliance affect SCADA system operation.

The SCADA system was created as an all-encompassing way to provide standard automation protocols access to various local control modules from various vendors. Large SCADA systems now work quite similarly to distributed control systems in reality. However, they use a variety of plant interfaces. They can manage large-scale operations that may span several locations and long distances.

Our vital infrastructure is under the supervision and control of SCADA systems. The United States has 15 essential infrastructures. Included in this list are the following: (i) energy; (ii) waste and water systems; (iii) telecommunications; (iv) transportation; (v) chemical; (vi) dams; (vii) emergency services; (viii) financial services; (ix) commercial facilities; (x) government facilities; (xi) critical manufacturing; (xii) defense; (xiii) food and agriculture; (xiv) healthcare and public health; and (xv) information technology (IT).

These vital infrastructures rely on one another. For instance, the energy sector would be significantly impacted by an assault on the telecom industry.

DOI: 10.1201/9781003394389-3

SCADA systems provide utilities in the electric power sector with useful information and capabilities to distribute electricity dependably and securely. A utility's most important and expensive distribution, transmission, and generating assets must be operated effectively, which depends on an efficient SCADA system.

Modern SCADA systems have incorporated new communication and network technologies to make the power grid smart and interactive. They are widely used to monitor and manage real-time electric power networks, especially those dealing with generation, transmission, and distribution. In the United States, electric power systems have been using SCADA systems for more than 50 years.

3.2.1 Benefits and Drawbacks

Implementing SCADA systems for electricity distribution has the following advantages:

i. Increases security and dependability through automation.
   Automation protects workers by allowing problem areas to be identified and addressed automatically.
ii. It also eliminates the need for manual data collection.
iii. Alarms and ongoing system monitoring enable operators to quickly identify and resolve issues.
iv. Operators can use powerful trending capabilities to detect future issues, improve routine equipment maintenance, and identify areas for improvement.
v. Information about the system that is current, accurate, and consistent.
vi. Quicker fault separation, system restoration, and reporting.
vii. Comprehensive reporting and statistical data are archiving.
viii. Centralized database and system parameter history.

The following are the main drawbacks of SCADA systems:

Utilizes proprietary programming and communications and is incompatible with goods from other suppliers.

i. Outdated software with well-known security flaws
ii. Any operator with access can completely shut the system off.

3.2.2 Applications

SCADA is utilized to produce petrochemicals, food, oil and gas, water, waste, and manufactured goods. In electrical power systems, it is commonly employed. The most important and expensive utility distribution, transmission, and generating assets must operate effectively; this is where SCADA solutions come into play. Power system components, including power plants, transmission facilities, and substations, are monitored and controlled remotely via SCADA systems. Power-producing facilities experience automation before transmission and distribution.

Network connectivity analysis, state estimation, load flow application, voltage VAR control, load shed application, fault management, system restoration, loss minimization through feeder reconfiguration, load balancing through feeder reconfiguration, operation monitoring, and distribution load forecasting are the main applications of SCADA in power systems. This text will carefully examine two of these applications.

The monitoring and management of multiple remote stations pose significant challenges in electric power distribution systems. In addition to maintaining the proper voltages, currents, and power factors, they provide capabilities like real-time visibility into the processes. They employ RTU to automatically monitor, safeguard, and regulate a variety of equipment in distribution systems.

Substation control: Substations are essential for sustaining the flow of electricity and managing load. The SCADA system continually assesses the condition of different substation components and then sends them the appropriate control signals. It carries out tasks including bus voltage regulation, bus load balancing, circulating current regulation, and overload regulation.

3.2.3 Challenges

Due to multiple vulnerabilities in the communications network, securing SCADA systems in a power utility context is difficult. To save money and advance technology, bigger SCADA systems are linked to other networks such as the Internet. Due to its interconnectedness, SCADA systems in the power grid are susceptible to various communication security challenges, including new types of threats and cyberattacks, including man-in-the-middle, denial-of-service, social engineering, and insider assaults. The attackers' objective is to undermine network security elements, including availability, authentication, confidentiality, or integrity.

The communication link between the SCADA server and RTU is the most probable target for an attack. Security devices for encryption and decryption of information exchange should be implemented at two locations to combat this hack.

Additionally, access permission is needed between the master systems and RTUs. In addition to password authorization, smart cards, firewalls, and intrusion detection systems are additional security measures. Vendors should include built-in security safeguards in their systems to address security issues.

SCADA system installation is costly and time-consuming in addition to providing security. Costs go up when SCADA software is updated and staff members training to utilize the new program. A system may have components from many vendors that communicate via various proprietary protocols.

3.3 SCADA SYSTEM AND ITS LEVELS

Looking at the overall structure of a SCADA system, there are four distinct levels within SCADA, these being (Figure 3.1):

i. Field instrumentation,
ii. PLCs and/or RTUs,
iii. Communications networks, and
iv. SCADA Host software.

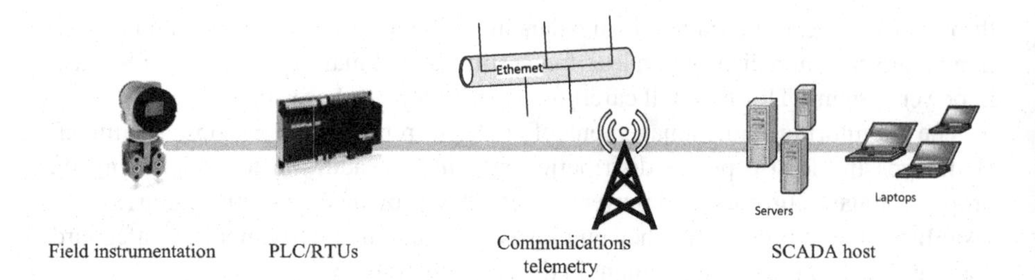

FIGURE 3.1 SCADA system levels.

3.4 BASIC FUNCTIONS OF THE SCADA SYSTEMS

3.4.1 Remote Supervision

This is one of the main functions of the SCADA system; the user can know the status of the different processes in the plant and make decisions depending on this information.

3.4.2 Remote Control of the Process

This function automatically enables or disables remote equipment (e.g., pumps, valves, compressors, transformers, and switches.) depending on the operator's decision. In the same way, it is possible to adjust parameters in the control loops and field devices.

3.4.3 Graphics Trends Presentation

It's referring the possibility of visualizing the process diagrams with information in real time.

3.4.4 Alarm Presentations

This function alerts the operator to abnormal conditions or events requiring intervention.

3.4.5 Storage of Historical Information

It also allows registration and stores operational information and alarms.

Other functions do not associate directly with the operations in a control room, but equally important, they are diagnostic of hardware and software, automated systems of preventive maintenance and corrective, integration with other operational systems, calibration of devices, and others.

3.4.5.1 Field Instrumentation

According to the old saying, "You can't manage what you can't measure," instrumentation is essential to a secure and effective control system. Historically, an operator would start/stop local pumps and manually open/close valves. Pumps and their related operating settings would also have been manually regulated. These instruments would have gradually, over time, been supplied with feedback sensors, such as limit switches, allowing communication for these wired devices into a local PLC or RTU, delivering data relay to the SCADA Host program.

The capacity to design, install, and maintain equipment today demands greater technical knowledge than in the past, but this is offset by the lower cost of automating operations and workers' better technical skill levels. Most field equipment, including valves, is now equipped with actuators so that a PLC or RTU can control the device instead of manually operating it. With this feature, the control system can respond more rapidly to unexpected occurrences and shut down or optimize production.

To guarantee that an electrical device has no unfavorable impacts on its surroundings or other electrical devices, instrumentation must also adhere to any EMC (electromagnetic compatibility) regulations that may be in place.

3.4.5.2 PLCs and RTUs

Formerly two very separate devices, PLCs and RTUs, are now quite similar. Technology has converged in this area as producers of these gadgets increased their capacities to satisfy consumer demand.

An RTU was a "dumb" telemetry unit for connecting field devices 30 years ago. The RTU had well-developed communication interfaces or telemetry but would only "relay" the data from the instruments to the SCADA Host without any processing or control. Equipment for remote telemetry (RTUs) are portable computers that are placed in strategic areas across the world. RTUs are neighborhood hubs for receiving sensor data and transmitting control instructions. The RTU's control programming was introduced in the 1990s, making it more PLC-like. On the other hand, PLCs have always been able to execute control programs, but they lack communication interfaces and data logging functionality, which have been somewhat added during the previous ten years.

Offering a specific application that could include several instruments and devices with an RTU/PLC and incorporate technology sets to provide an "off-the-shelf" approach to common process requirements, such as gas well production that includes elements of monitoring, flow measurement, and control that would extend as an asset into the SCADA Host, is a further development of devices in the field.

PLCs and RTUs must meet the same standards as instruments in terms of environmental and regulatory compliance since they work in the same setting. PLCs, however, have not always been as ecologically friendly as RTUs. This is mostly because PLCs were created to function in previously somewhat controlled environments, such as factory floors.

3.4.5.3 Remote Communications Networks

The SCADA Host, situated in the field office or central control center, receives data from distant RTU/PLCs out in the field or along the pipeline via the remote communication network. Communication is the SCADA system's "glue" or "linking component," and it is crucial to the system's functioning since assets are dispersed over a wide geographic region. A SCADA system's ability to effectively handle communication with distant assets is essential to its success.

In the past 10–15 years, many customers have converted to radio or satellite communications to save expenses and eliminate troublesome cabling difficulties. Twenty years ago, the communication network would have been leased lines or dial-up modems, which were highly costly to construct and operate. Cellular communications and upgraded radio equipment that offer faster transmission rates and better diagnostics are among the newer communication kinds that have become accessible. For contemporary, dispersed SCADA systems, the fact that these communication mediums are still prone to malfunction is a significant problem.

The protocols changed at the same time as the communication medium. PLCs and RTUs exchange data via protocols, which are electronic languages, either with other PLCs and RTUs or SCADA Host systems. Protocols have often been exclusive and the work of a particular company. As a further step, several manufacturers tended to stick with a single standard, MODBUS, but added proprietary components to satisfy certain functionality needs.

In recent years, true non-proprietary protocols such as DNP have emerged (distributed network protocol). These protocols were developed independently of any one manufacturer and have become more of an industry standard thanks to the participation of several people and businesses in their creation. These procedures haven't advanced sufficiently yet, to be widely applicable to the application procedure and regulatory requirements (Figure 3.2).

3.4.5.4 SCADA Host Software

Historically, the method for viewing graphical displays, alerts, and trends has been SCADA Host software. Only when control components for distant instruments were established was control from the SCADA Host itself possible. These systems belonged to operators, technicians, and engineers and were shut off from the outside world. Monitoring, maintaining, and engineering processes and SCADA components were within their purview. This is no longer the situation because of improvements in IT. The data that the SCADA Host software creates now has to be accessible in real time to a wide range of different stakeholders. With the help of data obtained from the SCADA system, accounting, maintenance management, and material purchase needs are performed, at least in part. There is a push for the SCADA Host to be an enterprise entity that provides data to various users and processes. This has prompted the adoption of standards and interfaces to support methods by SCADA

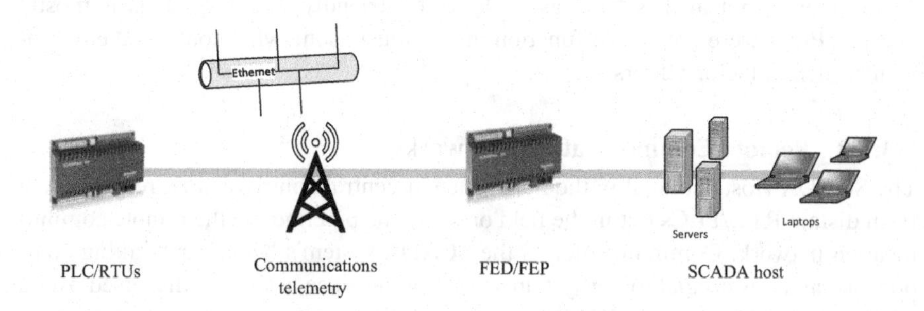

FIGURE 3.2 RTU/PLCs system connection.

Host software development. Additionally, it implies that IT, which has historically been kept apart from SCADA systems, is now actively engaged in maintaining networks, database interfaces, and user access to data.

Many of the first SCADA Host products weren't equipped with the telemetry communication characteristics needed by SCADA systems for geographically dispersed assets since they were created particularly for the industrial setting, where a SCADA system was housed in a single building or complex (Figure 3.3).

These first-generation SCADA Hosts sometimes require a hybrid PLC or RTU, often a front-end driver (FED) or front-end processor, to manage connections with remote devices (FEP). This had several problems since it cluttered communications and the need for specialist programming that couldn't be done on the SCADA Host platform. Several FED or FEP devices were utilized to alleviate part of this, but due to their specialized nature, their development and upkeep came with extra costs. Since contemporary SCADA software already provides telemetry features, these hybrid PLCs are no longer required for communications. They now use software programs called "drivers" integrated into the SCADA Host. Software drivers comprise the many protocols to connect to remote devices such as RTUs and PLCs.

As technology developed, software platforms for SCADA Host benefited from a wide range of new functionalities. These included developing integrated databases that could handle hundreds of changes per second for extremely large systems and were specifically designed with SCADA Host software requirements in mind.

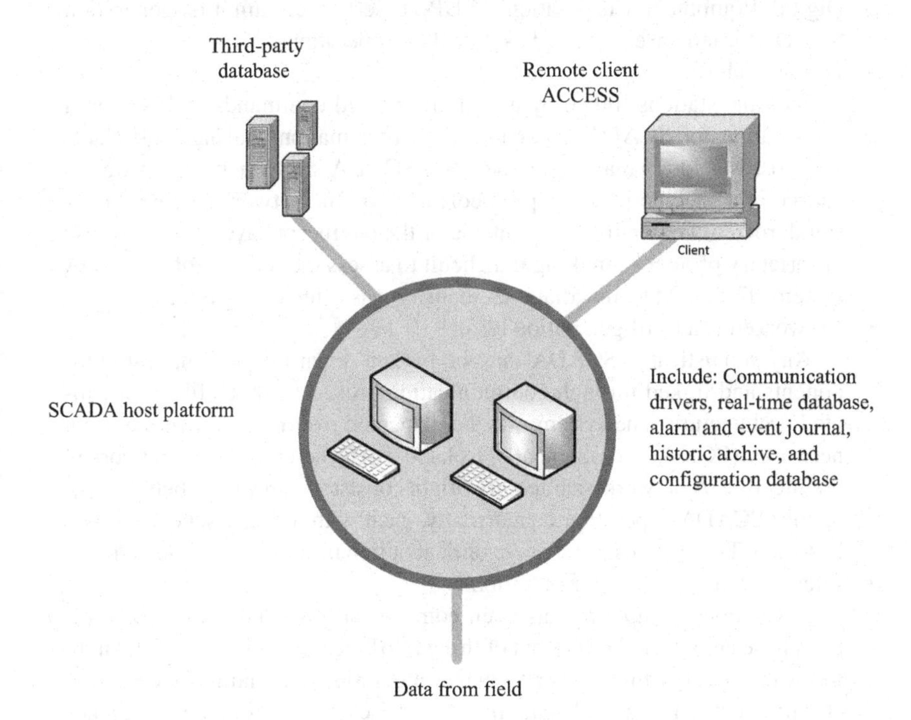

FIGURE 3.3 SCADA Host platform.

The SCADA Host program cannot offer data access to external databases without these criteria. Remote client access to the SCADA Host is another technology that has allowed users to operate and keep an eye on SCADA systems, whether on the go or in other locations.

The SCADA Host systems are under pressure to be operationally safe from the oil and gas industry. The 49 CFR 195.446 Control Room Administration guidelines analyze the SCADA Host software's operations, maintenance, and management. The degree of integration, use of open standards, and architecture of the SCADA system are also covered.

3.5 SCADA ARCHITECTURE DEVELOPMENT

The four generations of SCADA systems are as follows.

- The first generation is "monolithic."

  Large minicomputers were used in the early stages of SCADA system computing. When SCADA was created, there were no common network services. SCADA systems have no interface to other systems, making them stand-alone systems. At that time, the communication protocols were all proprietary. To establish redundancy for the first-generation SCADA system, a backup mainframe system linked to each RTU site was deployed if the primary mainframe system failed. On minicomputers such as the Digital Equipment Corporation's PDP-11 series, certain first-generation SCADA systems were created as "turnkey" operations.
- Distributed: second generation

  Multiple stations linked by a LAN processed commands and delivered information for SCADA. Near real-time information sharing took place. The cost was lower than first-generation SCADA since each station was in charge of a specific job. The protocols used on the networks were still not standardized. Few individuals outside of the engineers have access to these proprietary protocols, making it difficult to assess the security of a SCADA system. The SCADA installation's security was often disregarded.
- Networked is a third-generation term.

  Any sophisticated SCADA may be broken down into its smallest components and linked through communication protocols, much like a distributed architecture. The system may be dispersed over many process control networks (PCNs), geographically isolated LAN networks in a networked architecture. A network architecture might consist of many distributed architecture SCADAs operating concurrently, each with a single supervisor and historian. This enables a more economical option in very large-scale systems.
- Internet of things, fourth generation

  Since cloud computing has been commercially available, SCADA systems have embraced the Internet of things (IoT) technologies more and more to lower infrastructure costs and make integration and maintenance easier. Because of the horizontal scale provided by cloud environments, SCADA systems may now execute more complicated control algorithms than are realistically possible to perform on conventional PLCs. Furthermore, compared to the heterogeneous mix of proprietary network protocols typical of many decentralized SCADA implementations, the use of open network protocols such TLS (transport layer security), which is inherent in the IoT technology, provides a security boundary that is easier to understand and manage. One such use of this technology is a creative method for collecting rainwater via the use of real-time controls (RTC)

This data decentralization necessitates a different SCADA strategy than conventional PLC-based systems. The optimal approach for using a SCADA system locally is linking the graphics on the user interface to the information kept in certain PLC memory locations. The traditional 1-to-1 mapping, however, becomes troublesome when the data originates from a diverse mixture of sensors, controllers, and databases (which may be local or at several linked sites). Data modeling, an idea originating from object-oriented programming, provides a remedy for this.

The SCADA program creates a virtual representation of each device in a data model. These virtual representations (also known as "models") may include additional information (web-based data, database records, media files, etc.) that may be required by other SCADA/IoT implementation components in addition to the address mapping of the device they represent. As traditional SCADA becomes more "house-bound" due to the IoTs' increased complexity, and as communication protocols change to favor platform-independent, service-oriented architecture (such as OPC UA) [16], more SCADA software developers will probably incorporate data modeling.

3.6 SECURITY

In recent years, security for SCADA systems has become a significant and contentious issue. In the past, SCADA systems were solitary objects that belonged to operators, engineers, and technicians. As a result, platforms for SCADA Hosts were not always designed with secure connectivity to public networks. As a result, many SCADA Host systems were vulnerable to assault since they could not defend themselves.

Security has been a problem with distant assets communicating back to a SCADA Host for many years due to various recorded assaults on SCADA systems. However, an open standard has only recently made possible a safe encrypted and authenticated data transmission between distant assets and a SCADA Host platform.

Security solutions for SCADA Host and remote asset communications have very distinct needs. Security must be regarded holistically and in terms of the SCADA system. One wouldn't need to use the SCADA system to interrupt production, for instance, if one decided to do so. A trespasser might readily threaten a gas well head site or a monitoring station on a gas pipeline if they are located in a rural area. Other options that may or might not be included in the SCADA system would have to be considered if the asset is very significant, such as security camera monitoring.

The majority of unauthorized accesses to a SCADA system occur via the SCADA Host or PCs that connect to the system for maintenance or diagnostics, not from or at the distant assets themselves. For example, the recent Stuxnet virus assault was carried out by inserting a thumb drive into a computer that was used to access a SCADA system.

Many standards exist that outline how to protect a SCADA system, not only in terms of the technology used but also in practices and processes. This is crucial since the security solution for SCADA is a combination of behaviors and processes with technology solutions rather than a single technological panacea. These practices and procedures would include training materials, access to SCADA Hosts, and steps to take if SCADA security has been hacked. IT departments should be involved in building up processes and procedures and adopting technologies in contemporary SCADA systems since they are essential to developing and maintaining SCADA security for an organization.

3.7 FUTURE IMPLEMENTATIONS

A distributed control system, or DCS, performs real-time process control, while a SCADA system often refers to a system that coordinates but does not. The real-time control processes in SCADA systems incorporating DCS components.

The fourth generation of SCADA is entirely based on the IoT, which greatly lowers infrastructure costs and improves ease of maintenance and integration. IoT is the next big thing in the world. Consequently, SCADA systems can now provide real-time status reports, making them as good as their successor DCS.

SCADA systems have taken advantage of various technological advancements to advance their competence, from the introduction of actuators and transducers (which made monitoring of processes easier, more accurate, and less expensive) at the instrumentation level to the introduction of open standards (to improve the interchange of data between a SCADA system and other processes within an organization).

To provide operational benefits from the SCADA Host down to the instrumentation, not only in terms of controlling and retrieving data but also in terms of engineering, implementing, operating, and maintaining these assets, modern SCADA systems are driven to

- Provide instrumentation and RTUs/PLCs for the asset or process solutions that can be easily managed.
- Create and implement open standards utilizing best practices established by open organizations rather than a single manufacturing body to facilitate the integration of assets inside a SCADA system. As a result, the cost of owning SCADA will go down.
- Create safe environments for SCADA systems, their assets, and processes by implementing several practices and procedures in addition to technological solutions.

3.8 HARDWARE DEVICES

This section describes how to use the accompanying SCADA software and set a protective relay's address.

A setup contains several relays and must be assigned different addresses to ensure correct communication with the software.

Using Lucas-Nülle SCADA software for

- Starting the application.
- User interface.
- Settings/parameter changes.
- Generating trend graphs.
- Measuring signal length.

Table 3.1 lists all the devices used in the power system/SCADA networks. The essential elements of protection technology are described briefly in Table 3.2.

Before using the SCADA system for the first time, you must make a one-time hardware connection for mutual communication. In some cases, the various devices are connected via different interfaces to the PC. Refer to the subsequent overview to find out how individual devices should be linked to the PC. The configuration for each type of connection is described in the corresponding discussion of this section. Table 3.3 shows a multi-function relay. Table 3.4 shows an analog/digital multimeter.

The currents and voltages used with this instrument are to be transformed by a factor of 1:1000. In other words, 1 V in the model corresponds to 1 kV in real energy

TABLE 3.1
Lucas-Nülle GmbH Power System/SCADA Networks Devices

Device	Designation
	Adjustable three-phase power supply (0–400 V/2 A, 72 p.u.)
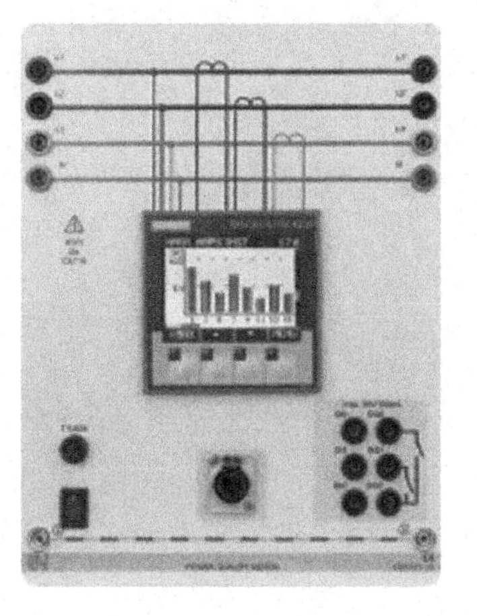	Three-phase power quality meters with display and long-term memory

(Continued)

TABLE 3.1 (*Continued*)
Lucas-Nülle GmbH Power System/SCADA Networks Devices

Device	Designation
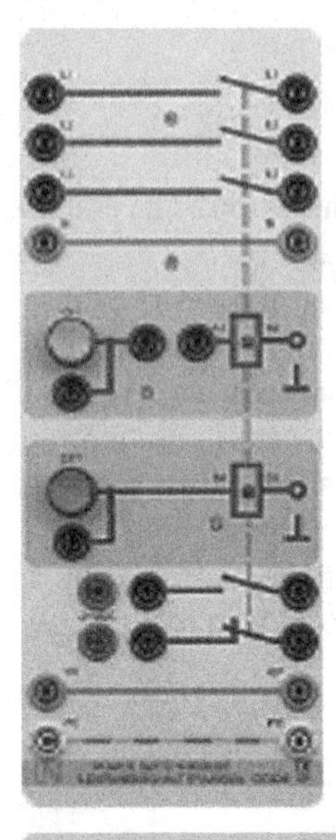	Circuit breaker (power-switch module)
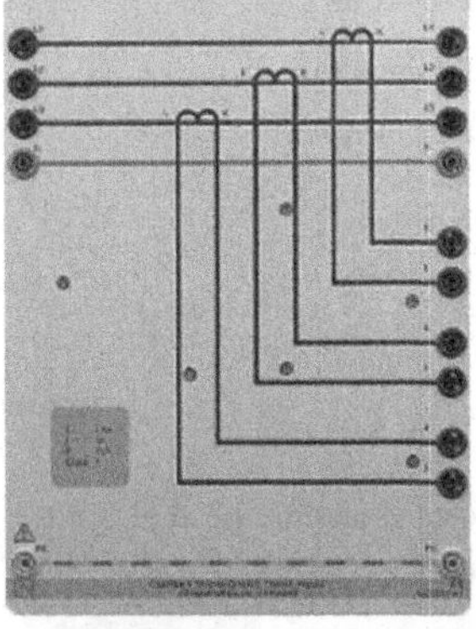	Current transformer
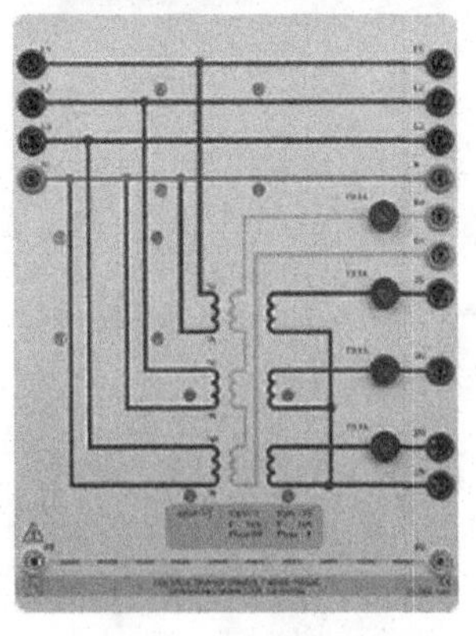	Voltage transformer

(*Continued*)

TABLE 3.1 (*Continued*)
Lucas-Nülle GmbH Power System/SCADA Networks Devices

Device	Designation
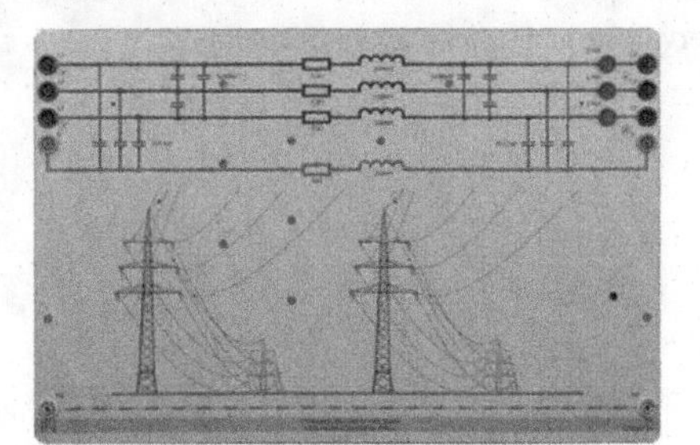	Transmission line model
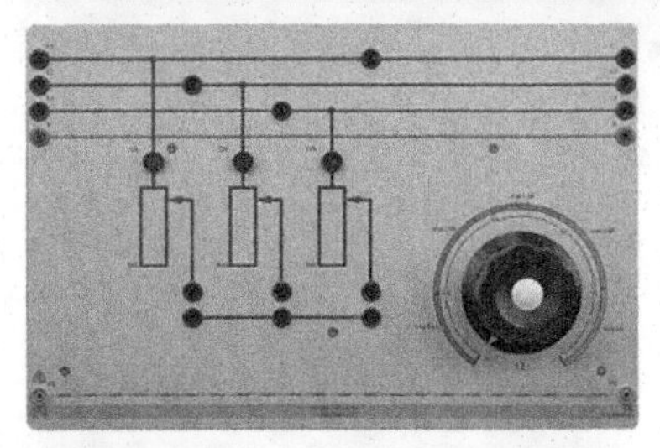	Resistive load
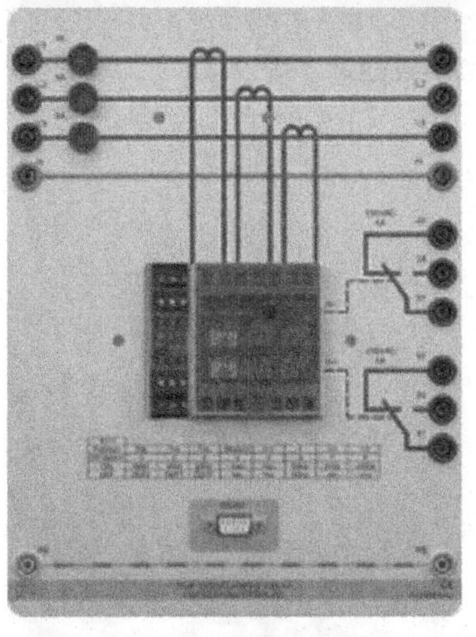	Overcurrent time protection relay
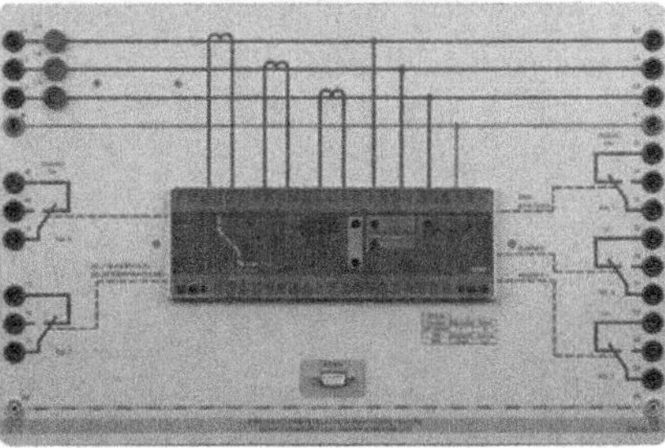	Time overcurrent relay with directional feature

(Continued)

TABLE 3.1 (*Continued*)
Lucas-Nülle GmbH Power System/SCADA Networks Devices

Device	Designation
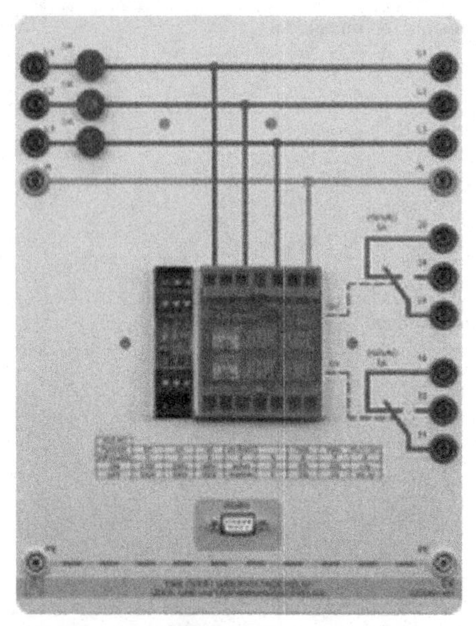	Overvoltage and undervoltage time relays
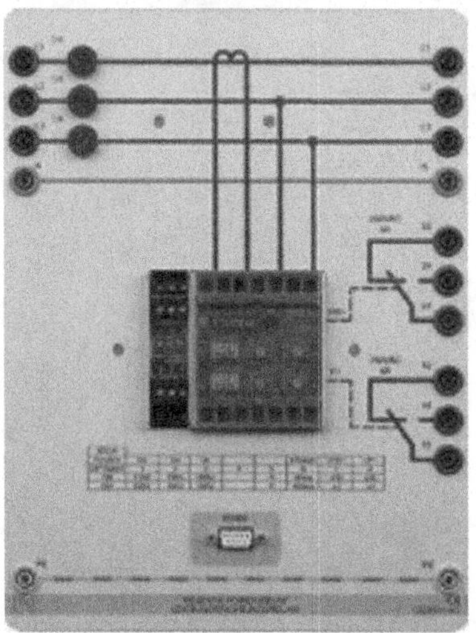	Directional power relay

(*Continued*)

TABLE 3.1 (*Continued*)
Lucas-Nülle GmbH Power System/SCADA Networks Devices

Device	Designation
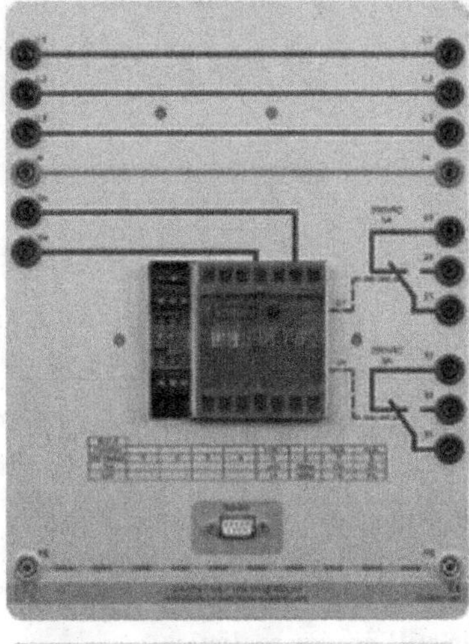	Earth-fault protection relay
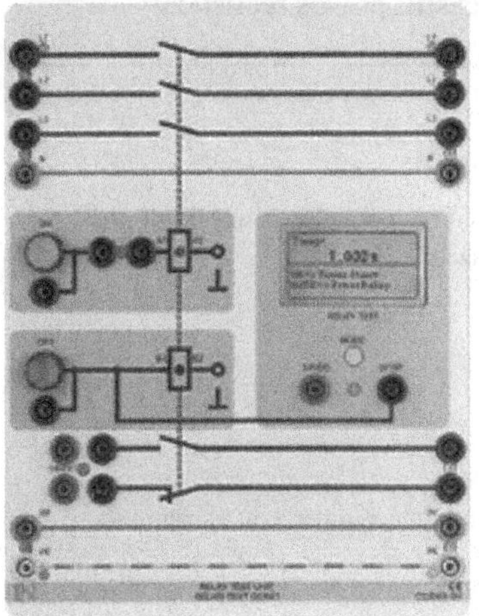	Relay test unit
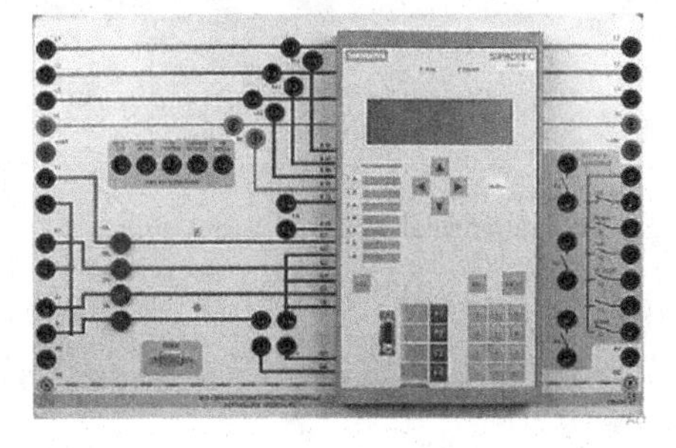	Distance protection relay

(*Continued*)

TABLE 3.1 (*Continued*)
Lucas-Nülle GmbH Power System/SCADA Networks Devices

Device	Designation
	Multimeter

transmission systems, 1 A, therefore, being equivalent to 1 kA. Regarding power, 1 W (or 1 VA) represents 1 MW (or 1 MVA).

The single lines employed here reproduce a 300-km (1864 miles) long, 380-kV overhead transmission line with the constants R', X', and C'. Due to the line-to-line voltage of 110 V or 220 V used in these experiments (corresponding to 110/220 kV on real lines), an overhead line with the same constants still proves realistic.

TABLE 3.2
Lucas-Nülle GmbH Protection Technology Elements

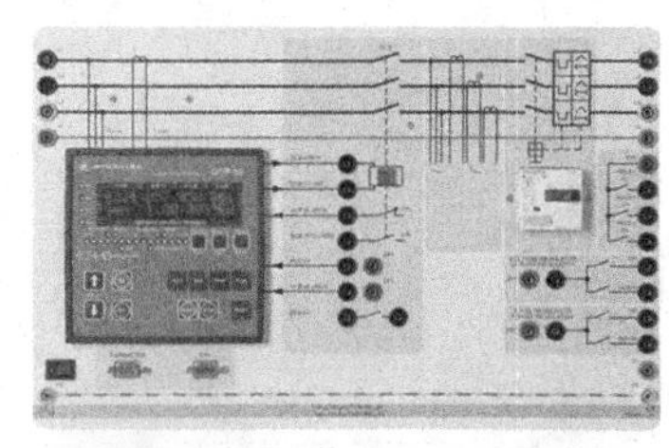	Multi-function relay, power controller, power factor controller, synchronization unit
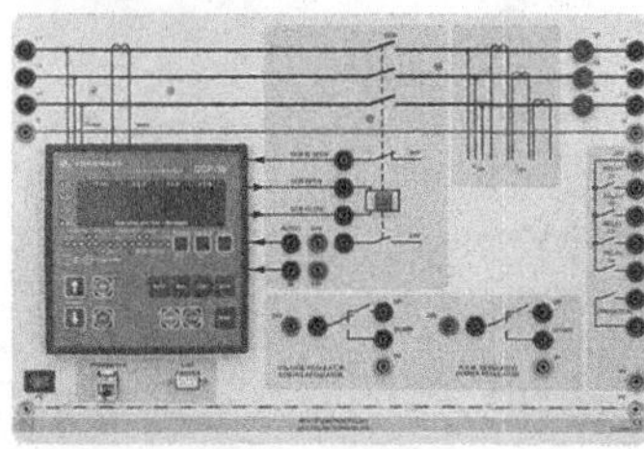	Multi-function relay, power controller, cos(φ) controller, synchronization unit
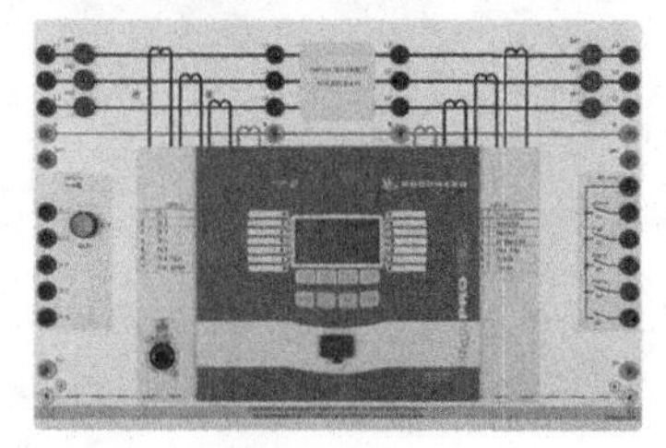	Transformer/generator differential protection relay

(*Continued*)

TABLE 3.2 (*Continued*)
Lucas-Nülle GmbH Protection Technology Elements

Generator HMI

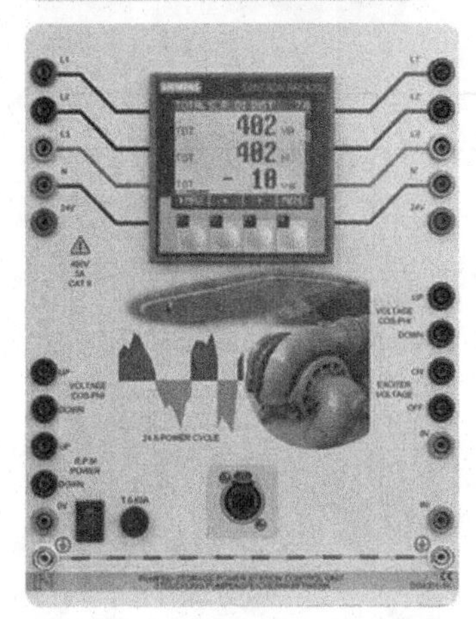

Pumped-storage power station control unit

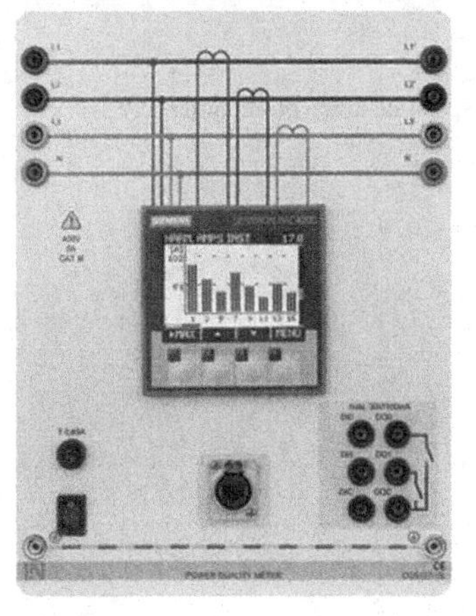

Power quality meter with graphic display and long-term storage

(*Continued*)

TABLE 3.2 (*Continued*)
Lucas-Nülle GmbH Protection Technology Elements

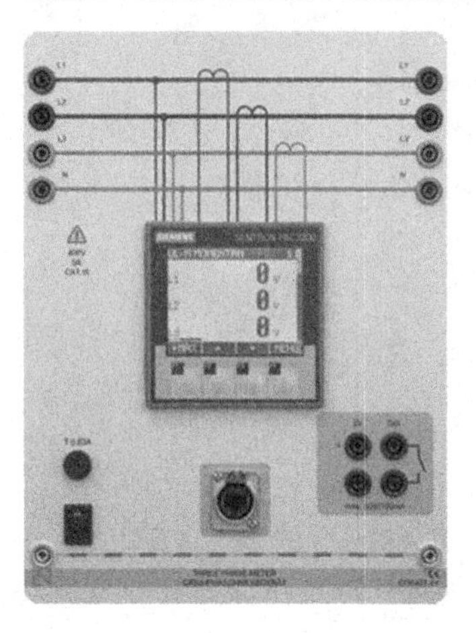	Three-phase meter

TABLE 3.3
Lucas-Nülle GmbH Multi-Function Relay

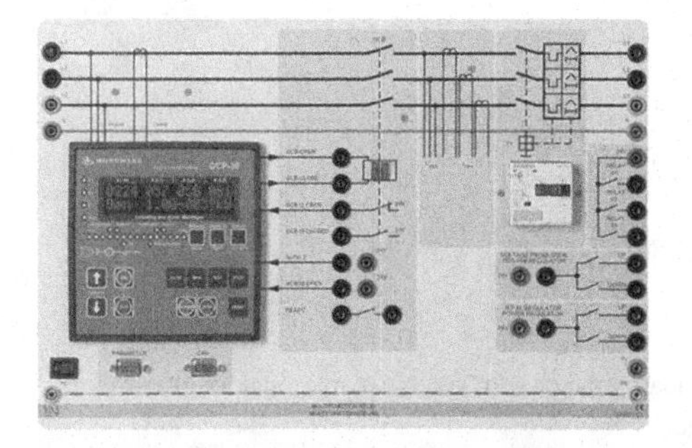	Multi-function relay, power controller, cos(φ) controller, synchronization unit
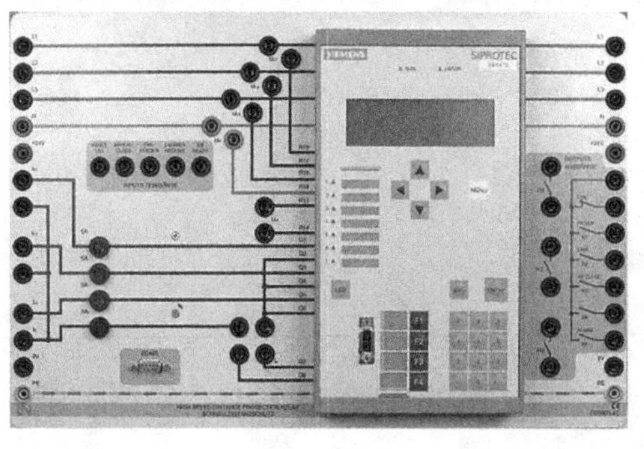	High-speed distance protection relay

(*Continued*)

TABLE 3.3 (*Continued*)
Lucas-Nülle GmbH Multi-Function Relay

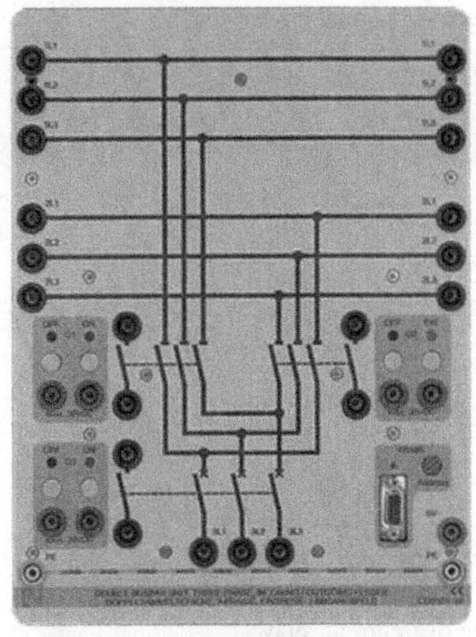	Double busbar, three-phase, incoming/outgoing feeder
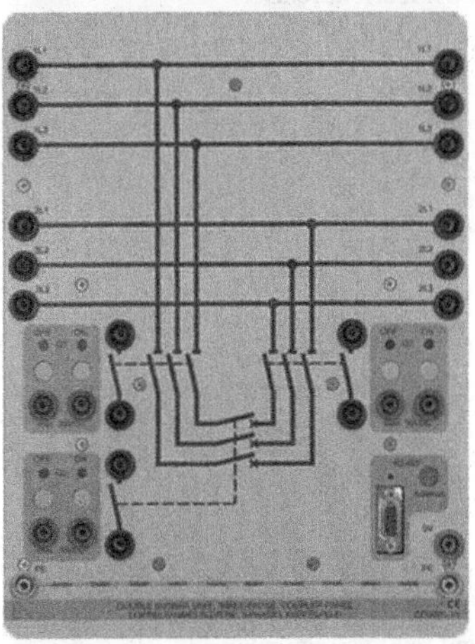	Double busbar, three-phase, coupler panel
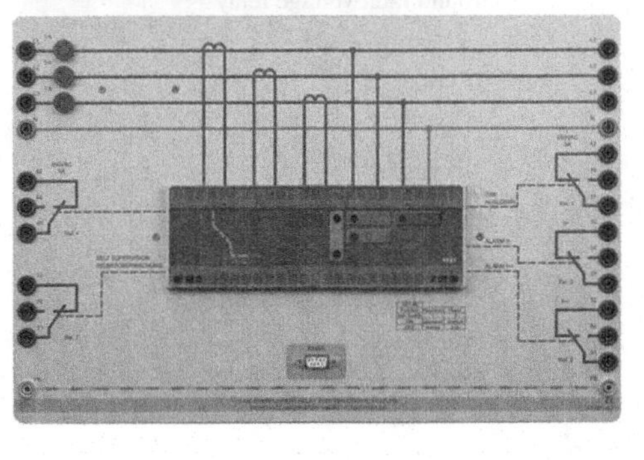	Directional time overcurrent relay

(*Continued*)

TABLE 3.3 (*Continued*)
Lucas-Nülle GmbH Multi-Function Relay

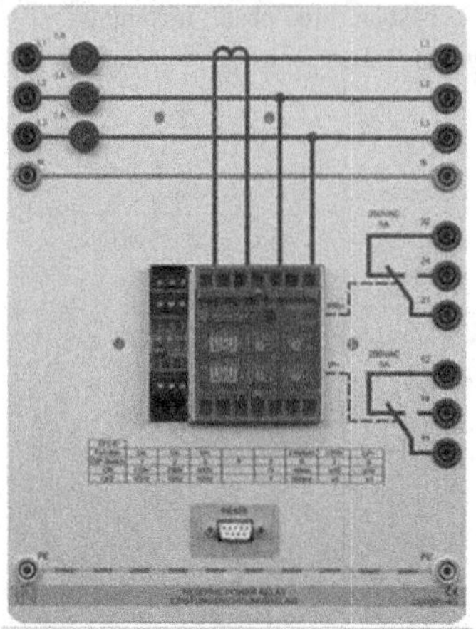	Power/directional power relay
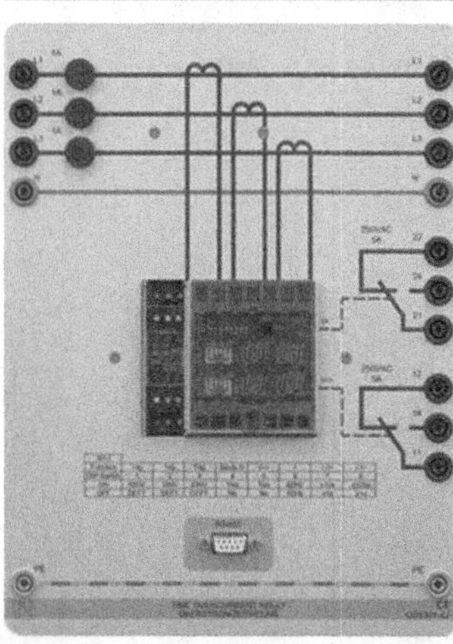	Time overcurrent relay
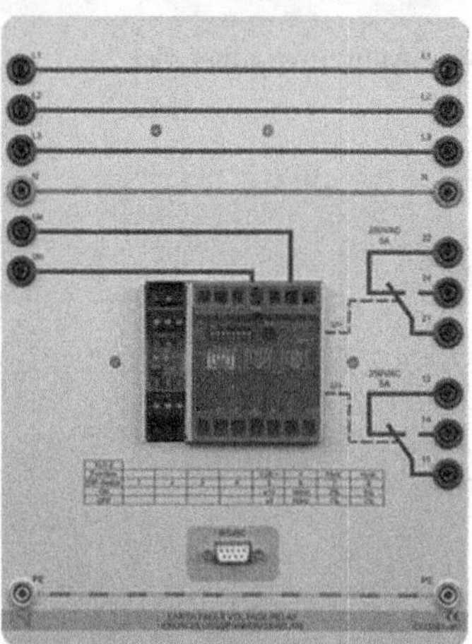	Ground fault voltage relay

(*Continued*)

TABLE 3.3 (*Continued*)
Lucas-Nülle GmbH Multi-Function Relay

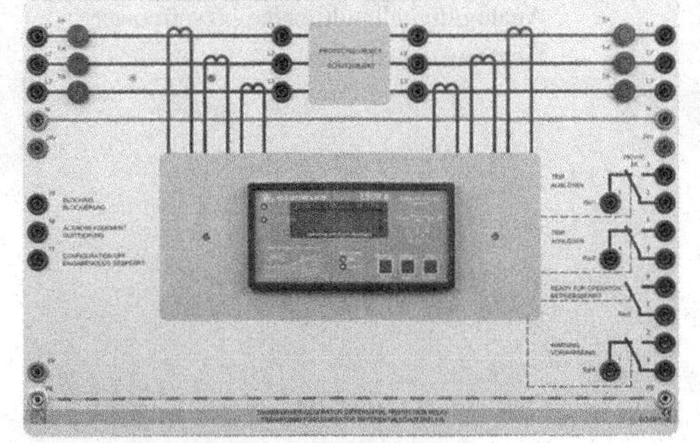	Time overvoltage/undervoltage relay
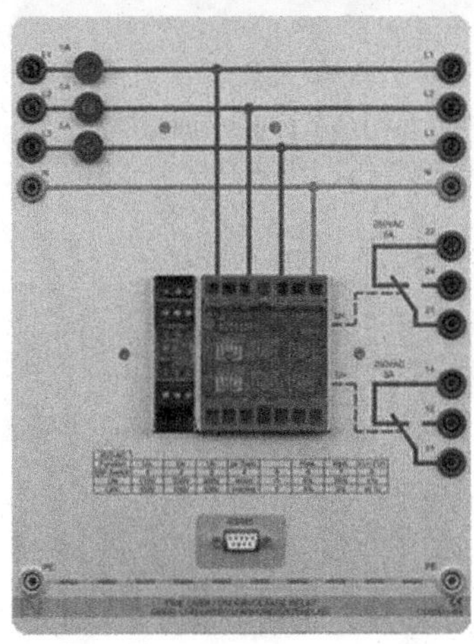	Transformer/generator differential protection relay

TABLE 3.4
Lucas-Nülle GmbH Analog/Digital Multimeter

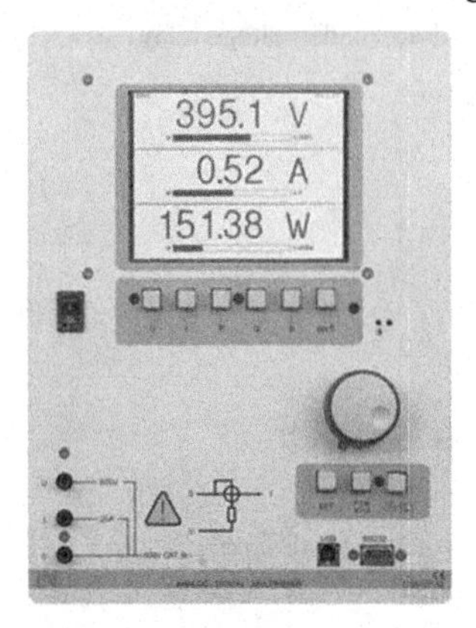	Analog/digital multimeter, power/power factor meter, software
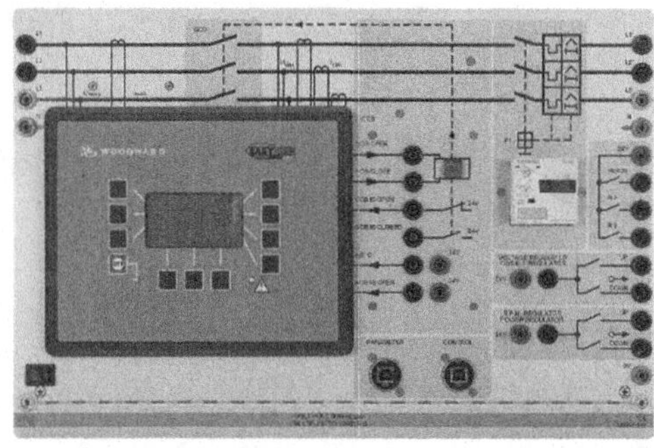	Multi-function relay, power controller, cos(φ) controller, synchronization unit

4 Faults Analysis

4.1 INTRODUCTION

Short circuits occur in the power system when equipment insulation fails due to system overvoltage caused by lightning or switching surges, insulation contamination (salt spray or pollution), or other mechanical causes. The resulting short circuit or "FAULT" current is determined by the synchronous machines' internal voltages and the system impedances between the machine voltage and the fault.

A large volume of network data must also be controlled and accurately handled. To assist the engineer in this, power-system planning, digital computers, and highly developed computer programs are used. Such programs include short circuits and transient programs.

The current chapter deals with fault analysis, introduction to the faults in power systems, transient phenomena, and three-phase short-circuit, unloaded synchronous machine. Short-circuit theory consists of balanced and unbalanced fault calculations in general and conventional methods for small systems. These fault types involve (single line-to-ground faults, line-to-line faults, and double line-to-ground faults). The last three unsymmetrical fault studies will require the knowledge and use of tools of symmetrical components.

4.2 FAULT CONCEPT

The great technical advances in the design and production of commercial and scientific general-purpose digital computers since the early 1950s have placed a powerful tool at the engineering profession's disposal. This advancement has made economically feasible digital computers' utilization for routine calculations encountered in everyday engineering work.

It has also provided the capability for performing more advanced engineering and scientific computations that were previously impossible because of their complex or time-consuming nature. These trends have increased the interest in digital computers immensely and have necessitated a better understanding of the engineering and mathematical bases for problem-solving. The development of computer technology has provided the following advantages to power system engineering:

- More efficient and economical means of performing routine engineering calculations are required to plan, design, and operate a power system.
- Better utilization of engineering talent by relieving the engineer from tedious hand calculations.
- The ability to perform more effective engineering studies.
- The capability of performing studies was impossible because of the volume of calculations involved.

DOI: 10.1201/9781003394389-4

Applying a computer to the solution of engineering problems involves several distinct steps (problem definition, mathematical formulation, selection of a solution technique, program design, programming, and program verification). The relative importance of each of these steps varies from problem to problem.

As electric utilities have grown and the number of interconnections has increased, planning for future expansion has become increasingly complex. The increasing cost of additions and modifications has made it imperative that utilities consider a range of design options and perform detailed studies of the effects on the system of each option based on several assumptions: normal and abnormal operating conditions, peak and off-peak loading, and present and future years of operation.

An essential part of a power supply network's design is calculating the currents, which flow in the components when faults of various types occur. In a fault survey, the fault is applied at various points in the network, and the resulting currents are obtained by digital computation. The magnitudes of the fault currents give the engineer the current settings for the protection to be used and the circuit breakers' ratings. In some circumstances, the effect of open circuits may need investigation.

When calculating short-circuit currents in high-voltage installations, it is often sufficient to work with reactances because they are generally much greater in magnitude than the effective resistances. The ratios of the rated system voltages are taken as the transformer ratios. Instead of the operating voltages of the faulty network, one works with the rated system voltage. It is assumed that the various network components' rated voltages are the same as the rated system voltage at their respective locations. The calculation is done with the aid of the %MVA system. The ohmic resistances of low-voltage cables are usually higher than their reactances.

With widely different ratios of the real and imaginary parts, these impedances are almost always present in a series arrangement. Complex calculation with impedance $Z = R + jX$ of the equipment is necessary; therefore, a complete short-circuit is always assumed when calculating the short-circuit currents. Other influences, particularly arc resistances, contact resistances, conductor temperatures, the inductances of current transformers, and the like, can reduce short-circuit currents. Since they are not available for calculation, they are allowed by a factor (C). The installation's apparatus and components must be designed for maximum dynamic and thermal short-circuit stress. On the other hand, short-circuit protection devices must respond to a lower short-circuit current. To reconcile these requirements, the term "*maximum short-circuit current*" and "*minimum short-circuit current*" have been introduced.

The maximum short-circuit currents with three-, two-, or single-phase faults are obtained with the short-circuit path's impedances at a conductor temperature of (20°C) and the factor ($C = 1.0$).

An electrical network under short-circuit conditions can be considered a network supplied by several sources (generators) with a single load connected to the node subjected to the short circuit. There are several reasons to have as accurate data as possible about short-circuiting currents and voltages in a system:

Each circuit breaker's interrupting capacity in every switching locality must be based upon the most severe short-circuit case.

The protective relaying system, intended to sense the fault and initiate selective switching, bases its operation upon the fault current's magnitude and directions.

The magnitude of bus voltages during short circuits determines the transient generator power outputs and the "transient stability."

Fault studies form an important part of power-system analysis. The problem consists of determining bus voltages and line currents during various faults. The three-phase balanced fault information is used to select and set phase relays, while the line-to-ground fault is used for ground relays. Fault studies are also used to obtain the rating of the protective switchgear.

For fault studies, generator behavior can be divided into three periods:

1. The subtransient period lasts only for the first few cycles.
2. The transient period covers a relatively long time.
3. The steady-state period.

Before about 1950, matrices were used only as research tools. They systematized the arrangement of materials and generally forced the research worker to be organized. Matrices reduced the computational effort; however, the absence of high-speed computers limited investigations to small sets of equations involving only very small matrices. The first generation of small-scale computers extended the use of matrices in solving network problems of limited size.

4.3 TYPES OF FAULTS

Faults in the three-phase system may be classified under the following headings:

i. Symmetrical three-phase faults.
ii. Single line-to-ground fault.
iii. Line-to-line faults.
iv. Double line-to-ground faults.

Note that three fault types involve line-to-ground (Earth fault). Most of these result from insulator flashovers for weather conditions and insulation quality. The balanced three-phase fault is the rarest in occurrence, accounting for above 5% of the total faults, and it is the least complex of all types of short-circuit studies as far as the calculations are concerned.

The types of faults commonly occurring in practice are illustrated in Figure 4.1, and the most common of these is the short circuit of a single conductor to earth. Often the path to earth contains resistance in the form of an arc, as shown in Figure 4.1f.

Although the single line-to-ground fault is the most common, calculations are frequently performed with the three-line, balanced short circuit (Figure 4.1d and e). This is the most severe fault and the most interesting calculation.

As well as fault current, fault MVA is frequently considered; this is obtained from the expression $\left\{\sqrt{3}\, V_L\, I_F \times 10^{-6}\right\}$ where V_L is the nominal line voltage of the faulted

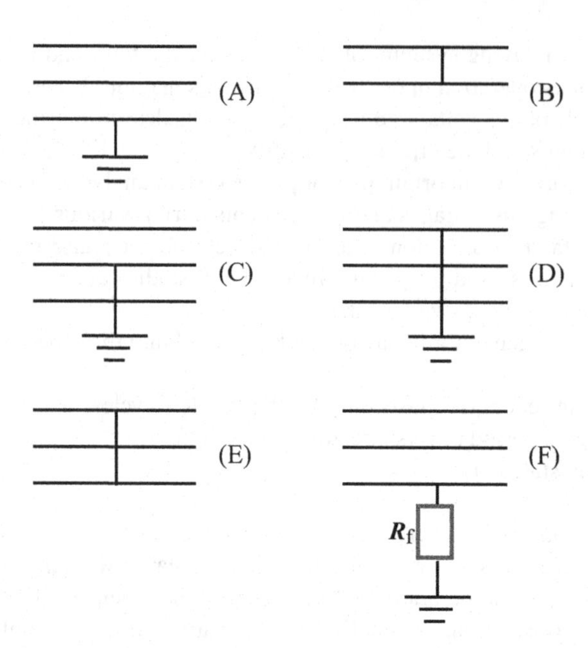

FIGURE 4.1 Common types of faults. (a) Single line-to-ground fault. (b) Line-to-line faults. (c) Double line-to-ground faults. (d) Three phase to ground fault. (e) Three phase fault. (f) Fault through resistor.

part before the fault. The MVA is often referred to as the *fault level.* The calculation of fault currents can be divided into the following two main types:

1. Faults short-circuiting all three phases when the network remains balanced electrically. Normal single-phase equivalent circuits may be used as in ordinary load-flow calculations for these calculations.
2. Faults other than three-phase short circuits when the network is electrically unbalanced.

The main objects of fault analysis may be enumerated as follows:

i. To determine the maximum and minimum three-phase short-circuit currents.
ii. To determine the unsymmetrical fault current for single and double line-to-ground, line-to-line, and open-circuit faults.
iii. To investigate the operation of protective relays.
iv. To determine rated rupturing capacity of breakers.
v. To determine fault-current distribution and busbar–voltage levels during faults.

4.4 SYMMETRICAL FAULT ANALYSIS

To make the correct choice in switchgear, it is necessary to perform short-circuit calculations. Such calculation enables the fault MVA due to a symmetrical three-phase fault to be determined at a point of interest. These calculations demand the reduction of the network to that of a single source feeding a single impedance.

The generator reactances are normally taken as their subtransient values to consider the worst conditions. The calculation may be performed by expressing impedance in ohms or per-unit values. When a transformer is involved, all impedance must be referred to as a selected voltage base. This could be either the primary or secondary of the transformer or any other selected voltage base.

4.4.1 Simplified Models of Synchronous Machines for Transient Analysis

For the salient pole, because of the nonuniformity of the air gap, the generator was modeled with direct axis reactance (X_d) and the quadrature axis reactance X_q. However, the circuit reactance is much greater under short-circuit conditions than the resistance. Thus, the stator current lags nearly $\pi/2$ radians behind the driving voltage, and the armature reaction m.m.f. is centered almost on the direct axis. Therefore, during a short circuit, the machine's effective reactance may be assumed only along the direct axis (i.e., only X_d).

The three-phase short circuit of the current decays from a very high initial value to a steady-state value. This is because of the machine reactance change due to the effect of the armature reaction.

A useful figure can be obtained by considering the field and damper windings as the transformer's secondary, whose primary is the armature winding, to determine the machine's equivalent circuits during normal steady-state and disturbance conditions.

There is no transformer action between the synchronous machine's stator and rotor windings during normal steady-state conditions. The resultant field produced by the stator and rotor both revolve with the same synchronous speed. This is similar to a transformer with open-circuited secondaries. For this condition, its primary may be described by the synchronous reactance X_d.

During the disturbance, the rotor speed is no longer the same as that of the revolving field produced by stator windings resulting in transformer action. Thus, field and damper circuits resemble much more nearly as short-circuited secondaries.

The equivalent circuit for this condition, referred to as the stator side, is shown in Figure 4.2.

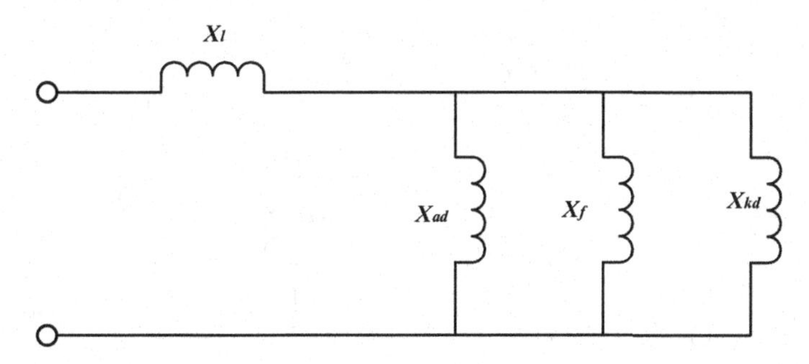

FIGURE 4.2 The equivalent circuit for the subtransient period.

Ignoring winding resistance, the equivalent reactance of Figure 4.2, known as the *direct axis subtransient reactance*, is:

$$X_d'' = X_l + \left(\frac{1}{X_{ad}} + \frac{1}{X_f} + \frac{1}{X_{kd}} \right)^{-1} \tag{4.1}$$

where X_l – Leakage reactance of armature.
X_{ad} – Reactance of armature reaction in the d-axis.
X_f – Reactance of field circuit.
X_{kd} – Reactance of damper circuit.

If the damper winding resistance R_k is inserted in Figure 4.2, and Thevenin's inductance seen at the terminals of R_k is obtained, the circuit time constant, known as the *direct axis short-circuit subtransient time constant,* becomes:

$$\tau_d'' = \frac{X_{kd} + \left(\frac{1}{X_l} + \frac{1}{X_f} + \frac{1}{X_{ad}} \right)^{-1}}{R_k} \tag{4.2}$$

For two-pole, turbo-alternator ($X_d'' = 0.07 - 0.12$p.u.) and for the water-wheel alternator ($X_d'' = 0.1 - 0.35$ p.u.).

where X_d'' is only used in calculations if the effect of the initial current is important, as, for example, when determining the circuit breaker short-circuit rating.

τ_d'' is very small, around 0.035 seconds, because the damper circuit has relatively high resistance. *Thus, this component of the current decays quickly.* It is then permissible to ignore the reactance of the damper circuit X_{kd}, and the equivalent circuit reduces to Figure 4.3.

Ignoring winding resistance, the equivalent reactance of Figure 4.3, known as the *direct axis short-circuit transient reactance*, is:

$$X_d' = X_l + \left(\frac{1}{X_{ad}} + \frac{1}{X_f} \right)^{-1} \tag{4.3}$$

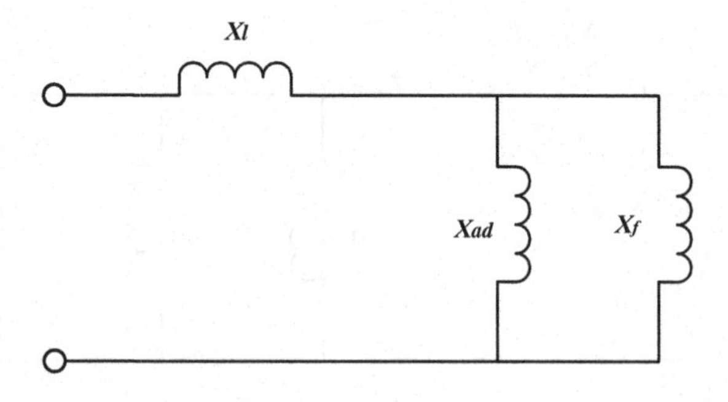

FIGURE 4.3 The equivalent circuit for the transient period.

If the field winding resistance R_f is inserted in Figure 4.3, and Thevenin's inductance seen at the terminals of R_f is obtained, the circuit time constant, known as the *direct axis short-circuit transient time constant,* becomes:

$$\tau_d' = \frac{X_f + \left(\frac{1}{X_l} + \frac{1}{X_{ad}}\right)^{-1}}{R_f} \tag{4.4}$$

where X_d' may lie between 0.1 and 0.25 p.u, and τ_d' is usually in the order of (1–2) seconds.

The field time constant, which characterizes the decay of transient with the armature open-circuited, is called the *direct axis open-circuit transient time constant.* This is given by

$$\tau_{do}' = \frac{X_f}{R_f} \tag{4.5}$$

where τ_{do}' it is around 5 seconds. τ_d' is related to τ_{do}' by:

$$\tau_d' = \frac{X_d'}{X_d}\tau_{do}' \tag{4.6}$$

Finally, when the disturbance is over, there will be no hunting of the rotor. Hence, there will not be any transformer action between the stator and the rotor, and the circuit reduces to Figure 4.4.

The equivalent reactance becomes the *direct axis synchronous reactance:* $X_d = X_l + X_{ad}$ is the same X_d which is obtained in a steady-state condition.

Similar equivalent circuits are obtained for reactances along the quadrature axis. These reactances are X_q'', X_q', and X_q. These reactances may be considered for cases when the circuit resistance results in a power factor appreciably above zero. The armature reaction is not necessarily totally on the direct axis, except that the machine's equivalent circuits are represented only by the direct axis's reactances.

The manufacturers provide synchronous machine reactances and time constants. A short-circuit test can obtain these values.

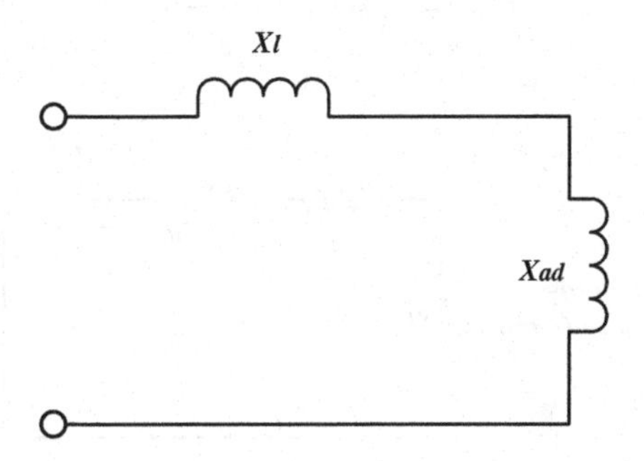

FIGURE 4.4 The equivalent circuit for the steady-state period.

4.4.2 Transient Phenomena

To understand the synchronous machine transient phenomena, we first study a simple R-L circuit's transient behavior, as shown in Figure 4.5.

The closing of the switch at $t = 0$ represents, to a first approximation, a three-phase short circuit at the terminals of an unloaded synchronous machine. The current is assumed to be zero before the switch closes.

$$v(t) = V_m \sin(\omega t + \alpha) \tag{4.7}$$

The angle α is the phase of the voltage wave at which the switch is closed

At $t = 0^+$, the instantaneous voltage equation for the circuit:

$$Ri(t) + L\frac{di(t)}{dt} = V_m \sin(\omega t + \alpha) \tag{4.8}$$

The solution for the current may be shown to be:

$$i(t) = I_m \sin(\omega t + \alpha - \theta) - I_m e^{-t/\tau} \sin(\alpha - \theta) \tag{4.9}$$

where

$$I_m = \frac{V_m}{Z}; \quad \tau = \frac{L}{R}; \quad \theta = \tan^{-1} \omega L/R;$$

$$Z = \sqrt{R^2 + (\omega L)^2} = \sqrt{R^2 + X^2}$$

The fault current $\{i(t)\}$ in Equation 4.9, called the *asymmetrical fault current,* consists of two components:

1. The first term is *the ac fault current* (also called symmetrical or steady-state fault current), which is a sinusoid and given by:

$$i_{ac}(t) = I_m \sin(\omega t + \alpha - \theta) \tag{4.10}$$

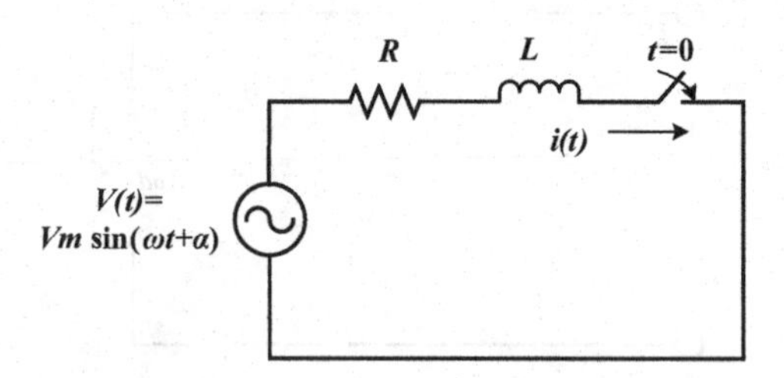

FIGURE 4.5 A simple series circuit with constant R and L.

2. The second term is *dc offset current,* which decays exponentially with a time constant $\tau = \frac{L}{R}$ And it is often referred to as a dc component since it is unidirectional and is given by:

$$i_{dc}(t) = -I_m e^{-t/\tau} \sin(\alpha - \theta) \tag{4.11}$$

At $(t=0)$, $i_{ac}(t)$ and $i_{dc}(t)$ are equal and opposite to satisfy the condition for zero initial currents.

The magnitude of the dc component, which depends on α (i.e., depends on the instant of application of the voltage to the circuit), varies from 0 when $\alpha = \theta$ to I_{max} when $\alpha = (\theta \pm \pi/2)$ radians.

Note that a short circuit may occur at any instant during a cycle of the ac source; that is, α can have any value. Since we are primarily interested in the largest fault current, we choose $\alpha = (\theta - \pi/2)$.

Then $i(t)$ becomes:

$$i(t) = I_m \sin(\omega t - \pi/2) + I_m e^{-t/\tau} \tag{4.12}$$

If $\omega L \gg R$, then $\theta \cong \pi/2$, so that circuit closer at voltage maximum $(\alpha = \pi/2, 3\pi/2, 5\pi/2,\ldots)$ would give no dc component, and closer at voltage zero $(\alpha = 0, \pi, 2\pi,\ldots)$ would cause the maximum dc component. See Figure 4.6.

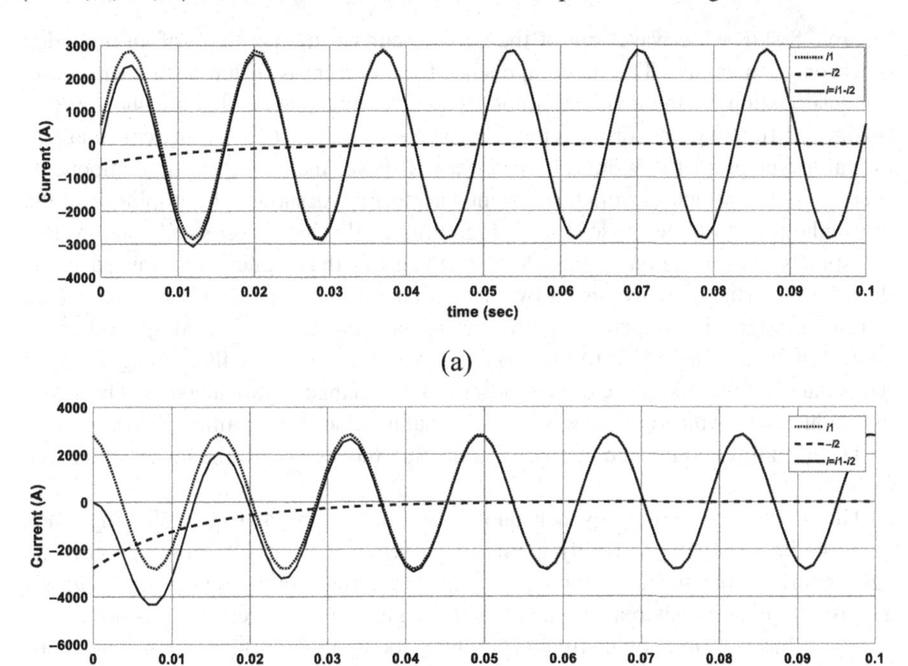

FIGURE 4.6 Current waveform. (a) With no dc offset. (b) With maximum dc offset.

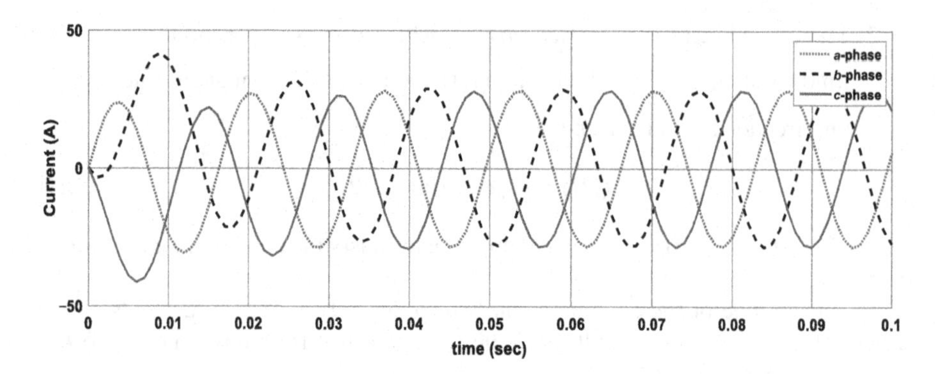

FIGURE 4.7 Steady-state current and transient current terms combine to form the resultant current.

A detailed study of Equation 4.9 will show the conditions which make for the maximum possible $i(t)$. The maximum of dc offset current is usually assumed to be the result of those conditions that make $\sin(\alpha - \theta) = 1$ or $\sin(\alpha - \theta) = -1$. See Figure 4.7, which shows that $i(t)$, $i_{ac}(t)$, and $i_{dc}(t)$ when $L > R$, $\theta = 85°$, $\sin(\alpha - \theta) = -1$, $\alpha = 355°$, at $t = 0$, $e \approx 0$.

4.4.3 Three-Phase Short-Circuit Unloaded Synchronous Machine

Figure 4.8 shows the waveform of the ac fault current in one phase of an unloaded synchronous machine. The dc component of the current is different in each phase. The justification for the different amounts of dc components is that the short circuit occurs at different points on each phase's voltage wave. Since, in a practical situation, we can never predict how much offset we will have; therefore, the dc components are removed (subtracted) from the current waveforms. As shown, the amplitude of the sinusoidal waveform decreases from a high initial value to a lower steady-state value.

A practical explanation for this phenomenon is that the magnetic flux caused by the short-circuit armature currents (or by the resultant armature m.m.f.) is initially forced to flow through high reluctance paths that do not link the field winding or damper circuits of the machine. This results from the theorem of constant flux linkages, which states that the flux linking a closed winding cannot change instantaneously. The armature inductance, which is inversely proportional to reluctance, is initially low.

The armature inductance increases as the flux moves toward the lower reluctance path.

The ac fault current in a synchronous machine is similar to that flowing when a sinusoidal voltage is suddenly applied to a series R-L circuit. However, there is one important difference: In the case of the R-L circuit, our reactance ($X = \omega L$) is a constant quantity, whereas, in the case of the generator, the reactance is not a constant one but is a function of time. Therefore, the ac fault current in a synchronous machine can be modeled by the series R-L circuit if a time-varying inductance $L(t)$ or reactance $X(t) = \omega L(t)$ is employed.

If we prefer to use r.m.s values, it is reasonable to divide the maximum values of the above figure by $\sqrt{2}$, as shown in Figure 4.9.

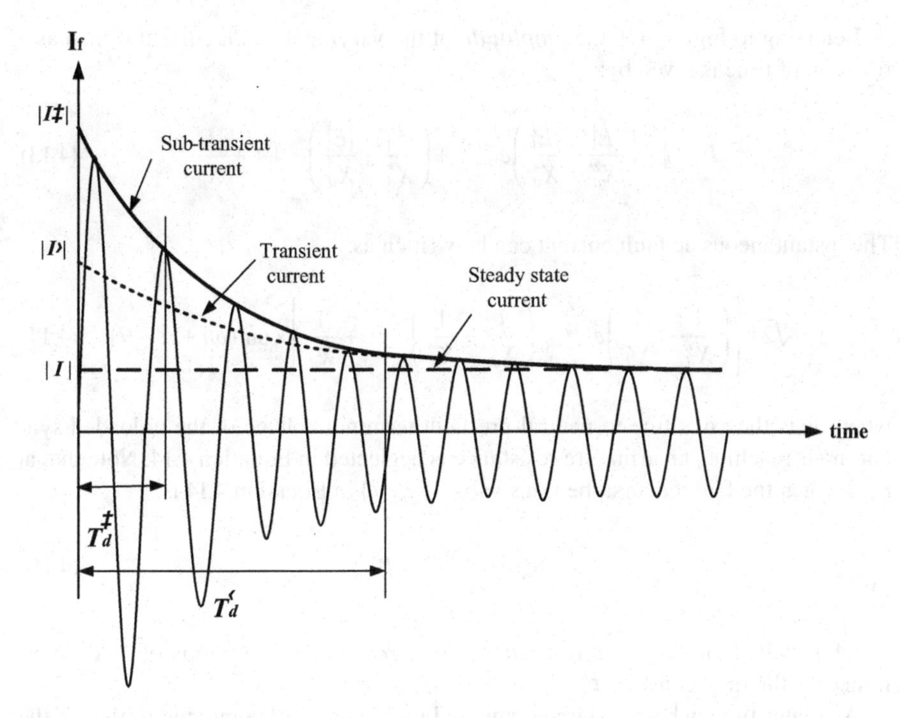

FIGURE 4.8 Fault current in one phase shows the subtransient, transient, and steady-state currents.

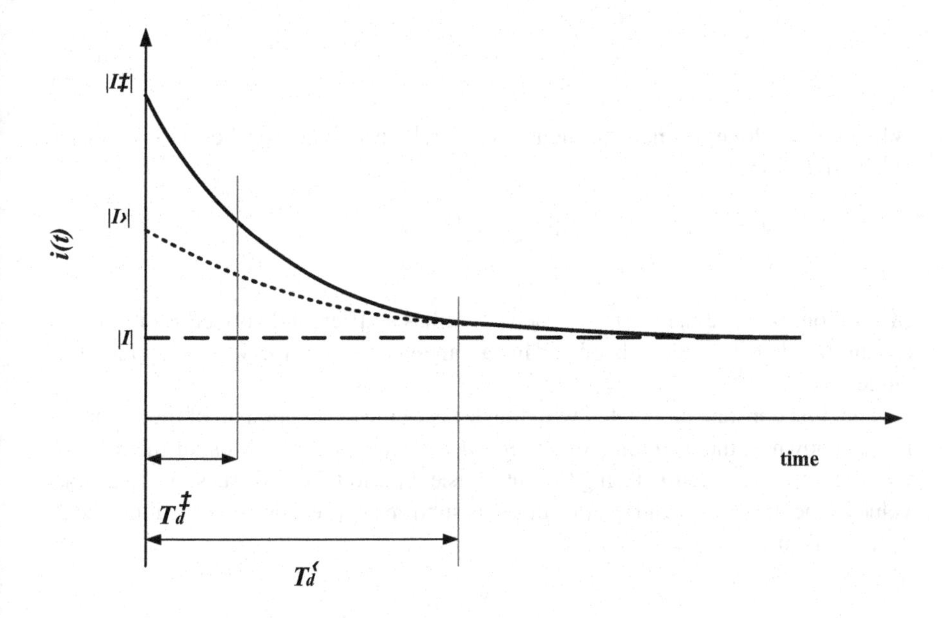

FIGURE 4.9 The subtransient, transient, and steady-state fault currents.

Referring to Figure 4.9, the *amplitude* of the varying ac fault current (r.m.s) as a function of time is given by:

$$|I_{ac}(t)| = \left(\frac{|E|}{X_d''} - \frac{|E|}{X_d'}\right)e^{-t/\tau_d''} + \left(\frac{|E|}{X_d'} - \frac{|E|}{X_d}\right)e^{-t/\tau_d'} + \frac{|E|}{X_d} \tag{4.13}$$

The instantaneous ac fault current can be written as:

$$i_{ac}(t) = \sqrt{2}E\left[\left(\frac{1}{X_d''} - \frac{1}{X_d'}\right)e^{-\frac{t}{\tau_d''}} + \left(\frac{1}{X_d'} - \frac{1}{X_d}\right)e^{-\frac{t}{\tau_d'}} + \frac{1}{X_d}\right] \times \sin(\omega t + \alpha - \theta) \tag{4.14}$$

where E is the r.m.s line-to-neutral pre-fault terminal voltage of the unloaded synchronous machine, an armature resistance is neglected in Equation 4.14. Note that at $t=0$, when the fault occurs, the r.m.s value of $i_{ac}(t)$ in Equation 4.14 is:

$$I_{ac}(0) = \frac{E}{X_d''} = I_d'' \tag{4.15}$$

which is called the r.m.s *subtransient fault current, I_d''*. The duration of I_d'' is determined by the time constant τ_d''.

At a later time, when t is large compared to τ_d'' but small compared to the τ_d', the first exponential term in Equation 4.14 has decayed almost to zero, but the second exponential has not decayed significantly. The r.m.s ac fault current then equals the r.m.s *transient fault current,* given by:

$$I_d' = \frac{E}{X_d'} \tag{4.16}$$

When t is much larger than τ_d', the r.m.s ac fault current approaches its steady-state value, given by:

$$I_{ac}(\infty) = \frac{E}{X_d} = I_d \tag{4.17}$$

In addition to the ac fault current, each phase has a different dc offset. As in the R-L circuit, the dc offset depends on the instantaneous value of the voltage applied (i.e., angle α).

The time constant associated with the decay of the dc component of the stator current is known as the *armature short-circuit time constant* (τ_a). Most of the decay of the dc component occurs during the subtransient period. For this reason, the average value of the direct axis and quadrature axis subtransient reactance is used for finding τ_a. It is given by:

$$\tau_a = \frac{X_d'' + X_q''}{2\omega R_a} \tag{4.18}$$

And it is *approximately* given by:

$$\tau_a = \frac{X_d'' + X_q''}{2R_a} \tag{4.19}$$

Typical value of τ_a is around 0.05–0.17 seconds.

In comparison, in Equation 4.11, the r.m.s of the dc component for phase (a) is given by:

$$I_{\text{dc}}(t) = \sqrt{2}\frac{E}{X_d''}e^{-t/\tau_a}\sin(\alpha-\theta) \tag{4.20}$$

$$I_{\text{dc}}(t) = \sqrt{2}\,I_d''\,e^{-t/\tau_a}\sin(\alpha-\theta) \tag{4.21}$$

$$I_{\text{dc(max)}}(t) = |I_{\text{dc}}(t)| = \sqrt{2}\,I_d''0\,e^{-t/\tau_a} \tag{4.22}$$

$$I_{\text{dc(max)}}(0) = \sqrt{2}\,I_d'' \tag{4.23}$$

The waveform of asymmetrical fault current is a superposition of dc and ac components

$$i_{\text{asy}}(t) = \sqrt{2}E\left[\left(\frac{1}{X_d''}-\frac{1}{X_d'}\right)e^{-\frac{t}{\tau_d''}} + \left(\frac{1}{X_d'}-\frac{1}{X_d}\right)e^{-\frac{t}{\tau_d'}} + \frac{1}{X_d}\right]$$

$$\times\sin(\omega t+\alpha-\theta) - \sqrt{2}\frac{E}{X_d''}e^{-\frac{t}{\tau_a}}\sin(\alpha-\theta) \tag{4.24}$$

In Equation 4.24, the degree of asymmetry depends upon the point of the voltage cycle at which the fault occurs, and if $\omega L \gg R$, then $\theta \cong \pi/2$.

The r.m.s value of $i_{\text{asy}}(t)$ is of interest. Since $i_{\text{asy}}(t)$ in Equation (4.31) is not strictly periodic. Its r.m.s value is not strictly defined. However, treating the exponential term (dc component) as a constant, we stretch the r.m.s concept to calculate the r.m.s asymmetrical fault current with maximum dc offset, as follows:

$$I_{\text{asy(rms)}}(t) = \sqrt{[I_{\text{ac}}(t)]^2 + [I_{\text{dc}}(t)]^2} \tag{4.25}$$

where $I_{\text{ac}}(t)$ is the magnitude of Equation 4.13 and $I_{\text{dc}}(t)$ is the magnitude of Equation 4.22.

Therefore, the maximum r.m.s current at the beginning of the short circuit $I_{\text{asy(rms)}}(0)$ is:

$$I_{\text{asy(max)}}(0) = \sqrt{\left(\frac{E}{X_d''}\right)^2 + \left(\sqrt{2}\frac{E}{X_d''}\right)^2} = \sqrt{(I_d'')^2 + \left(\sqrt{2}I_d''\right)^2}$$

$$I_{\text{asy(max)}}(0) = \sqrt{3}I_d'' \tag{4.26}$$

In practice, the *momentary duty* of a *circuit breaker (CB)* is given in terms of the asymmetrical short-circuit current, that is, mean:

$$I_{\text{momentary of CB}}(t) = I_{\text{asy(rms)}}(t) \quad (4.27)$$

The factor is normally taken as 1.6 instead of $\sqrt{3}$ in assessing the momentary current for the CB above 5 kV. This factor was reduced for voltage under 5 kV.

The *momentary duty* is not to be confused with *the interrupting capacity* of the CB.

To calculate the interrupting capacity of the CB, the subtransient currents are used. The calculation depends on the CB's speed, the ratio of X to R in the circuit, the distance between the fault and the generating station, and so forth. If X/R is small, the dc component will decay quickly; therefore, a smaller multiplying factor must be used. Of course, the faster the breaker, the higher the multiplying factor.

Typical values of multiplying factor for CBs of different speeds are as provided subsequently.

Circuit Breaker Speed	Multiplying Factor
One cycle	1.6
Two cycles	1.4
Three cycles	1.2
Five cycles	1.1
Eight cycles	1.0

Example 4.1

A bolted short circuit occurs in the series R-L circuit, with $V=18$ kV, $X=10\ \Omega$, $R=0.5\ \Omega$, and dc offset. The CB opens three cycles after fault inception. Determine:

1. The r.m.s ac fault current.
2. The r.m.s momentary current at $\tau = 0.75$ cycle passes through the breaker before it opens.
3. The r.m.s asymmetrical fault current that the breaker interrupts.

Solution

1. $I_{\text{ac(max)}} = \dfrac{V_m}{Z}$, and $I_{\text{ac(rms)}} = \dfrac{V_{\text{rms}}}{Z}$

$$\therefore \quad I_{\text{ac(rms)}} = \frac{20\times10^3}{\sqrt{(8)^2+(0.8)^2}} = \frac{18\times10^3}{10.0124} = 1.797\text{ kA}$$

2. $I_{\text{momentary}} = I_{\text{asy(rms)}}(t) = \sqrt{[I_{\text{ac}}(t)]^2 + [I_{\text{dc}}(t)]^2}$

$$= \sqrt{\left(\frac{V_{\text{rms}}}{Z}\right)^2 + \left(\sqrt{2}\frac{V_{\text{rms}}}{Z}e^{-t/\tau}\right)^2}$$

$$= \frac{V_{\text{rms}}}{Z}\sqrt{1+2e^{-2t/\tau}} = I_{\text{ac(rms)}}\sqrt{1+2e^{-2t/\tau}}$$

$$\tau = \frac{L}{R} = \frac{X}{\omega R} = \frac{X}{2\pi f R} = \frac{10}{2\pi f(0.5)} = \frac{20}{2\pi f}$$

$$t = \frac{\tau}{f}$$

where τ is time in cycles

$$\therefore \quad t = \frac{0.75}{f}$$

$$\frac{t}{\tau} = \frac{0.75/f}{1/(\pi \cdot f)} = 0.75\pi$$

Therefore,

$$I_{\text{momentary}} = 1.797 \times \sqrt{1+2e^{-2\times 0.75\times\pi}} = 1.813\ \text{kA}$$

$$3 - I_{\text{asy(rms)}}(3\ \text{cycle}) = 2.488\ \sqrt{1+2e^{-2\pi(3)}} \approx 2.488\ \text{kA}$$

Example 4.2

A 600 MVA 18 kV, 60 Hz synchronous generator with reactances $X_d'' = 0.25$, $X_d' = 0.34$, $X_d = 1.2$ p.u. and time constants $\tau_d'' = 0.025$, $\tau_d' = 2.1$, $\tau_a = 0.23$ seconds is connected to a CB. The generator operates at 6% above-rated voltage and no load when a bolted three-phase short circuit occurs on the breaker's load side. The breaker interrupts the fault three cycles after fault inception. Determine:

1. The subtransient fault current in per-unit and kA r.m.s.
2. Maximum dc offset as a function of time.
3. r.m.s asymmetrical fault current, which the breaker interrupts, assuming maximum dc offset.

Solution

1. The no-load voltage before the fault occurs $E = 1.06$ p.u. The subtransient fault current that occurs in each of the three phases is:

$$I_d'' = \frac{E}{X_d''} = \frac{1.06}{0.25} = 4.24\ \text{p.u.}$$

The generator base current is:

$$I_{base} = \frac{S_{rated}}{\sqrt{3}\ V_{rated}} = \frac{600}{(\sqrt{3})(18)} = 19.245\ \text{kA}$$

$$\therefore I_d'' \text{ in p.u.} = (4.24)(19.245) = 81.6\ \text{kA}$$

2. The $I_{dc(max)}(t)$ offset that occurs in any one phase is:

$$I_{dc(max)}(t) = \sqrt{2}(81.6)e^{-\frac{t}{0.23}} = 115.4\, e^{-t/0.23}\ \text{kA}$$

3. To calculate $I_{asy(rms)}(t)$ must determine $|I_{ac}(t)|$ and $|I_{dc}(t)|$

$$t = \frac{\tau}{f} = \frac{3\,\text{cycle}}{60} = 0.05\,\text{seconds.}$$

$$|I_{ac}(0.05)| = 1.06\left[\left(\frac{1}{0.25} - \frac{1}{0.34}\right)e^{-\frac{0.05}{0.025}} + \left(\frac{1}{0.34} - \frac{1}{1.2}\right)e^{-\frac{0.05}{2.1}} + \frac{1}{1.2}\right]$$

$$= 3.217\ \text{p.u.}$$

$$= (3.217)(19.245) = 61.91\,\text{kA}$$

$$|I_{dc}(0.05)| = \sqrt{2}I_d''\, e^{-0.05/\tau_a} = \sqrt{2}\,(81.6)\,e^{-\frac{0.05}{0.23}} = 92.852\,\text{kA}$$

$$I_{asy(rms)}(0.05) = \sqrt{\left[I_{ac}(0.05)\right]^2 + \left[I_{dc}(0.05)\right]^2}$$

$$= \sqrt{(61.91)^2 + (92.852)^2}$$

$$= 111.6\,\text{kA}$$

4.4.4 Effect of Load Current

If the fault occurs when the generator delivers a pre-fault load current, two methods might be used to solve three-phase symmetrical fault currents.

1. Use of internal voltages behind reactances:
   When there is a pre-fault load current, three *fictitious* internal voltages E'', E', and E may be considered to be effective during the subtransient, transient, and steady-state periods, respectively, as shown in Figure 4.10.

$$E'' = V + jX_d''I_L \tag{4.28}$$

$$E' = V + jX_d'I_L \tag{4.29}$$

$$E = V + jX_dI_L \tag{4.30}$$

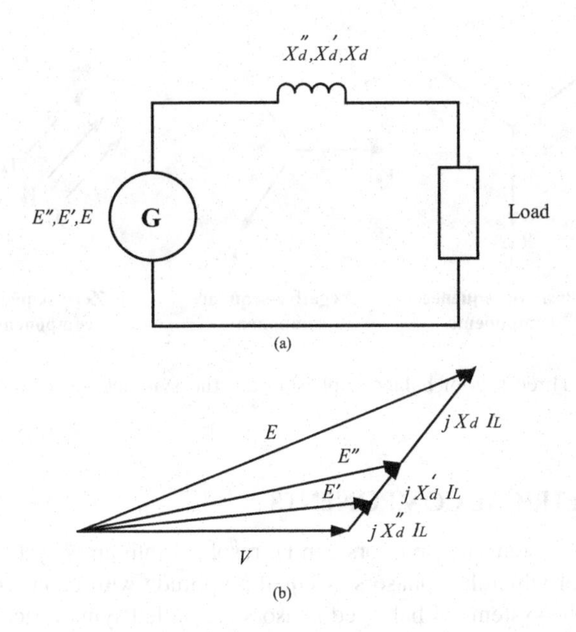

FIGURE 4.10 One-phase diagram showing the subtransient, transient, and steady-state currents. (a) Circuit diagram. (b) Phasor diagram.

2. Using Thevenin's theorem and superposition with load current:

   The fault current is found in the absence of the load by obtaining Thevenin's equivalent circuit to the point of fault. The total short-circuit current is then given by superimposing the fault current with the load current.

4.5 UNSYMMETRICAL FAULTS ANALYSIS

As stated earlier, most system faults occur in practice and are unbalanced or unsymmetrical.

An unbalanced power system may be defined as one in which the current or voltages are unbalanced. It will be evident that such an imbalance may be brought about if either the alternator voltage is unsymmetrical or the system's circuit is so.

Most of the faults on power systems are unsymmetrical faults, consisting of unsymmetrical short circuits, unsymmetrical faults through impedances, or open conductors.

Unsymmetrical faults occur as single line-to-ground faults, line-to-line faults, or double line-to-ground faults. The path of the fault current from line to line or line to the ground may or may not contain impedance. One or two open conductors result in unsymmetrical faults, either through the breaking of one or two conductors or through the action of fuses and other devices that may not open the three phases simultaneously.

Since any unsymmetrical fault causes unbalanced currents to flow in the system, it is shown below the general network's sequence components and Thevenin's equivalent circuits.

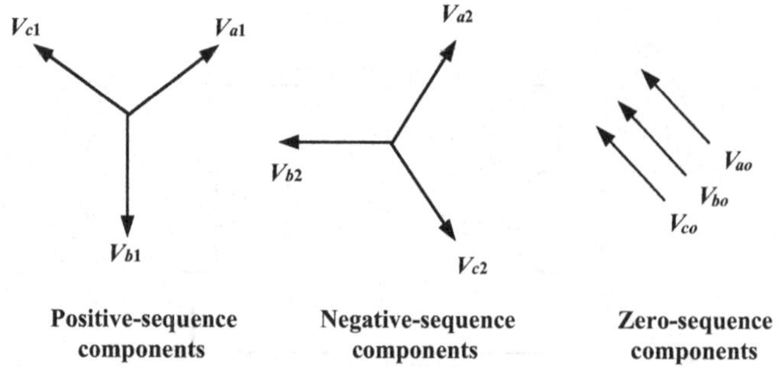

V_{c1} V_{a1} V_{b1} V_{a2} V_{b2} V_{c2} V_{ao} V_{bo} V_{co}

Positive-sequence components **Negative-sequence components** **Zero-sequence components**

FIGURE 4.11 Three sets of balanced phasors are the symmetrical components of three unbalanced phasors.

4.6 SYMMETRICAL COMPONENTS

An unbalanced system of *N* phasors can be resolved into an *N* system of balanced phasors, each of which has phasors of equal amplitude with equal angles between each phasor. The systems of balanced phasors are called symmetrical components. The vector sum of the symmetrical components equals the original system of unbalanced phasors. The symmetrical components of a three-phase system, illustrated in Figure 4.11, are as follows.

4.6.1 Positive-Sequence Components

These are three phasors of equal magnitude, offset from each other by 120° and rotating in the same direction as the original phasors, usually signified counterclockwise in a phasor diagram.

4.6.2 Negative-Sequence Components

Three phasors of equal magnitude, offset from each other 120° in phase, and rotating in the opposite direction to the original phasor, make up the negative-sequence components. Normally the rotation of the phasors is considered to be in the same direction as the original phasors, but two of the phase vectors, *b* and *c*, and reversed, as shown in Figure 4.11; thus, the negative-sequence components phase rotation is *a c b* instead of *a b c*.

4.6.3 Zero-Sequence Components

These are three phasors of equal magnitude, with an in-phase offset between them. They are in phase. The zero-sequence components are often non-rotating, like a dc voltage. However, zero-sequence impedance is complex, so the zero-sequence voltage is treated as a single-phase source for calculating zero-sequence impedance in relaying problems.

The defining equations for the symmetrical components, where V_a, V_b, and V_c are the original unbalanced phasors, are:

$$V_a = V_{a1} + V_{a2} + V_{a0}$$
$$V_b = V_{b1} + V_{b2} + V_{b0} \tag{4.31}$$
$$V_c = V_{c1} + V_{c2} + V_{c0}$$

The phasor diagram is shown in Figure 4.12.

The operator is used to indicate a 120° phase shift. The three phasors' relative angular position for a three-phase voltage can be expressed as the product of the phasor's amplitude and the operator (*a*), referring to Figure 4.11.

$$V_{b2} = a\ V_{a2},\quad a = 1\ \angle 120° = -0.5 + j0.866$$

$$V_{b1} = a^2 V_{a1},\quad a^2 = 1\ \angle 240° = -\ 0.5 - j0.866$$

$$V_{c2} = a^2\ V_{a2},\quad 1 + a + a^2 = 0$$

$$V_{c1} = a\ V_{a1},\quad V_{c0} = V_{b0} = V_{a0}$$

Therefore, it can use previous relations to final V_a, V_b, and V_c;

$$V_a = V_{a1} + V_{a2} + V_{a0}$$
$$V_b = a^2\ V_{a1} + a\ V_{a2} + V_{a0} \tag{4.32}$$
$$V_c = a\ V_{a1} + a^2\ V_{a2} + V_{a0}$$

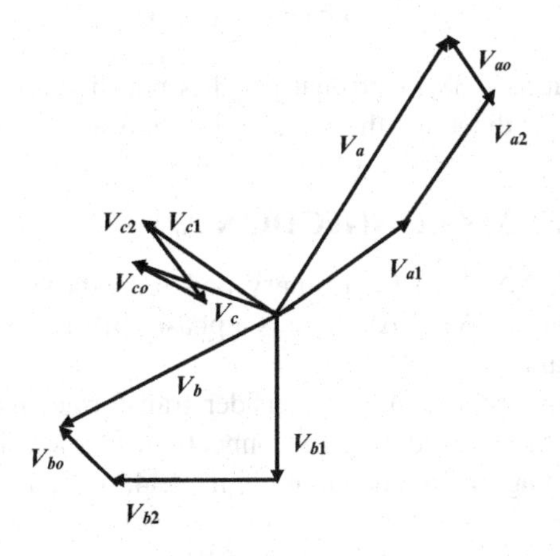

FIGURE 4.12 Three unbalanced phasor diagrams.

It rewrites Equation 4.32 as

$$\begin{bmatrix} V_a \\ V_b \\ V_c \end{bmatrix} = \begin{bmatrix} 1 & 1 & 1 \\ 1 & a^2 & a \\ 1 & a & a^2 \end{bmatrix} \begin{bmatrix} V_{a0} \\ V_{a1} \\ V_{a2} \end{bmatrix} \tag{4.33}$$

And these transformations can do it for currents, too, as

$$\begin{bmatrix} I_a \\ I_b \\ I_c \end{bmatrix} = \begin{bmatrix} 1 & 1 & 1 \\ 1 & a^2 & a \\ 1 & a & a^2 \end{bmatrix} \begin{bmatrix} I_{a0} \\ I_{a1} \\ I_{a2} \end{bmatrix} \tag{4.34}$$

4.7 EFFECT OF SYMMETRICAL COMPONENTS ON IMPEDANCE

To find the effect of the symmetrical component on impedance, we assumed that the start relation is:

$$V_{abc} = [Z_{abc}] \cdot I_{abc} \tag{4.35}$$

where $[Z_{abc}]$ is a matrix of dimension (3×3) that gives self- and mutual impedance in phases and between them.

$$[A] \cdot [V_{012}] = [Z_{abc}][A][I_{012}] \tag{4.36}$$

$$[V_{012}] = [A]^{-1}[Z_{abc}][A][I_{012}] \tag{4.37}$$

$$[Z_{012}] = [A]^{-1}\,[Z_{abc}]\,[A] \tag{4.38}$$

$$[V_{012}] = [Z_{012}] \cdot [I_{012}] \tag{4.39}$$

The important Equation 4.38 found that $[Z_{abc}]$ is not diagonal but somewhat symmetrical, while $[Z_{012}]$ is diagonal; these make the analysis very easy.

4.8 PHASE SHIFT Δ/Y CONNECTION Δ/Y

Type of connection (ΔΔ, YY) for primary and secondary causes no phase shift between them. While for (ΔY, YΔ), there is a phase shift between the transformer's primary and secondary.

To understand this problem, let us consider transformer has (Y-Δ) connection, as shown in Figure 4.13, considering (Δ-connection) is low-voltage side, then from studying the phasor diagram, as shown in Figure 4.14, we can see that,

$$V_{AB} = V_{HV}\ \angle 30^\circ \tag{4.40}$$

$$V_{ab} = V_{LV}\ \angle 0^\circ \tag{4.41}$$

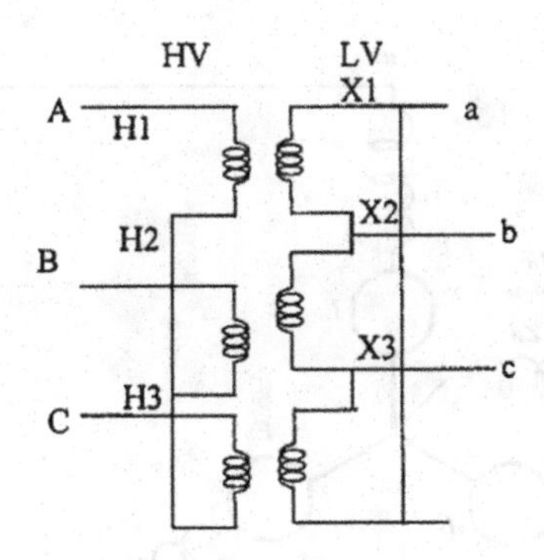

FIGURE 4.13 The transformer has a (**Y-Δ**) connection.

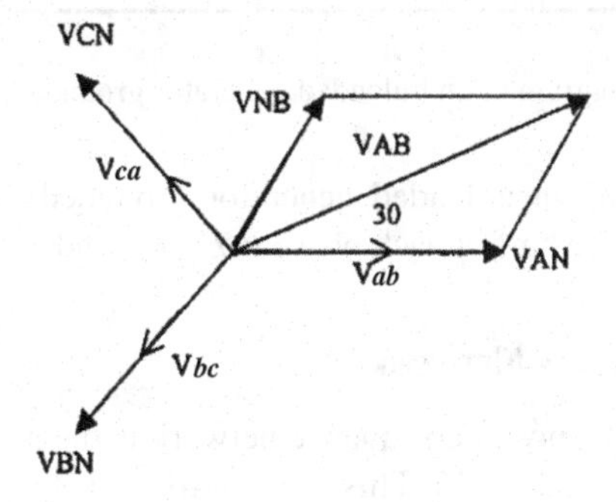

FIGURE 4.14 The phasor diagram.

All of the high-voltage magnitudes lead to low voltages by (30°), and from many tries, it can say that "for each star delta, delta star named the phasors as make positive sequence magnitudes in high voltage side lead more than positive sequence magnitudes in low voltage side by (30°) and vice versa according to negative sequence magnitudes."

4.9 SEQUENCE NETWORK OF UNLOADED GENERATOR

When a fault occurs at the generator's terminals, currents I_a, I_b, and I_c flow in the lines; if the fault involves ground, the current flowing into the neutral of the generator is designated (I_n).

One or two of the line currents may be zero, but the currents can be resolved into symmetrical components regardless of how unbalanced they may be.

Drawing the sequence network is simple. The generated voltages are of positive sequence only since the generator is designed to supply balanced three-phase voltages. The positive-sequence network is composed of an emf in series with a positive-sequence impedance of the generator. The negative and zero-sequence networks contain no emf but include the generator's impedance to negative and zero-sequence currents.

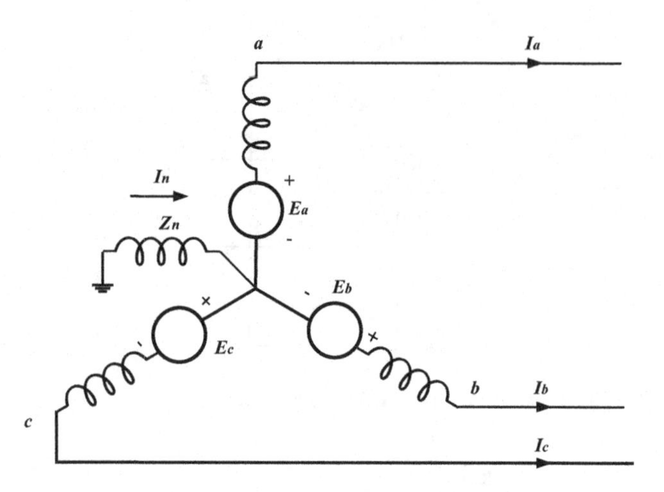

FIGURE 4.15 Circuit diagram of an unloaded generator grounded.

A circuit diagram of an unloaded generator grounded through a reactance is shown in Figure 4.15. The emf of each phase is E_a, E_b, and E_c.

4.9.1 Positive-Sequence Network

The generated emf in the positive-sequence network is the no-load terminal voltage to neutral, as shown in Figure 4.16. This is equal to the voltage behind transient and subtransient reactances and the voltage behind synchronous reactance since the generator is not loaded. The positive-sequence network's reactance is the subtransient, transient, or synchronous reactance depending on whether subtransient, transient, or steady-state conditions are being studied. From Figure 4.16b

$$V_{a1} = E_a - I_{a1}Z_1 \tag{4.42}$$

E_a is the positive sequence no-load voltage to neutral.
Z_1 is the positive-sequence impedance of the generator.

4.9.2 Negative-Sequence Network

The reference bus for positive and negative-sequence networks is the neutral of the generator. As far as positive and negative-sequence components are concerned, the generator's neutral is at ground potential since only zero-sequence currents flow in the neutral and ground impedance. Figure 4.17 shows the path for phase currents of the negative sequence in the generator and the corresponding sequence networks. From Figure 4.17b.

$$V_{a2} = -I_{a2}Z_2 \tag{4.43}$$

where Z_2 is the negative-sequence impedance of the generator.

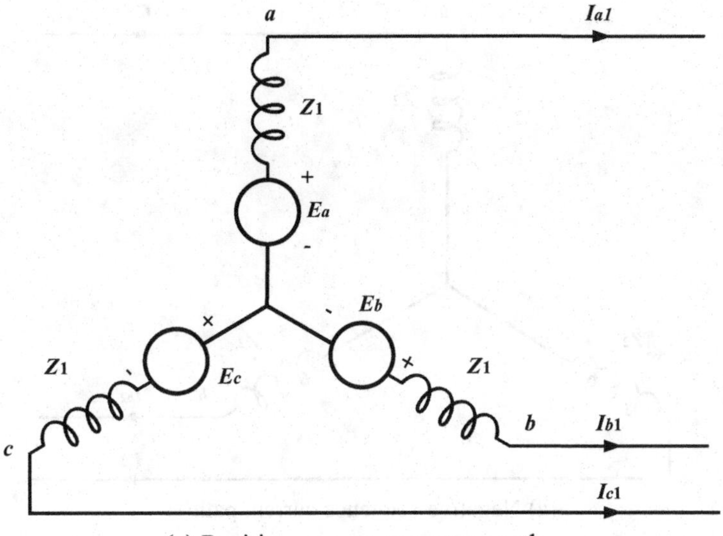

(a) Positive-sequence current path

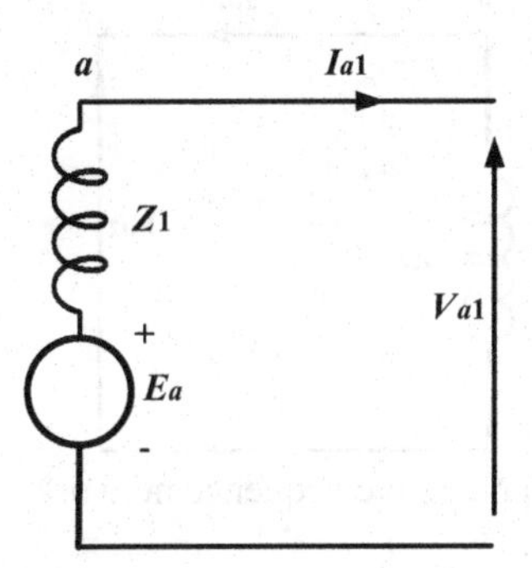

(b) Positive-sequence network

FIGURE 4.16 The path for phase current of positive sequence in the generator (a) and the corresponding sequence networks (b).

4.9.3 Zero Sequence

The current flowing in the impedance Z_n between neutral and ground is $(3I_0)$. By referring to Figure 4.18a, we see that the voltage drop of zero sequences from point a to ground is $(-3I_{a0}\,Z_n - I_{a0}\,Z_{g0})$, where Z_{g0} is the zero-sequence impedance per phase of the generator.

The zero-sequence network, a single-phase circuit assumed to carry only the zero-sequence current of phase, must, therefore, have an impedance of $(3Z_n + Z_{g0})$, as shown in Figure 4.18b, The total zero-sequence impedance through which Iao flows is:

$$Z_0 = 3Z_n + Z_{g0} \tag{4.44}$$

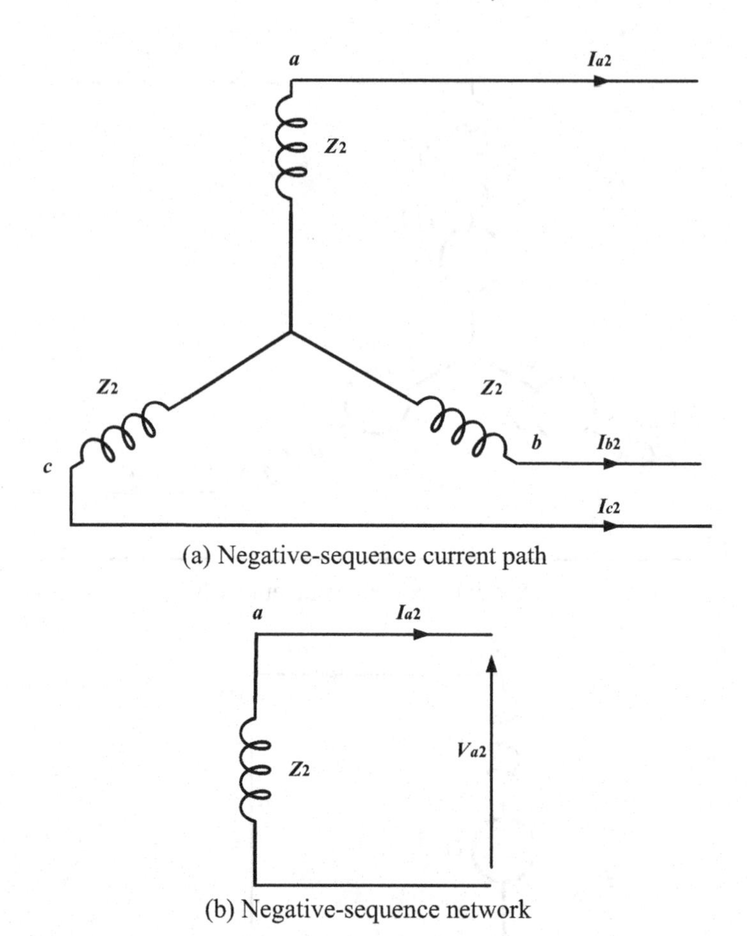

FIGURE 4.17 The path for phase current of the negative sequence in the generator (a) and the corresponding sequence networks (b).

The equations for the components of voltage drop from point (a) of phase (a) to the reference but (or ground) are, as may be deduced from Figure 4.18b,

$$V_{a0} = -I_{a0}Z_0 \tag{4.45}$$

where Z_0 is the zero-sequence impedance defined by Equation (4.44).

4.10 ANALYSIS OF UNSYMMETRICAL FAULTS USING THE METHOD OF SYMMETRICAL COMPONENT

The fault occurs at any point in the power-system network represented by the interconnection between sequence components at the fault position. Therefore, the important thing that must be determined accurately is the way of connecting these networks, and they must follow these steps for analysis:

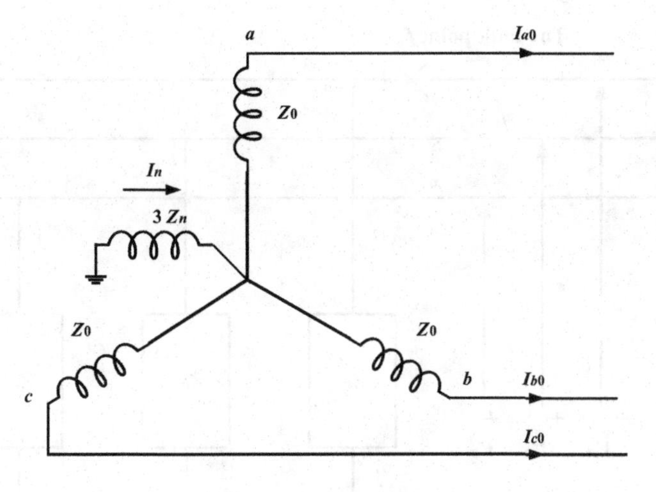

(a) Zero-sequence current path

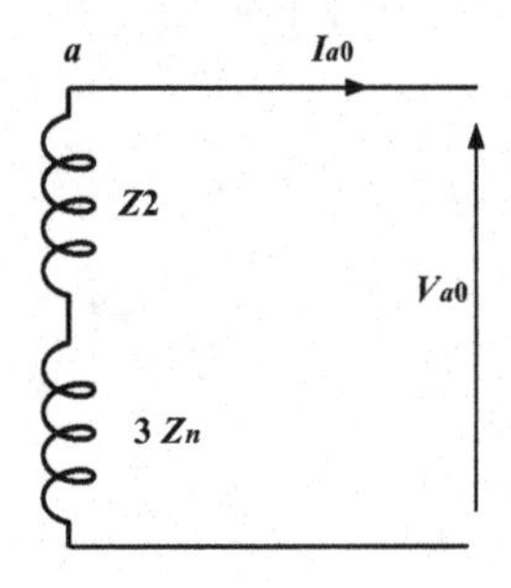

(b) Zero-sequence network

FIGURE 4.18 The path for phase current of zero sequence in the generator (a) and the corresponding sequence networks (b).

1. Make a detailed diagram for the circuit and determine all phases connected to the fault position, as shown in Figure 4.19.
2. Write down all circumstances of fault dealing with phase current and phase voltage.
3. Transform magnitudes of phase current and phase voltage (*a-b-c*) that are determined in item (2) above to sequence components (0 1 2) by using [*A*] or $[A]^{-1}$.
4. Determine the connection of sequence network terminals (*N*, *F*) using the information of sequence currents determined in item (2).
5. Determine the connection of recent sequence components terminals of the network and all impedances concerning the fault using items (3, 4).

$$\begin{bmatrix} V_{a0} \\ V_{a1} \\ V_{a2} \end{bmatrix} = \begin{bmatrix} 0 \\ E_a \\ 0 \end{bmatrix} - \begin{bmatrix} Z_0 & 0 & 0 \\ 0 & Z_1 & 0 \\ 0 & 0 & Z_2 \end{bmatrix} \begin{bmatrix} I_{a0} \\ I_{a1} \\ I_{a2} \end{bmatrix} \tag{4.46}$$

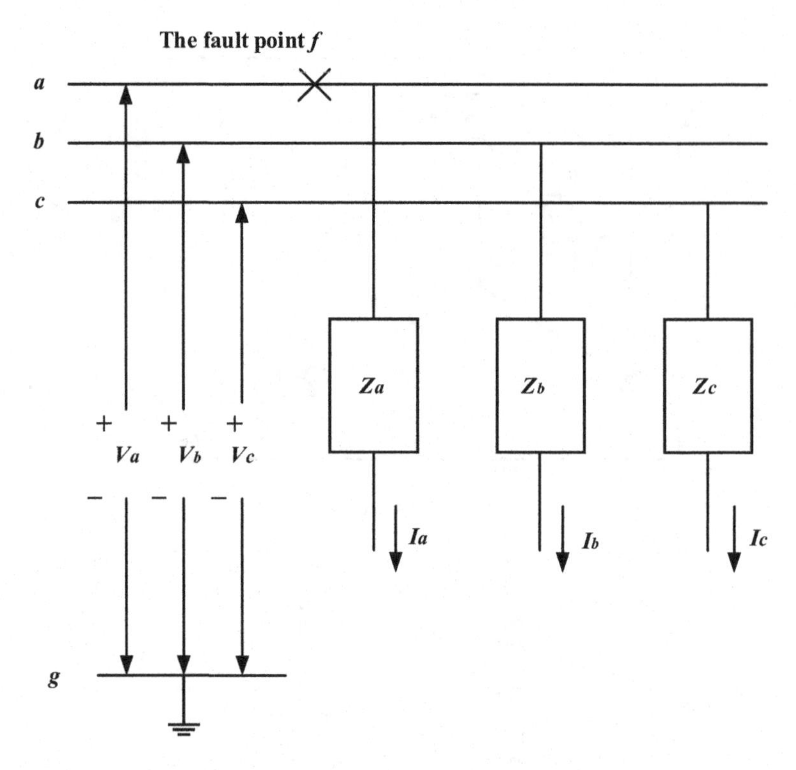

FIGURE 4.19 Circuit diagram at fault point (F).

4.10.1 Single Line-to-Ground Fault

Consider a single line-to-ground fault from phase a to ground at the general three-phase bus shown in Figure 4.20. For generality, we include a fault impedance Z_f in case of a bolted fault, $Z_f = 0$, whereas, for an arcing fault, Z_f is the arc impedance. In the case of a transmission line insulator flashover, Z_f includes the total fault impedance between the line and ground, including the arc and transmission tower's impedances and the tower footing if there are no neutral wires.

Fault condition in the phase domain

$$I_b = I_c = 0 \tag{4.47}$$

Single line-to-ground

$$V_{ag} = Z_f I_a \tag{4.48}$$

As

$$V_a = Z_f I_a \tag{4.49}$$

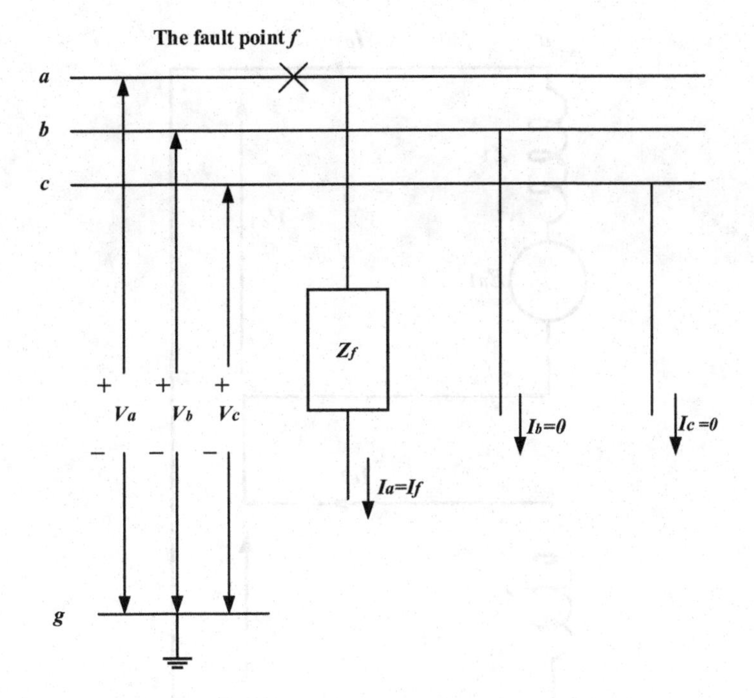

FIGURE 4.20 General three-phase bus.

Therefore,

$$V_a = 3Z_f I_{a1} \tag{4.50}$$

$$V_{a0} + V_{a1} + V_{a2} = 3Z_f I_{a1} \tag{4.51}$$

Then from Figure 4.21, the sequence components of the fault current are:

$$I_{a0} = I_{a1} = I_{a2} = \frac{V_f}{Z_0 + Z_1 + Z_2 + (3Z_f)} \tag{4.52}$$

4.10.2 Line-to-Line Fault

Consider a line-to-line fault from phase b to c, shown in Figure 4.22.

Fault conditions in the phase domain

$$I_a = 0 \tag{4.53}$$

line-to-line fault current

$$I_c = -I_b \tag{4.54}$$

$$V_{bg} - V_{cg} = Z_f I_b \tag{4.55}$$

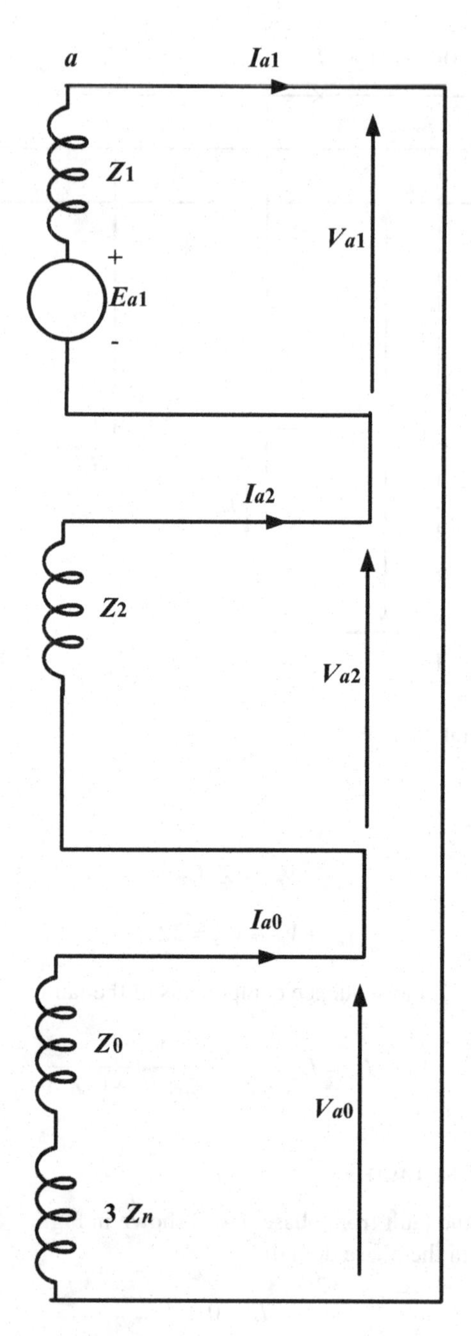

FIGURE 2.21 Interconnected sequence networks for SLG fault.

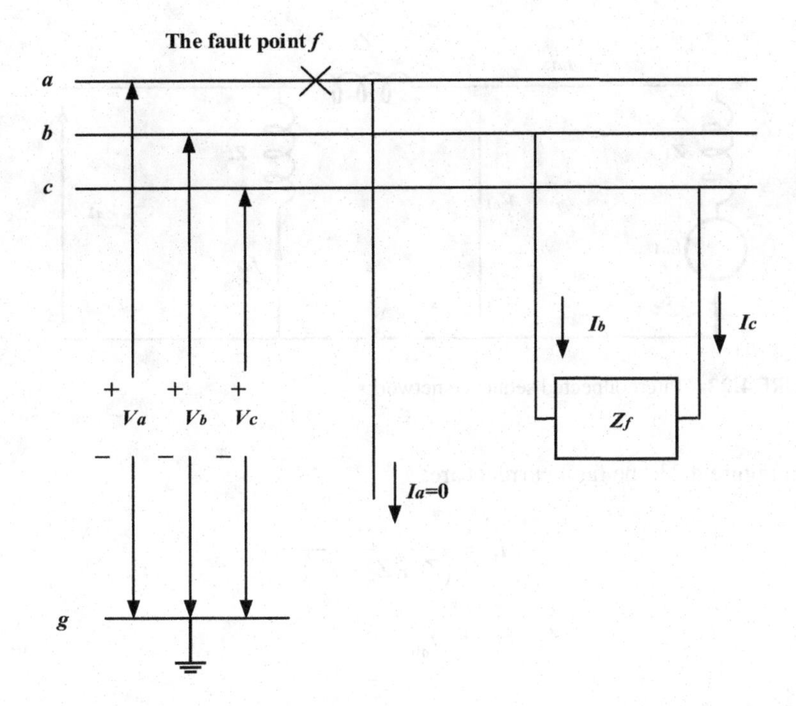

FIGURE 4.22 General three-phase bus.

From using the transformation equation:

$$I_{a012} = A^{-1} I_{abc} \tag{4.56}$$

and

$$I_{a0} = 0, \quad I_{a2} = -\ I_{a1} \tag{4.57}$$

$$Z_f I_{a1} = V_{a1} - V_{a2} \tag{4.58}$$

Fault conditions in sequence domain

$$I_a = 0 \tag{4.59}$$

line-to-line fault

$$I_{a1} = -I_{a2} \tag{4.60}$$

$$V_{a1} - V_{a2} = Z_f I_{a1} \tag{4.61}$$

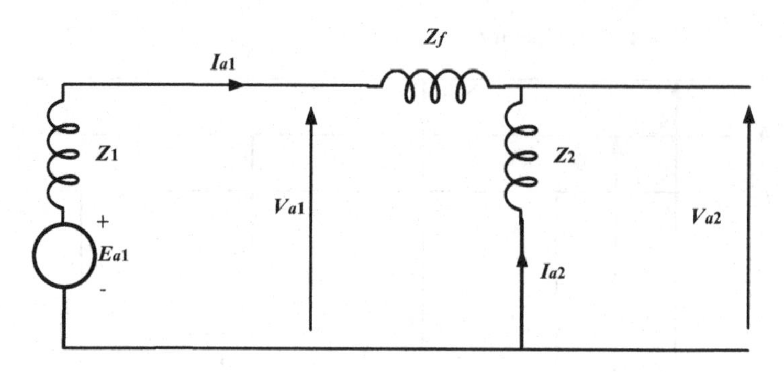

FIGURE 4.23 Interconnected sequence networks.

From Figure 4.23, the fault currents are:

$$I_{a1} = \frac{V_f}{\left(Z_1 + Z_2 + Z_f\right)} \tag{4.62}$$

$$I_{a0} = 0 \tag{4.63}$$

4.10.3 Double Line-to-Ground Fault

A double line-to-ground fault from phase b to phase c through impedance Z_f to the ground is shown in Figure 4.24. From this figure we observe that:

Fault conditions in the phase domain

$$I_a = 0 \tag{4.64}$$

Double line-to-ground fault

$$V_{cg} = V_{bg} \tag{4.65}$$

$$V_{bg} = Z_f\left(I_b + I_c\right) \tag{4.66}$$

Fault conditions in sequence domain

$$I_{a0} + I_{a1} + I_{a2} = 0 \tag{4.67}$$

Double line-to-ground fault

$$V_{a0} - V_{a1} = \left(3Z_f\right)I_{a0} \tag{4.68}$$

$$V_{a1} = V_{a2} \tag{4.69}$$

$$I_{a0} + I_{a1} + I_{a2} = 0 \tag{4.70}$$

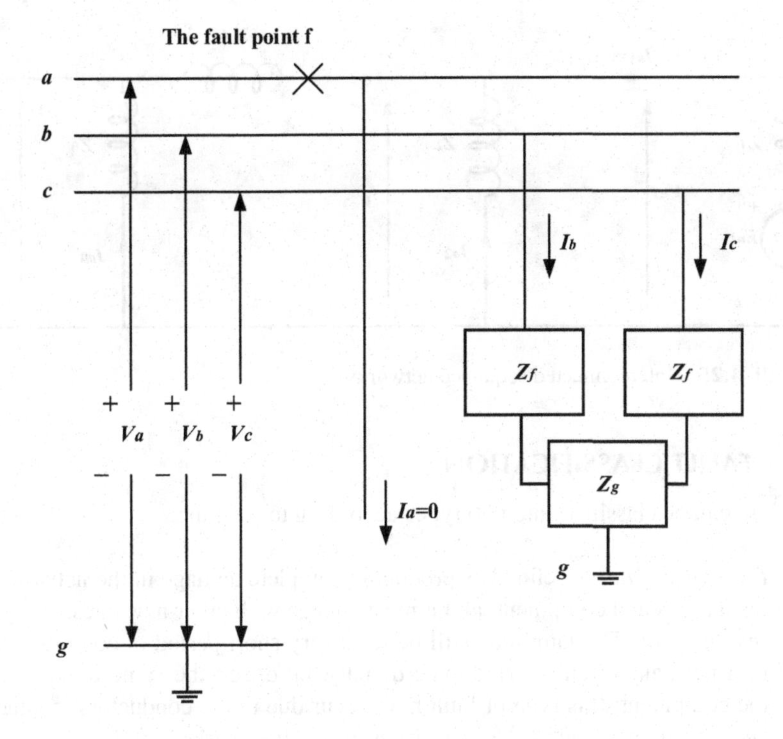

FIGURE 4.24 General three-phase bus.

$$V_b = V_{a0} + a^2 V_{a1} + aV_{a2} \tag{4.71}$$

$$V_c = V_{a0} + aV_{a1} + a^2 V_{a2} \tag{4.72}$$

The domain is satisfied from fault conditions in sequence by connecting the zero, positive, and negative-sequence networks in parallel at the fault terminal; additionally, ($3Z_f$) is included in the zero-sequence network series. This connection is shown in Figure 4.25 from this figure; the positive-sequence fault current is

$$I_{a1} = \frac{V_f}{Z_1 + \left[\frac{Z_2(Z_0 + 3Z_f)}{Z_2 + Z_0 + 3Z_f}\right]} \tag{4.73}$$

Using the current division in Figure 4.25, the negative and zero-sequence fault currents are

$$I_{a2} = (-I_{a1})\left[\frac{Z_0 + 3Z_f}{Z_0 + 3Z_f + Z_2}\right] \tag{4.74}$$

$$I_{a0} = (-I_{a1})\left[\frac{Z_2}{Z_0 + 3Z_f + Z_2}\right] \tag{4.75}$$

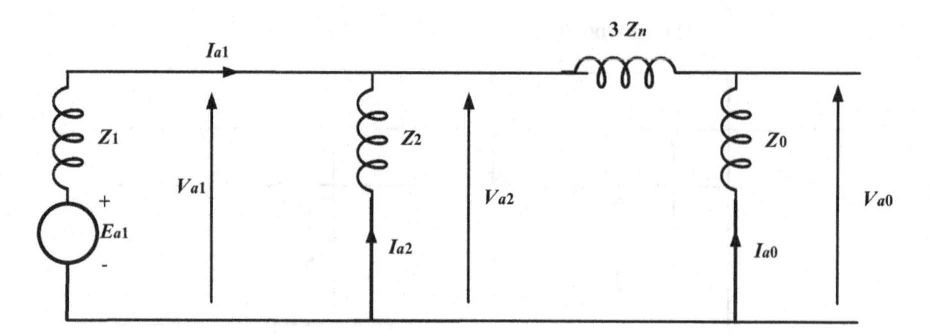

FIGURE 4.25 Interconnected sequence networks.

4.11 FAULT CLASSIFICATION

The fault can be classified into two types according to the time:

1. *Permanent fault* is defined as producing complete damage in the network's insulations and equipment, and it must change with other new ones.
2. *Instant fault:* The fault may still be for a very short period of compression to a permanent fault period. The damage produced the same damage in the equipment; this type of fault may occur due to the conductors' contact because of heavy wind. Then it causes an electric spark.

This type of fault is classified into two subtypes:

- *Recursion fault:* It occurs due to the contact between the conductors and comes back to normal operation and may contact conductors, and hence, these cases may occur due to heavy wind.
- *Not recursion fault*: It occurs due to contact between the conductors, then to return to normal operation and remove the contact.

Thus, the first part of this presentation will deal with the balanced and unbalanced fault calculation in general and conventional methods for small systems. The second part will deal with a large system using a digital approach.

4.12 ASSUMPTIONS AND SIMPLIFICATIONS

In the system data presentation, it is assumed that the system is balanced, and only one phase in the three-phase system is considered. The impedances represent the per-phase impedance of transmission lines and transformers. It is considered that the impedance to the flow of positive, negative, or zero-sequence currents is the same in each phase except at points of the system unbalanced, and these unbalances require special treatment.

The rest of the network comprises balanced impedances such that the per-phase representation is possible. The generators have different impedance to the flow of positive, negative, and zero-sequence currents; however, a general simplifying assumption is made as follows:

1. All load currents are negligible.
2. All generated voltages are equal in phase and magnitude to the positive-sequence pre-fault voltage.
3. The positive and negative-sequence networks are identical.
4. The networks are balanced except at fault points.
5. All shunt admittance (line charging susceptance, etc.) is negligible.

4.13 FAULT VOLTAGE-AMPS

Voltage – ampere is always called fault level, which it will produce by multiplying the fault current by reference voltage (V_r). In the case of a ground fault, the voltage is represented as phase voltage, while in the case of a phase fault, the voltage is line voltage. Now we discuss how we can compute voltage – ampere directly from the equivalent circuit of the fault; if we use per-unit system, let (Z_t) represent the equivalent circuit of the fault, then

$$Z_t = Z_1 + Z_2 + Z_0 + 3Z_f \text{ (Ground fault)} \tag{4.76}$$

$$Z_t = Z_1 + Z_2 + Z_0 \text{ (Phase fault)} \tag{4.77}$$

Let (V_r) be the reference phase voltage for the faulted part of the system,

$$\text{VA 3 ph(base)} = 3V_r \cdot I_r = \frac{3V_r^2}{Z_r} \tag{4.78}$$

$$\left(\text{Fault VA}\right) = \frac{3V_r\, E_{an}}{Z_t}\left(\text{For ground fault}\right) \tag{4.79}$$

Then, in the per-unit system,

$$\left(\text{Fault VA}\right)\text{p.u.} = \frac{3V_r\, E_{an}}{\left[\dfrac{3V_r^2}{Z_r}\right]Z_t} = \frac{\dfrac{E_{an}}{V_r}}{\dfrac{Z_t}{Z_r}} = \frac{(E_{an})\,\text{p.u.}}{(Z_t)\,\text{p.u.}} \tag{4.80}$$

The same thing happened for phase fault:

$$\left(\text{Fault VA}\right) = \frac{\sqrt{3}V_r\, \sqrt{3}E_{an}}{Z_t} \tag{4.81}$$

Then in the per-unit system:

$$(\text{Fault VA})\ \text{p.u} = \frac{3V_r E_{an}}{\left[\dfrac{3V_r^2}{Z_r}\right] Z_t} = \frac{\dfrac{E_{an}}{V_r}}{\dfrac{Z_t}{Z_r}} = \frac{(E_{an})\ \text{p.u.}}{(Z_t)\ \text{p.u.}} \tag{4.82}$$

then

$$\left(\text{Fault VA}\right)\text{p.u.} = \frac{1}{Z_t} \tag{4.83}$$

Example 4.3

Draw the positive, negative, and zero sequences for the power system shown in Figure 4.26 (Figures 4.27–4.29)

Solution

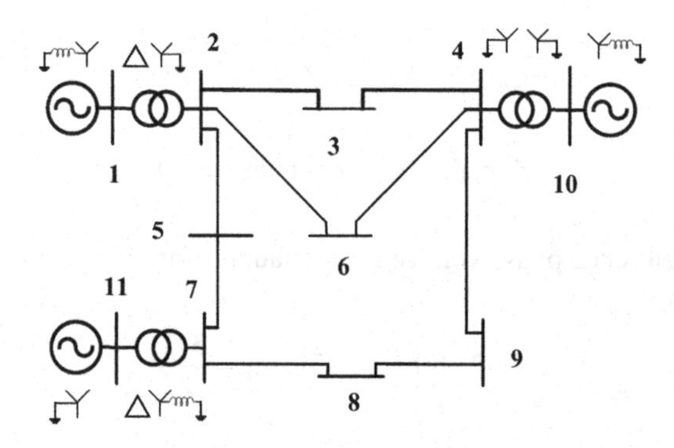

FIGURE 4.26 The one-line diagram for Example 4.1.

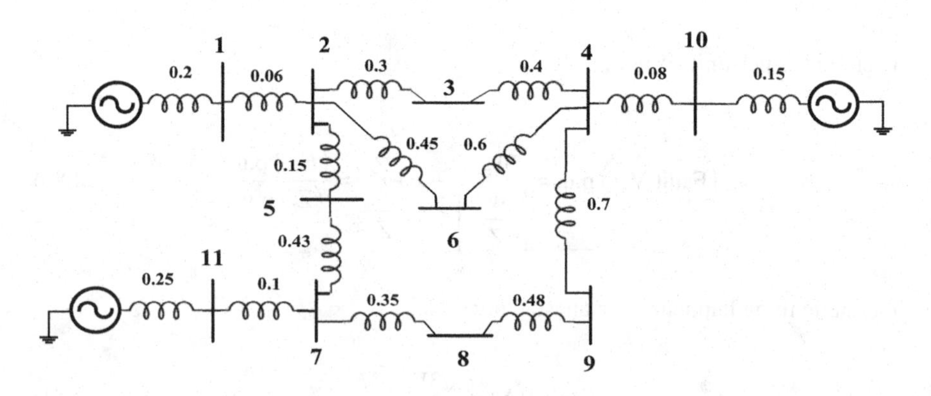

FIGURE 4.27 Positive-sequence network for Example 4.1.

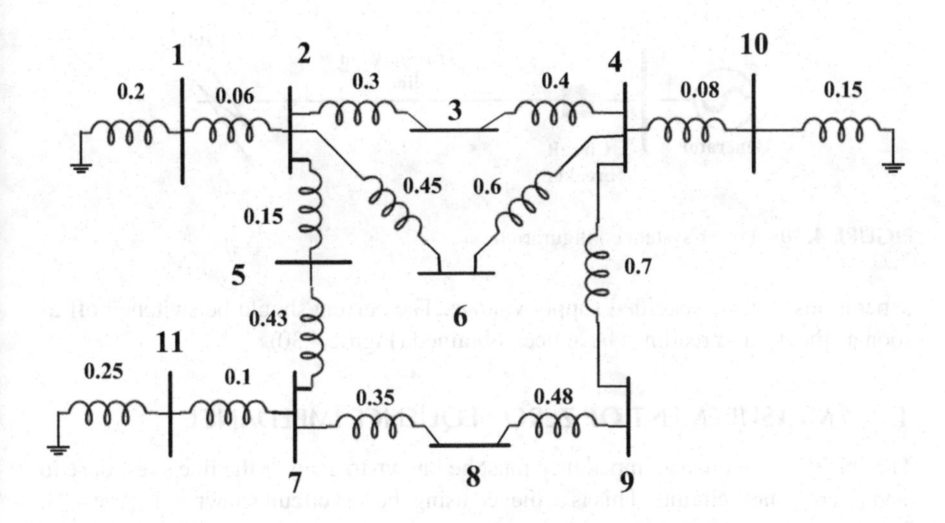

FIGURE 4.28 Negative-sequence network for Example 4.1.

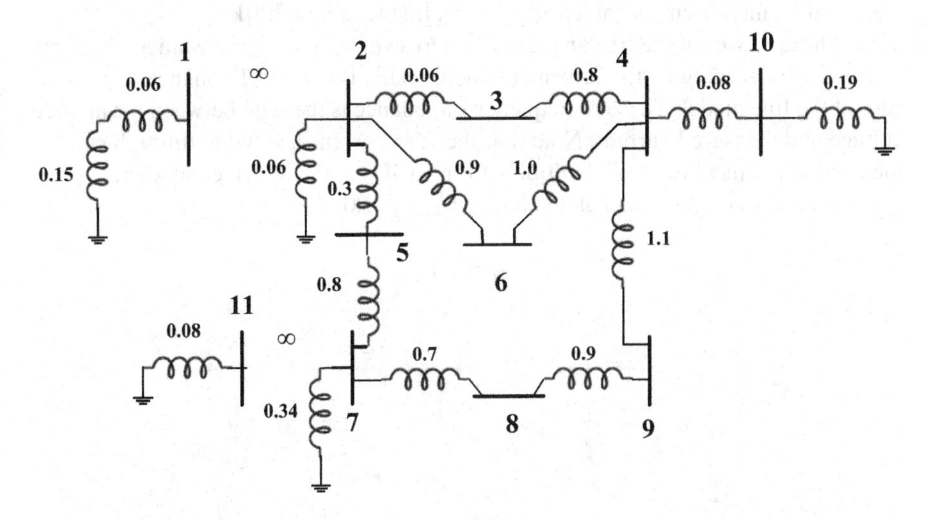

FIGURE 4.29 Zero-sequence network for Example 4.1.

4.14 FAULT ANALYSIS BY THE SCADA SYSTEM

Because of the high currents occurring in the event of a failure, the line model is operated at a rated voltage of 110 V during the experiments. At this rated voltage, too, the model realistically reproduces the values R, L, and CB. The measurements are carried out at a line length of 300 km. Two multimeters can simultaneously measure all voltages and currents at both lines' ends.

Increase the voltage of the feeding transformer at the start of the line in small increments to the specified value, ensuring that the maximum permissible amperage (2.5 A) is not exceeded anywhere. All short-circuit tests should be carried out

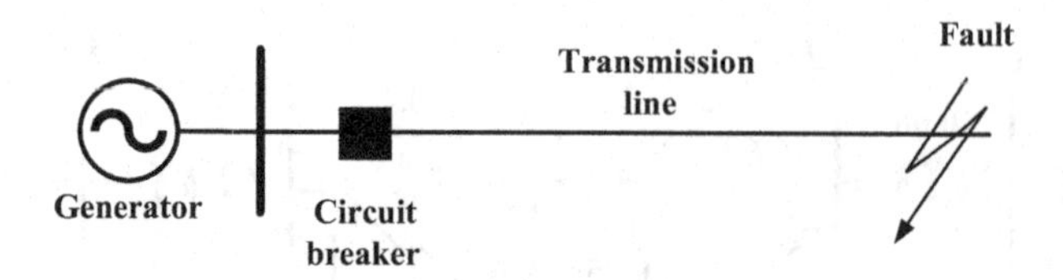

FIGURE 4.30 Power-system configuration.

expeditiously at the specified supply voltage. The current should be switched off as soon as the desired readings have been obtained (Figure 4.30).

4.15 MEASUREMENT OF ZERO-SEQUENCE IMPEDANCE

The line's zero-sequence impedance must be known to analyze the line's response to asymmetric short circuits. This is achieved using the test circuit shown in Figure 4.31. The negative-sequence impedance need not be determined separately, equal to the positive-sequence impedance in the case of static systems such as transmission lines. The conduct measurements at a transmission line length of 300 km.

All three phases of the line are connected to a single alternating voltage, the earth or earth wire serving as the return conductor. This is a neutral conductor N in the case of the line model. The zero-sequence impedance is the ratio between the applied voltage and measured current. Note that the zero current I_0 flowing through each of the conductors increases by a multiple of three $(3I_0)$ in the return conductor.

Conduct the measurement at a voltage of $V_{LN} = 100$ V.

$$I_a = 1.1$$

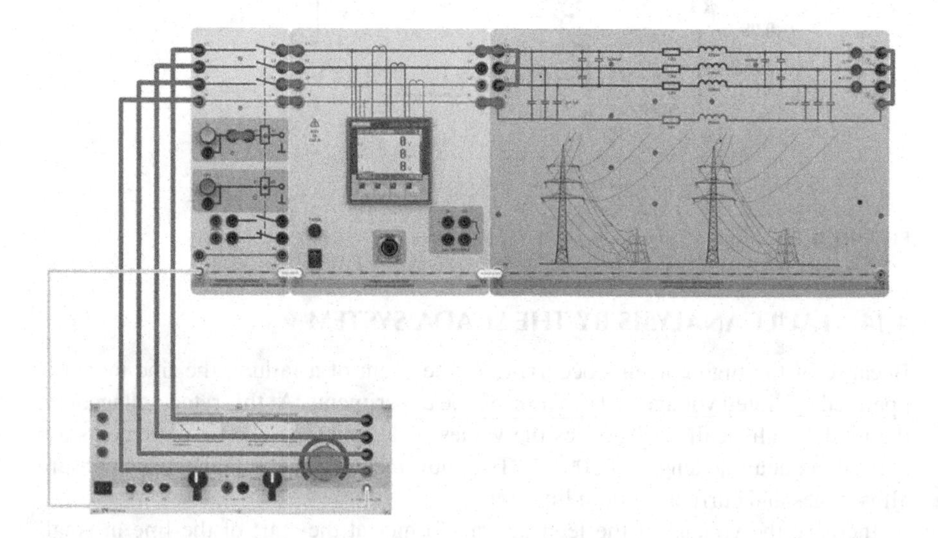

FIGURE 4.31 Lucas-Nülle GmbH test circuit for determining zero-sequence impedance emulator at PVAMU.

$$I_n = I_a/3 = 1.1/3 = 0.36$$

$$P = 34\text{ W},\quad Q = 314\text{ VAR},\quad S = 316\text{ VA}$$

Consequently, the value of the zero-sequence impedance

$$Z_0 = V_0/I_0 = 100/0.36 = 272\ \Omega.$$

The power measurement can be used to determine the active component of Z_0.

The active power consumed by the whole line was measured. Accordingly, $P_0 = 13$ W are distributed proportionally among the conductors and the return conductor. The resistor R_0 is determined from the equation

$$R_0 = P_0/I_0^2 = 13/0.362 = 100\ \Omega,$$

and the reactive component

$$X = \sqrt{Z_0^2 - R_0^2} = \sqrt{272^2 - 100^2} = 252.95\ \Omega.$$

The result is to be compared with a calculated value:

For

$$Z_{\text{TL}} = (7.2 + j\ 86.7)\ \Omega, \text{ and}$$

$$Z_E = (15 + j\ 75.4)\Omega \text{ are used}$$

$$\text{So } Z_0 = Z_m + 3 \cdot Z_E$$

$$\text{and, then } Z_0 = (52.2 + j\ 312.9)\Omega.$$

4.16 SYMMETRIC (THREE-POLE) SHORT CIRCUIT

For comparison, a three-pole short circuit is analyzed first. Set up the circuit as shown in Figure 4.32; connect the three outer conductors and the neutral conductor after the right-hand multimeter.

Raise the line-to-line voltage at the start of the line to 110 V, and then measure the parameters listed subsequently at a line length of 300 km.

$$\text{Voltage } L_1 - L_2 (\text{at line start}) = 0\text{ V}$$

$$\text{Current } L_1 = 1.1\text{ A}$$

$$\text{Active power } P \text{ per phase} (\text{at line start}) = 13\text{ W}$$

$$\text{Reactive power } Q \text{ per phase} (\text{at line start}) = 100\text{ VAR}$$

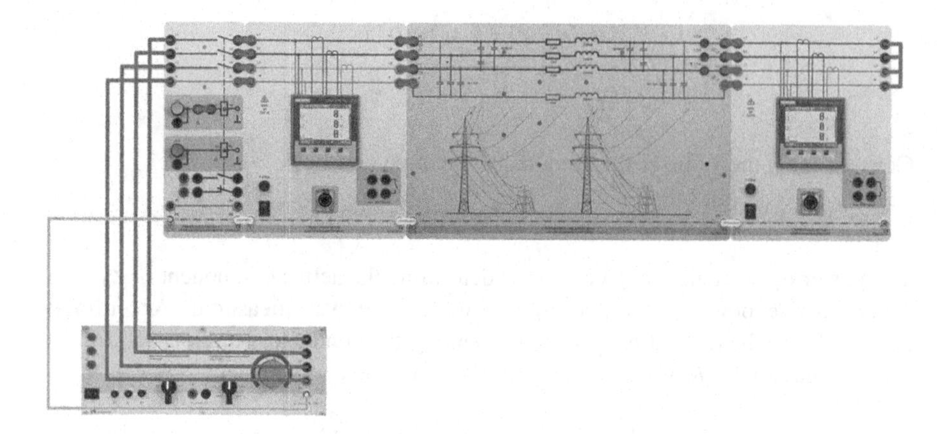

FIGURE 4.32 Lucas-Nülle GmbH test circuit for measurements in the event of a three-pole short-circuit emulator at PVAMU.

The power measurement reveals that the reactive power component predominates in the event of a short circuit. This is easily understandable because any existent load resistances are bridged during the short circuit, and a high-voltage line's reactance is always much higher than its active resistance.

When the connection to the neutral conductor is removed, and because a symmetric short circuit is involved, the measured values do not change. In the case of a three-pole short circuit, the return conductor remains de-energized (except for minor asymmetries attributable to the components).

4.17 ASYMMETRIC SHORT CIRCUITS

To determine currents and voltages in the case of unbalanced faults, the circuit only needs to be modified slightly. The same fault types specified in the theoretical part are simulated here.

4.17.1 Single-Pole Short Circuit (Earth Fault)

After the second multimeter shown in Figure 4.33, the outer conductor L_1 connects to the neutral conductor.

Raise the line-to-line voltage at the start of the line to 110 V and then measure the parameters listed next.

- Voltage $L_1 - L_2$ (at line start) = 177 V
- Current L_1 = 0.42 A
- Current L_2 = 0.17 A
- Current L_3 = 0.17 A
- The voltage of outer conductor L_2 concerning earth = 142 V
- The voltage of outer conductor L_3 concerning earth = 138 V
- Relationship for the short-circuit current: $I_{\text{sc single-pole}} = 3 \cdot E''/\left(Z_m + Z_g + Z_0\right)$

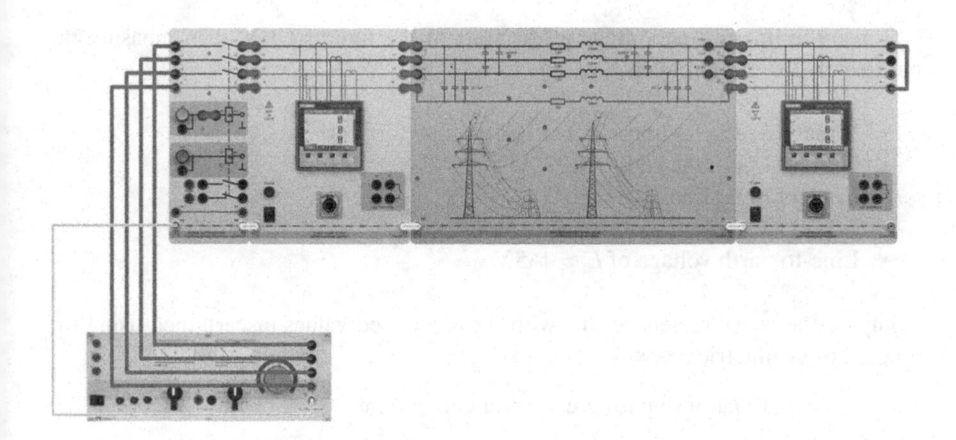

FIGURE 4.33 Lucas-Nülle GmbH test circuit for measurements in the event of a single-pole short-circuit emulator at PVAMU.

A value of $110\,\text{V}/\sqrt{3}$ is to be used for the driving voltage E''. Moreover, $Z_m = Z_g = (7.2 + j\ 86.7)\Omega$ and $Z_0 = (52.2 + j\ 312.9)\Omega$.

Accordingly, $Z_m + Z_g + Z_0 = (66.6 + j\ 486.3)\Omega$; value: 490.9 Ω.

The single-pole short-circuit current (value) is $I_{\text{sc single-pole}} = 0.39$ A.

A detailed calculation of the intact line-to-earth voltages is dispensed here. According to the equations in the previous section, these voltages have the following values: $V_2 = 84$ V and $V_3 = 87$ V.

4.17.2 Two-Pole Short Circuit with Earth Fault

After the right-hand multimeter shown in Figure 4.34, the outer conductors L_2 and L_3 connect to the neutral conductor. Use an additional ammeter to measure the sum $I_{\text{sc two-pole}}$ of the two currents.

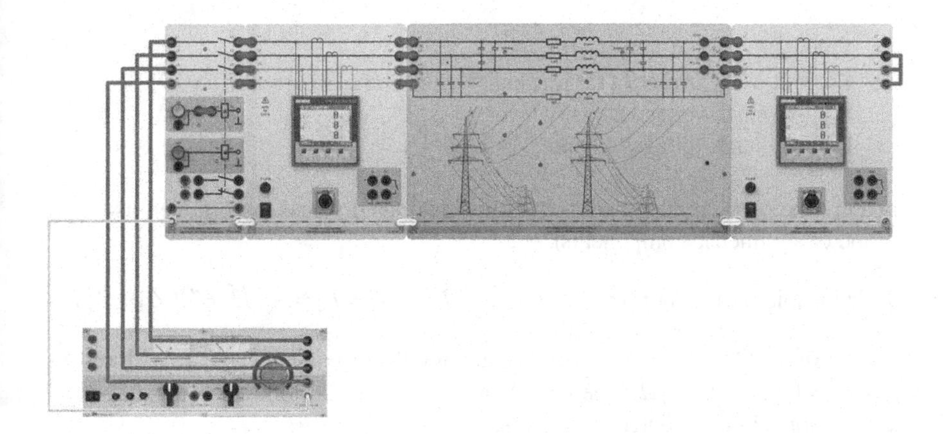

FIGURE 4.34 Lucas-Nülle GmbH test circuit for measurement in the event of a two-pole short-circuit emulator at PVAMU.

Raise the line-to-line voltage at the start of the line to 110 V, then measure the parameters listed subsequently.

- Voltage $L_1 - L_2$ (at line start) = 168 V
- Current L_2 (at line end) = 0.93 A
- Current L_3 (at line end) = 0.93 A
- In = 0.33 A
- Line-to-earth voltage of L_1 = 145 V

Compare the measurement results with the calculated values (ascertained using the method of symmetric components):

Relationship for the short-circuit current:

$$I_{\text{sc two-pole}} = j \cdot 3 \cdot E'' \cdot Z_g / (Z_m \cdot Z_g + Z_m \cdot Z_0 + Z_0 \cdot Z_g).$$

The impedances Z_m, Z_g, and Z_0 calculated previously result in the following denominator for the fraction discussed earlier:

$$N = (Z_m \cdot Z_g + Z_m \cdot Z_0 + Z_0 \cdot Z_g) = (-60{,}977 + j\ 14{,}807)\Omega^2.$$

Accordingly, the value of the two-pole short-circuit current is: $I_{\text{sc two-pole}} = 0.26$ A.

A detailed calculation of the two partial currents and intact line-to-earth voltage is dispensed here. The previous equations result in the following values: I_2 $I_2 = 0.69$ A, $I_3 = 0.66$ A, $V_1 = 94$ V.

4.17.3 Two-Pole Short Circuit without Earth Fault

In this case, simply disconnect outer conductors L_2 and L_3 from the neutral conductor. Raise the line-to-line voltage at the start of the line to 110 V and measure the parameters listed subsequently.

- Voltage $L_1 - L_2$ (at line start) = 167 V
- Current L_2 = Current L_3 (at line end) = Short-circuit current I_{sc} = 0.91 A
- Line-to-earth voltage L_1 = 109 V
- Line-to-earth voltages L_2 and L_3 = 52 V

Compare the measurement results with the calculated values (ascertained using the method of symmetric components):

Relationship for the short-circuit current: $I_2 = -I_3 = I_{sc} = -j\sqrt{3}\ E''/(Z_m + Z_g)$.

A value of 110 V/$\sqrt{3}$ is to be used again for the driving voltage E". Moreover, $Z_m = Z_g = (7.2 + j\ 86.7)\Omega$. The calculation results in the following amperages for the two short-circuited outer conductors:

$$I_2 = I_3 = I_{sc} = 0.63\,\text{A}$$

A detailed calculation of the three line-to-earth voltages is dispensed here. The equations in the previous section result in the following values for these voltages: $V_1 = 64$ V, $V_2 = V_3 = 32$ V.

4.18 EARTH FAULTS AND THEIR COMPENSATION

Although in practice, high-voltage systems are not operated with an isolated neutral point at nominal voltages = 110 kV. The line model also serves as a good demonstration of responses to earth faults with and without a quenching coil if such voltage levels are simulated.

4.18.1 Earth-Fault Compensation

Modify the circuit slightly by connecting the earth-fault quenching coil to the neutral conductor between the power switch and line model.

Figure 4.35 will use a line-to-line voltage of 100 V at the end of the line before the occurrence of an earth fault. Successively set all inductance values between 1.0 and 2.0 H on the coil, and measure the corresponding earth-fault currents. L_2 and L_3 voltages are 187 Vph.

4.18.2 Earth Fault with an Isolated Neutral Point

Set up the circuit as shown subsequently, leaving the neutral conductor disconnected from the transformer. After the right-hand multimeter produces an earth fault between the outer conductor L_1 and N (Figure 4.36).

Raise the line-to-line voltage at the end of the line, and measure the parameters listed next

- Voltage $L_1 - L_2$ (end of the line) = 219 V
- Current L_1 (end of the line) = 0.37 A

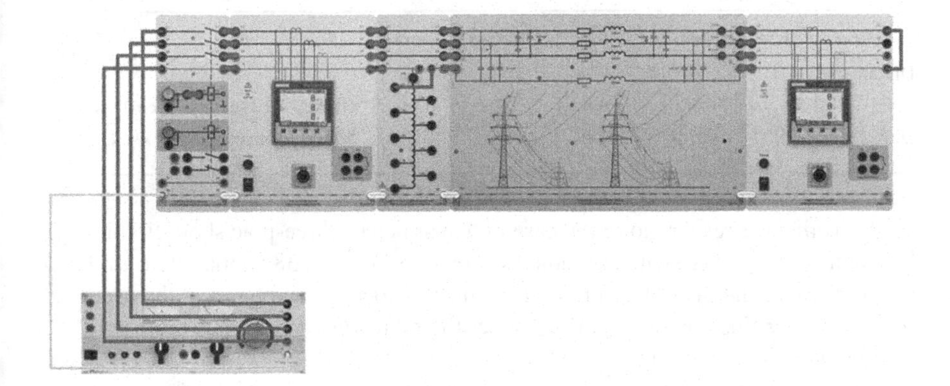

FIGURE 4.35 Lucas-Nülle GmbH test circuit for measurements with an earth-fault quenching coil fault emulator measurement at PVAMU.

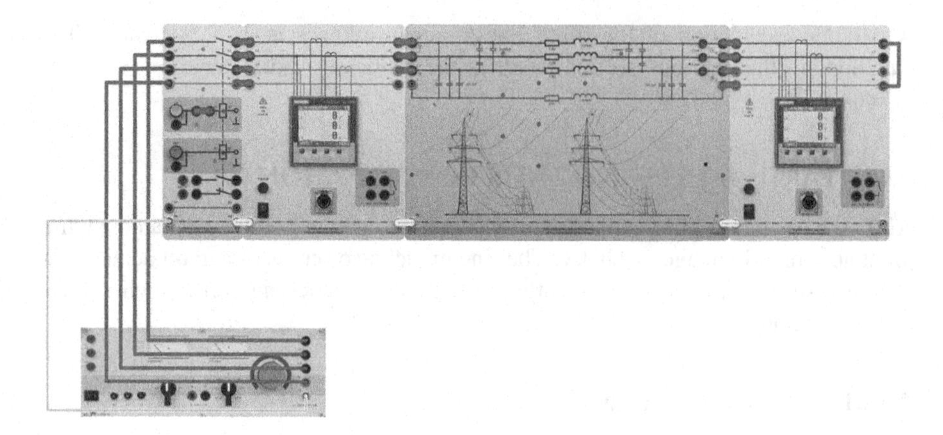

FIGURE 4.36 Lucas-Nülle GmbH test circuit for measurements in the event of an earth-fault emulator measurement at PVAMU.

- Line-to-earth voltages L_2 and $L_3 = 220\,\text{V}$
- Value of the earth-fault current: $I_{\text{E}} = \sqrt{3} \cdot U_N \cdot \omega \text{CE}$. Using a 2.2 μF for the line-to-earth capacitance results in $I_E = 0.16$ A.

4.19 OVERCURRENT TIME PROTECTION

The overhead transmission line receives a three-phase power supply and is loaded symmetrically at the receiving end. Located before the transmission line is a CB for disconnecting the line from the power supply in the event of a fault. The time overcurrent relay measures the current in each phase via a current transformer.

Figure 4.37 shows the circuit diagram and layout plan.

Using the LN SCADA software system, Set the load to its lowest level and set the relay's DIP switches as indicated in the table provided subsequently (active setting = green background).

Function	**Trip**	**Trip**	**Trip**	**Block $I_>$**	**Block $I_{>>}$**	**f**	**$t\,I_>$**	**$t\,I_>$**
DIP switch	1	2	3	4	5	6	7	8
ON	Inverse	Strong i.	Extreme i.	Yes	Yes	60 Hz	× 10 seconds	× 100 seconds
OFF	DEFT	DEFT	DEFT	No	No	50 Hz	× 1 seconds	× 1 seconds

To determine the reset ratio in the case of a symmetric, three-pole short circuit:

Connect the power-switch module as shown in Figure 4.38 so that the right-hand side is bridged and the left-hand side is connected to all three phases at the overhead line's end. For this connection, disconnect the relay output from power switch 1 to prevent premature tripping.

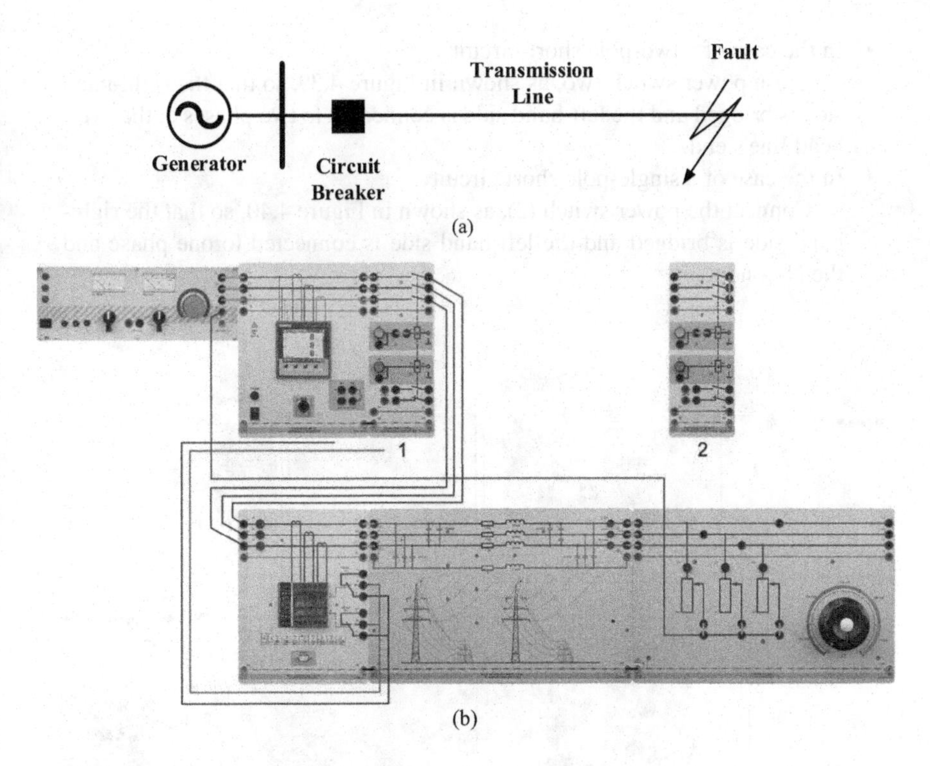

FIGURE 4.37 Lucas-Nülle GmbH test circuit for measurements in the event of an overcurrent time protection. (a) Power-system configuration. (b) Layout plan.

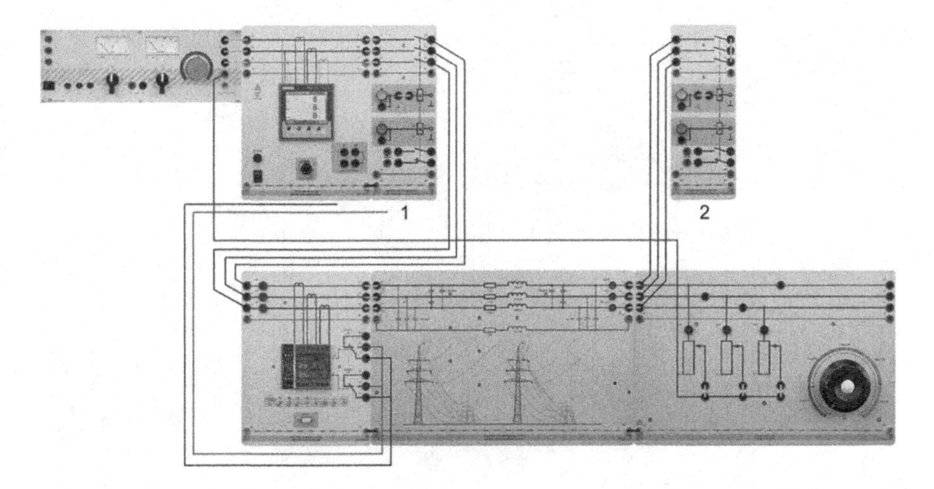

FIGURE 4.38 Lucas-Nülle GmbH three-pole short-circuit emulator at PVAMU.

- In the case of a two-pole short-circuit

  Close power switch two, as shown in Figure 4.39, so that the right-hand side is bridged and the left-hand side is connected to two phases at the overhead line's end.
- In the case of a single-pole short circuit

  Connect the power switch (2), as shown in Figure 4.40, so that the right-hand side is bridged and the left-hand side is connected to one phase and the N-conductor.

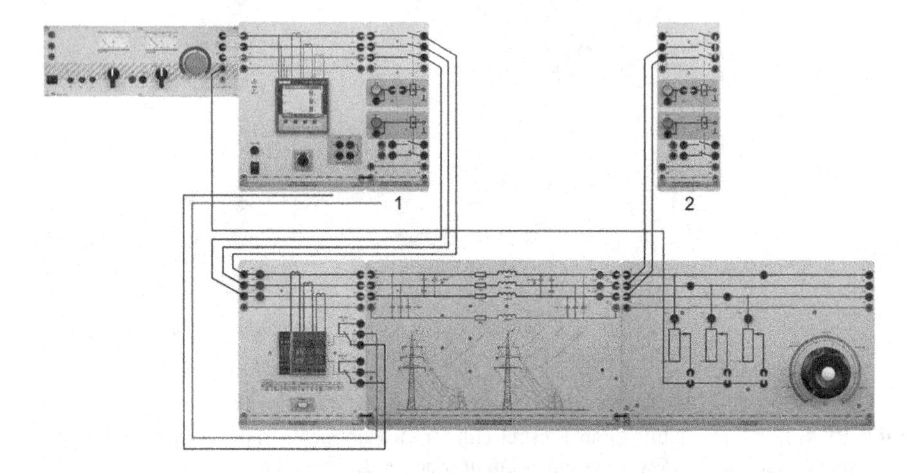

FIGURE 4.39 Lucas-Nülle GmbH two-pole short-circuit emulator at PVAMU.

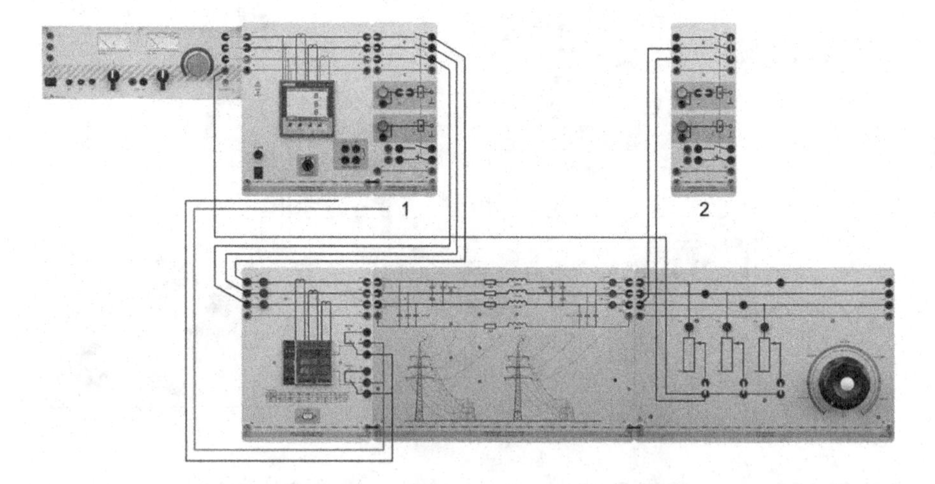

FIGURE 4.40 Lucas-Nülle GmbH single-pole short-circuit emulator at PVAMU.

PROBLEMS

4.1. A single-line diagram of a power system is given in Figure 4.41. Positive-sequence, negative-sequence, and zero-sequence reactances in per unit are also given. The synchronous generator operates at a rated MVA and is 5% above the rated voltage. The neutrals of the generator and Δ-Y transformers are solidly grounded. The motor neutral is grounded through a reactance $X_n = 0.03$ per unit on the motor base. Pre-fault voltage is $V_F = 1.05\angle 0°$ per unit. The pre-fault load current and Δ-Y transformer phase shifts are neglected.

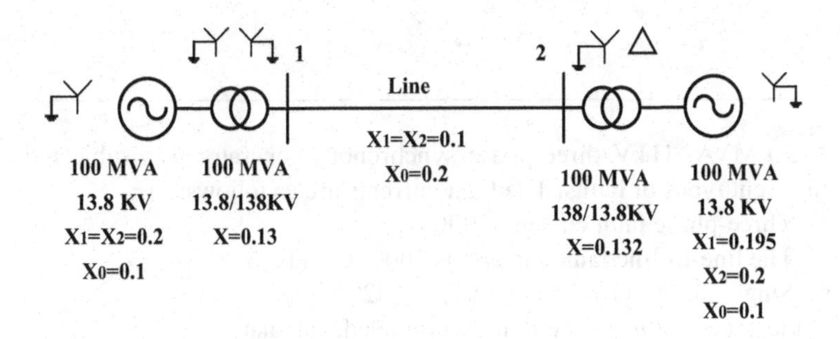

FIGURE 4.41 Power-system diagram for Problem 4.1.

a. Draw the system's per-unit zero-sequence, positive-sequence, and negative-sequence networks on a 100 MVA, 14.8-kV base.
b. Reduce the sequence networks found in part (a) to their Thevenin equivalents, as viewed from bus 2. Draw Thevenin equivalent circuits.
c. Calculate the subtransient fault current per per-unit for a bolted single line-to-ground short circuit from phase a to ground at bus 2.

4.2. A simple three-phase power system is shown in Figure 4.42. Assume that the ratings of the various devices in this system are as follows:

- Generator G_1: 100 MVA, 14.8 kV, $X_1 = X_2 = 0.2$ p.u., $X_0 = 0.25$ p.u.,
- Generator G_2: 100 MVA, 20.0 kV, $X_1 = X_2 = 0.25$ p.u., $X_0 = 0.25$ p.u.,
- $X_n = 0.05$ p.u.
- Transformer T_1: 100 MVA, 14.8/138 kV, $X = 0.10$ p.u.
- Transformer T_2: 100 MVA, 20.0/138 kV, $X = 0.15$ p.u.
- Each Line: $X_1 = X_2 = 40\ \Omega$, $X_0 = 100\ \Omega$.

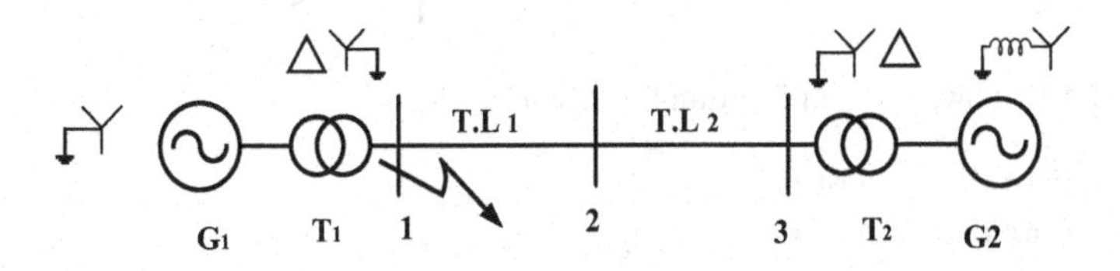

FIGURE 4.42 Power-system diagram for Problem 4.2.

Fill table for the fault at bus no. 1.

Types Required \ Fault	Three-Phase Fault	SingleLine-to-Ground Fault	Line-to-Line Fault	Double Line-to-Ground Fault
I_f				
I_a				
I_b				
I_c				
V_a				
V_b				
V_c				

4.3. A 50 MVA, 11 kV three-phase synchronous generator was subjected to different types of faults. The fault currents are as follows:

- Three-phase fault current: 2000 A,
- The line-to-line fault current: is 2600 A, and
- Single line-to-ground fault current: 4200 A.

If the generator neutral is solidly grounded, calculate

i. Positive-sequence reactance of the generator.
ii. Negative-sequence reactance of the generator.
iii. Zero-sequence reactance of the generator. Neglect the resistance.

4.4. In a three-phase, four-wire power system, current in *A*, *B*, and *C* lines under abnormal conditions of loading are:

$$I_A = 100\angle 50^\circ \text{ A}, I_B = 120\angle -30^\circ \text{ A, and } I_C = 70\angle -10^\circ \text{ A}$$

A. Calculate A-line's zero, positive, and negative phase sequences.
B. Calculate the return current in the neutral connection.

4.5. For the power system shown in Figure 4.43. All values are given in per unit based on the same base. Assume the pre-fault voltage is 1.0 p.u. with zero phase angle (Table 4.2).

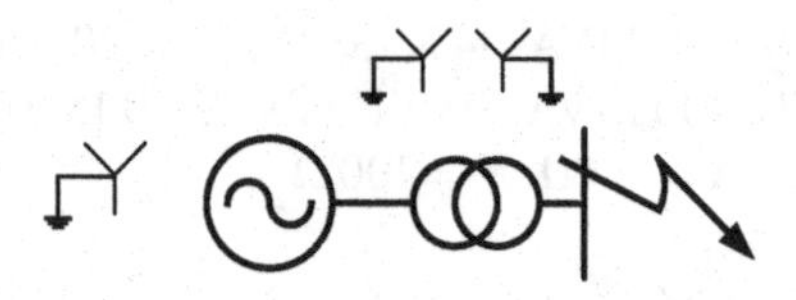

FIGURE 4.43 Power-system diagram for Problem 4.5.

A. Find the ratio of $I_{f\,three\text{-}phase}/I_{f\text{-}SLG}$.
B. Find the ratio of $I_{f\,three\text{-}phase}/I_{f\text{-}DLG}$.

Item	Connection	X_1	X_2	X_0
Generator	Grounded star	0.1	0.1	0.05
Transformer	Grounded star for both sides	0.2	0.2	0.2

4.6. In a three-phase power system, the voltage in phases A, B, and C lines under abnormal conditions are:

$$V_A = 100\angle 50°\ \text{V},\ V_B = 120\angle -30°\ \text{V, and}\quad V_C = 70\angle -10°\ \text{V}$$

A. Calculate the zero, positive, and negative voltage phase sequences in line A.
B. Calculate the line voltage V_{AB}, V_{BC}, and V_{CA}

4.7. For the power system shown in Figure 4.44. All values are given in per unit based on the same base. Assume the pre-fault voltage is 1.0 p.u. with zero phase angle.

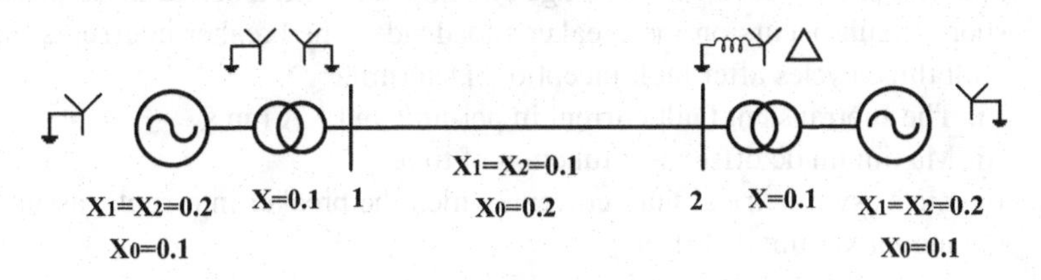

FIGURE 4.44 Power-system diagram for Problem 4.7.

Draw the zero, positive, and negative-sequence system and find the fault current for

A. The three-phase fault occurs on bus 2.
B. The line-to-line fault occurs on bus 2.
C. Double line-to-ground fault occurs on bus 2.

4.8. The synchronous generator in Figure 4.45 operates at rated MVA, 0.9 p.f and lagging, and at 5% above-rated voltage when a three-phase short circuit occurs at bus 2. Calculate the per unit and the actual values of (a) subtransient fault current, (b) subtransient generator and motor currents, neglecting pre-fault current, and (c) subtransient generator and motor currents, including pre-fault.

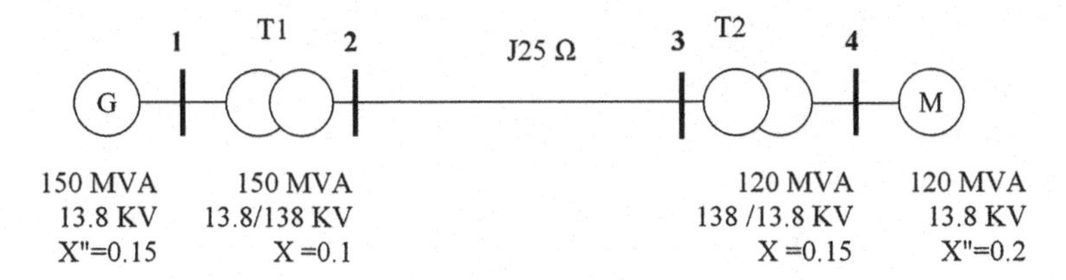

FIGURE 4.45 Power-system diagram for Problem 4.8.

4.9. For the power system shown in Figure 4.34 (in *Problem 4.8*), a bolted double line-to-ground fault occurred at bus 2. Use $S_b = 150$ MVA, $V_b = 138$ kV in the transmission line circuit, assume that $X'' = X_2$, $X_0 = 0.05$ p.u. for the generator and motor, $X_0 = 3X_1$ and $X_1 = X_2$ for the transmission line, Transformers are the star to ground connected. Calculate in p.u.

i. The subtransient fault current in each phase.
ii. The neutral fault current.
iii. The motor contribution to the fault current.

4.10. A bolted short circuit occurs in the R-L circuit, with $V = 22\,\text{kV}$, $X = 12\ \Omega$, $R = 0.55\ \Omega$, and dc offset. The circuit breaker opens 3.5 cycles after fault inception. Determine:
   i. The r.m.s ac fault current.
   ii. The r.m.s momentary current at $\tau = 0.5$ cycles passes through the breaker before it opens.
   iii. The r.m.s asymmetrical fault current that the breaker interrupts.

4.11. A 600 MVA 18 kV, 60 HZ synchronous generator with reactances $X_d'' = 0.15$, $X_d' = 0.35$, $X_d = 1.25$ p.u. and time constants $\tau_d'' = 0.015$, $\tau_d' = 1.8$, $\tau_a = 0.25$ seconds is connected to a circuit breaker. The generator operates at 6% above-rated voltage and no load when a bolted three-phase short circuit occurs on the breaker's load side. The breaker interrupts the fault three cycles after fault inception. Determine:
   i. The subtransient fault current in per-unit and KA r.m.s
   ii. Maximum dc offset as a function of time
   iii. r.m.s asymmetrical fault current, which the breaker interrupts, assuming maximum dc offset.

5 Fuses and Circuit Breakers

5.1 INTRODUCTION

A fuse is a device linked in series with conductors attached to the current-carrying load device. The oldest and most operational system for safety fuses was introduced in 1890 when Thomas Edison invented electric fuses. There are many kinds of fuses, and every type of fuse has a purpose. The various types of fuses, their design, function, and applications are discussed in this chapter.

The fuse is a level detector consisting of a sensor and an intermediate sensor. A fuse element will respond to a high current flow when mounted in series with the system being protected and/or triggered. The melting time of the fuses is inversely proportional to the current flowing through the fuse. During interrupting the current flow, the fuse connection is broken. For delivering multiple shots, there can be mechanical arrangements of fuse configuration.

Fuses may only be capable of interrupting currents up to the maximum short-circuit ratio or limiting the size of the short-circuit current by interrupting the flow until the maximum value is reached. This current-limiting action is a significant feature with applications in many industrial applications and low-voltage installations.

The description of fuses and circuit breakers, types, and specifications are discussed in this chapter. The chapter includes an introduction to the construction and working of a fuse, the characteristics of a fuse, and fuses applications. Also, the chapter discusses high-voltage circuit breakers.

5.2 LOAD AND FUSE CURRENT

Under normal conditions, the load current is less than the fuse's ampere rating, and the fuse remains intact to connect the load to the source. The current through the link will increase when an overload or short circuit occurs. Once the link has completely dissolved, there is an open circuit in the fuse, and the circuit will remain open until the fuse is replaced. Fuses are a "one-time" use device. Figure 5.1 shows describes three kinds of fuses.

When the load current is lower than the fuse's rated current under normal conditions, the fuse remains intact to connect the load to the source. The current through the connection will rise when an overload or short circuit occurs. Once the connection is fully dissolved, an open circuit is present in the fuse, and the circuit stays open until replaced. Fuses in a system are for "one-time" use only.

The fuse link resistance is extremely low during normal service when the fuse runs less than the constant current rate. The link's temperature will eventually reach

DOI: 10.1201/9781003394389-5

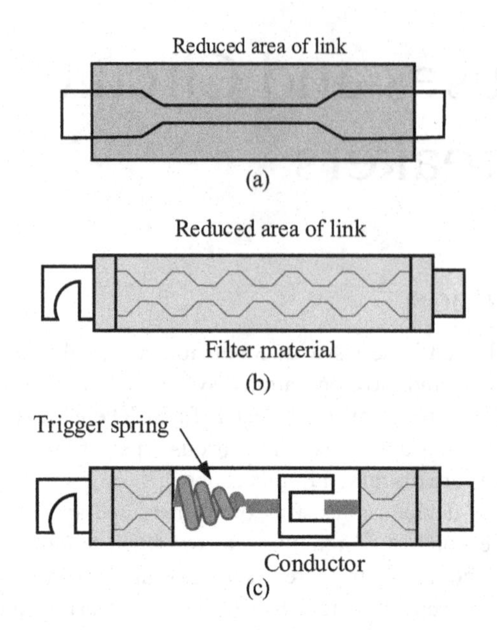

FIGURE 5.1 Short-circuit elements. (a) Non-current limiting, non-time-delay fuse; (b) current-limiting non-time-delay fuse; and (c) current-limiting time-delay fuse.

a level that causes the bound portion of the link to melt if an overload current occurs and lasts for more than a short time. The results of this overload lead to

i. A gap creation, and an electric arc formation.
ii. The gap width increases as the arc cause the connecting metal to burn.
iii. Eventually, arc resistance reaches a degree so high that the arc cannot continue and is extinguished.
iv. The electricity would be cut off entirely because of the failure of the fuse.

Usually, fuse requirements depend on the following four variables (Figure 5.2):

i. *DC rating*: Without melting and washing, the fuse can hold RMS current indefinitely.
ii. *Voltage rating*: This RMS voltage defines the fuse's ability to suppress the internal arc after the fuse connection's melting. Its voltage level must be able to survive a blown fuse. Most low-voltage fuses have a 250–600 V rating, and the fuse rate of medium voltage varies from 2.4 to 34.5 kV.
iii. *Rated interrupt current*: The fuse can safely cut the biggest RMS asymmetric current. For medium voltage, the normal interrupting rate is 65, 80, and 100 kA.
iv. *Response time*: A fuse's melting and winding time depend on the overcurrent magnitude defined by the curve of "current time."

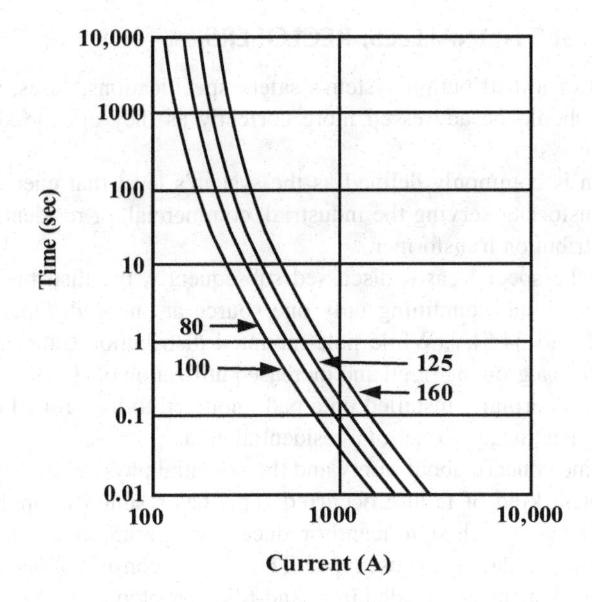

FIGURE 5.2 Current versus time–fuse characteristics.

Technically, the fuses form the protective relay background, especially for radial feeders such as distribution lines or auxiliary systems of power stations. The two main drawbacks of fuses are the following.

i. *Delay*: One-shot feature requires replacing a blown fuse before restoring service. This means delay, the need for the correct backup fuses, and the qualified maintenance personnel who have to go in and replace fuses in the field. The multi-shot feature can be provided by installing several parallel fuses and a mechanical trigger mechanism. The blowing of one of the fuses automatically shifts its position.
ii. A single-phase to a ground fault would cause a single fuse to explode in a three-phase circuit, de-energizing only one phase and allowing connected equipment – such as motors – to remain connected to the remaining phases, with subsequent increased heating and vibrations due to the unbalanced voltage supply.

The protective relays were developed as logic components separate from the circuit interrupt function to overcome these drawbacks. Relays require low-level inputs (voltages, currents, or contacts). They derive their input from power transformers, such as current or voltage and switching contacts. They are only fault detection devices and require an associated interrupt device – a circuit breaker – to clear the fault. One of the most important developments was the separation of the fault detection function and the outage function, giving the relay designer the ability to design a protection system that matched the power system's needs.

5.3 FUSES, SECTIONALIZES, RECLOSERS

In the context of a distribution system's safety specifications, fuses, sectionalizes, and reclosers should be addressed more correctly as they are the key protection devices for that system.

Distribution is commonly defined as the system's level that energizes the final step-down transformer serving the industrial, commercial, or residential consumer, that is, the distribution transformer.

Except for the special cases discussed subsequently, the distribution system is almost entirely radial, containing only one source at one end. Operating voltage ranges from 2.4 to 34.5 kV. While pole-mounted distribution transformers remain prevalent, there is a growing trend and increased utilization of (URD) systems. URD transformers are normally installed on a pad, mounted to the ground on a concrete pad, generally at a distant corner of a residential area.

There is some concern about safety and the potential physical damage that would occur if a violent kind of failure occurred. This has resulted in applying current-limiting (CL) fuses, which significantly reduces the "permissible" power of a high current fault compared to other fuse types. The CL fuse consists of one or more silver wires or ribbon elements suspended in a sand-filled envelope.

When operating against a high current, the fusible element instantly melts along its length. The resulting arc quickly loses its thermal energy to the surrounding sand. The rapid thermal energy loss limits the current to a small value known as the "let" current. The most used protection device in the distribution circuit is the fuse. Fuses differ widely in properties from one manufacturer to another. The fuses' characteristic current time curves are given in the minimum melt and total clearance times, as shown in Figure 5.3. Minimum melting is the time between the initiation of a current large enough to cause the melting of the element responding to the current and the moment when arcing occurs.

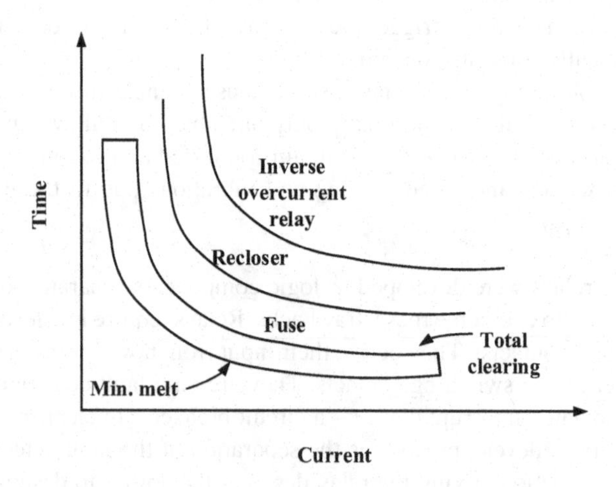

FIGURE 5.3 Overcurrent device coordination.

Total time of clearance (TCT) is the total time elapsed from the onset of overcurrent to terminal circuit interruption, i.e.,

$$\text{TCT} = \text{Minimum melting limit} + \text{bending time}.$$

It is important not to mix fuse types such as ANSI "K" or "T." Operational characteristics differ enough that there may be a lack of cohesion. While implementing other safety devices, such as overcurrent relays, the same precaution must be taken. In general, while the fuses or relays may be identical, there are variations in output and subtle differences in operating characteristics that will trigger coordination, current level, and capture mode difficulties.

Fuses have various load-carrying capabilities with various fusion curves, which must be considered. Most manufacturers' application tables display three load current values: continuous capture, hot load, and cold load capture. The continuous load is the highest estimated current that will not affect the fuse for three hours or more. The sum of that, a hot load, is without inflating; it can be carried continuously, interrupted, and reactivated instantly. A 30-minute break follows the cold load and is elevated due to loss of diversity when service is restored. Because during this time, the fuse will also cool down, capturing the cold and hot loads that will approach similar values.

International standards and requirements have been established for ease of replacement and installation due to the large size of the distribution system's equipment. Unique distribution systems-related equipment such as potheads (transition conductors and insulators) monitor the expense and alternative inventory rather than the specific application. Besides the fuse itself, the interrupters are sectional devices and reclosers. It is difficult to interrupt the partition by error.

It "counts" the times it "sees" a fault current and turns on after a predetermined number while the circuit de-energizes. The recloser cannot interrupt the fault and re-shutdown automatically in a programmed sequence.

Figure 5.4 shows the three locations of a fault

i. The branch fuse must be removed in location A, leaving the service unobstructed to the mainline and other branches.
ii. The splitter should remove the fault in location B, but since the splitter cannot cut the fault, the passenger makes the actual clearing. However, the divider "sees" the fault current and records a single number. The divider

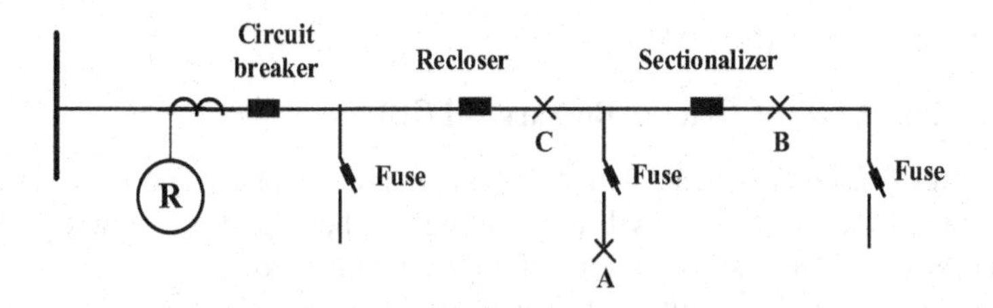

FIGURE 5.4 Radial power system.

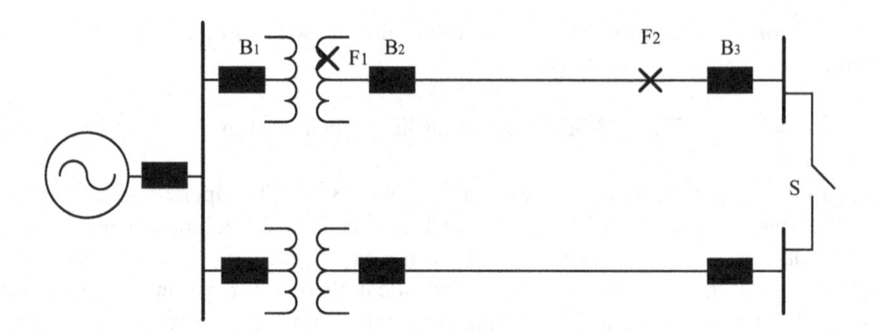

FIGURE 5.5 Feeder to feeder switching.

will experience the second count of the fault current if the fault remains. The recloser will get stuck again, allowing section opening and error clearing; the recloser will successfully reset and restore service until broken.

iii. For fault in location C, the recloser jogs and re-closes as programmed. The divider does not see the error and does not calculate.

The relays in a distribution station must recognize these potential short-circuit current differences in the same way as in a grid system.

Switching from feeder to feeder provides a backup source if a substation transformer is out of service or part of the distribution system must be deactivated.

In Figure 5.5, switch *S* is normally open, and each terminal transformer feeds its load. In the event of a permanent failure at F_1, the sectional parts or re-locking devices on the transformer side will automatically open. The line must then be de-energized by opening the circuit breaker, closing switch *S*, and transferring the remaining load to the other transformer. For the transformer fault at F_2, the breakers of stations (B_1) and (B_2) are opened, and the entire load can be fed from the other station.

Switching from one substation to another changes the magnitude and path of the fault current, and when applying and tuning line defense systems, this must be considered. Similarly, in the distribution system, the cogenerator's use adds another source of electricity, both separate and far from the utility substation, which influences the fault current's size and direction. Usually, this move is performed manually by assigning workers to various locations. If the transition could be performed automatically, a substantial saving of time and resources would be possible.

5.4 ELCB, MCB, AND MCCB

5.4.1 Earth Leakage Circuit Breaker (ELCB)

This breaker is used to detect earth leakage faults. Once phase and neutral are connected to the ELCB, a current will flow through a phase, and the same current will have to return to neutral so that the resulting current is zero.

Once there is an earth fault on the load side, the phase current will pass directly through the earth and will not return to neutral through the ELCB. This means that it

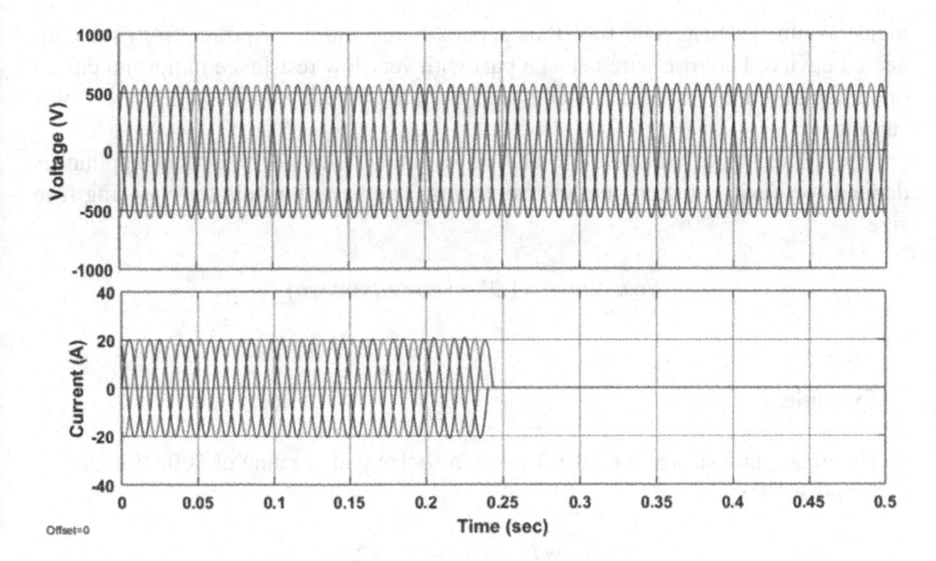

FIGURE 5.6 Voltage and current waveforms at generator-transmission line terminals.

will protect the other circuits from defective loads once the sideways current travels and will not return to the same point due to this difference in the current ELCB. If the neutral current is not earthed, the fault current will be high, the full fault current will return through the ELCB, and there will be no difference in current.

5.4.2 Miniature Circuit Breaker (MCB)

It is thermal, operated, and used for short-circuit protection in small current rating circuits. Normally it is used where the normal current is <100 A.

5.4.3 Molded Case Circuit Breaker (MCCB)

It is thermal operated for overloading current and magnetic operations for instant trip in a short-circuit condition. Under-voltage and under-frequency may be inherent. Normally it is used where the normal current is more than 100 A (Figure 5.6).

5.5 CONSTRUCTION AND WORKING OF A FUSE

A fuse consists of a low-resistance wire of metal covered by a material that is not combustible. On the other hand, electrical appliances can be affected without a fuse and circuit breaker because they cannot handle the overcurrent according to their rated limits and connect and mount in series with a circuit and a system that requires a short circuit or overcurrent protection.

The working theory of the fuse is based on the "heating effect of current," that is, when a short circuit happens, a rise in current or a mismatched load is connected because of the heat from the heavy current flowing through it, where the thin wire melts within the fuse. The fuse thus separates the source of power from the connected device. The fuse wire is just a part with very low resistance in normal circuit operation and does not affect the device's normal operation connected to the power supply.

The choice of a suitable fuse and its estimated size for electrical appliances depends on various factors and environments. The basic formula for choosing fuse size

$$\text{Fuse rating} = 1.25 \times (\text{power/voltage})$$

Example 5.1

Find the right fuse size for a 10 A two-pin socket with a rating of 500 W socket supply a 110 V.

$$(500\,\text{W}/120\,\text{V}) \times 1.25 = 5.2\,\text{A}$$

The power rating can be controlled through the two-pin socket and the main supply voltage in single-phase 120 V AC in the United States.

But you should go for the max, i.e., 6 A fuse rating instead of 5.2 A, for safe and reliable operation of the circuit.

5.6 CHARACTERISTICS OF A FUSE

Different types of fuses can be categorized according to the characteristics as follows:

i. Current rating and current-carrying capacity of fuse
ii. Voltage rating of fuse
iii. Breaking capacity of a fuse
iv. I^2t value of fuse
v. Response characteristic
vi. Rated voltage of the fuse
vii. Packaging size

The above categories are briefly described as follows.

5.6.1 Fuse Current-Carrying Capacity

The current-carrying capacity is the amount a fuse can easily conduct without interrupting the circuit.

5.6.2 Breaking Capacity

The maximum current value that can safely be interrupted by the fuse is called breaking capacity and should be higher than the prospective short-circuit current.

5.6.3 Rated Voltage of Fuse

The fuse can comfortably accommodate a maximum voltage rating, anticipating the current power of the current. For example, if a fuse is built for 32 V and cannot be used with 120 V, different insulation is needed in the different fuses operating at different voltage levels. Each fuse has a maximum allowable voltage rating. Voltage rating rules, HV (high voltage), LV (low voltage), and miniature fuses can be fused.

5.6.4 I^2t Value of Fuse

The I^2t terms related to fuses are normally used in short-circuit conditions. The quantity of energy dissipated is cleared when the fuse element clears the electrical fault.

The operation time of the fuse at high current levels is inversely proportional to the square of the current during the pre-arcing stage and proportional to the voltage during the arcing stage.

For any conductor, its temperature rise depends on the I^2t factor. The empirical formula can calculate this factor:

For copper conductors

$$I^2t = 11.5 \times 10^4 \ A^2 \log_{10} \frac{273 + \theta_m}{273 + \theta_o}$$

For aluminum conductors

$$I^2t = 5.2 \times 10^4 \ A^2 \log_{10} \frac{273 + \theta_m}{273 + \theta_o}$$

I = Short-circuit current (A)
t = Duration of the short circuit (seconds)
A = Net cross-sectional area of the conductor (mm^2)
θ_o = Initial temperature of the conductor (°C)
θ_m = Final temperature of the conductor

If the pre-arcing I^2t is not exceeded, there will be no fuse performance deterioration. This is taken into account when discrimination is required between fuses. If the total I^2t of the smaller fuse is less than the pre-arcing I^2t of the larger fuse, then the smaller fuse would operate without causing any deterioration of the larger fuse.

Example 5.2

It is proposed to use a No.30 AWG copper wire as a fuse element. If its initial temperature is 50°C, calculate the following:

a. The I^2t needed to melt the wire (copper melt at 1083°C).
b. The time needed to melt the wire if the short-circuit currency is 30 A.

Solution

$$30\,\text{AWG} = 0.0507\ \text{mm}^2$$

a. $I^2t = 11.5\times10^4\ A^2 \log_{10}\dfrac{273+\theta_m}{273+\theta_o}$

$$= 11.5\times10^4\ \times 0.0507^2 \times \log_{10}\frac{273+1083}{273+50}$$

$$= 184\ A^2\ \text{seconds.}$$

b. For a current of 30 A:

$$30^2\ t = 184$$

$$t = 0.2\ \text{seconds}$$

Example 5.3

It is proposed to use a No.30 AWG copper wire as a fuse element. If its initial temperature is 60°C, calculate the following:

i. Tha I^2t needed to melt the wire (aluminum melt at 1000°C).
ii. The time needed to melt the wire if the short-circuit currency is 25 A.

Solution

From fuse tables of wires 30 AWG = 0.0507 mm².

i. For aluminum conductors

$$I^2t = 5.2\times10^4\ A^2 \log_{10}\frac{273+\theta_m}{273+\theta_o}$$

$$= 5.2\times10^4\ \times 0.0507^2 \times \log_{10}\frac{273+1000}{273+60}$$

$$= 77.844\ A^2\ \text{seconds.}$$

ii. For a current of 25 A:

$$25^2 t = 77.844$$

$$t = 0.125\,\text{seconds}$$

5.6.5 Response Characteristic of a Fuse

The speed at which a fuse blows depends on how much current passes through the wire.

The more current passes through the wire, the quicker the reaction time.

The response characteristic indicates the response time to an occurrence that is overcurrent.

Ultra-fast fuses or rapid fuses are called fuses that react rapidly to an overcurrent state.

Most semiconductor devices are fused because semiconductor devices are too easily destroyed by excessive current.

Another fuse is called a slow-burning fuse, and switching fuses do not respond to an overcurrent event immediately but rather detonate several seconds after the overcurrent occurs. These fuses have found their application in motor control electronic systems because they take in much more current in starting.

5.7 CLASSIFICATION OF FUSES

Fuses can be categorized according to the usage of various applications:

i. Just fuse for one time,
ii. Restored feature,
iii. Limiting and present, and
iv. Non-CL fuses.

Once fuses are used, the consumer needs to manually replace these fuses; switch fuses are inexpensive and commonly used in most electronics and electrical systems. If an overcurrent, overload, or unmatched load link event occurs.

On the other hand, after the service, the resettable fuse resets automatically when the device faults.

The present limiting fuse generates a high resistance for a short time. In contrast, in high current flow, the non-CL fuse produces an arc to disrupt the corresponding circuit's current.

5.8 TYPES OF FUSES

Different fuses are available in the market and can be categorized based on different aspects. Fuses are used in AC as well as DC circuits.

Fuses can be divided into two main groups according to the type of supply voltage:

i. AC fuses
ii. DC fuses

There is a slight difference between AC and DC fuses used in the power systems, which have been reviewed as follows.

5.8.1 DC Fuses

In the case of a DC system, when the metallic wire melts due to the heat generated from the excess current, the electric arc is produced, and it is very difficult to extinguish this arc due to the constant value of the DC. The DC fuse is slightly larger than the AC fuse to reduce the arc by increasing the distance between the electrodes to reduce this arc.

5.8.2 AC Fuses

In the case of an AC system, the voltage at a frequency of 50 or 60 Hz changes the capacitance from 0 to 50 or 60 times every second so that the arc can be easily mutated compared to DC. Therefore, AC fuses are a bit small compared to DC fuses. The fuses can be classified based on one-time or multiple operations.

Depending on the fuse current rating, the fuses can be one of the following types for LV applications:

1. Semi-enclosed fuse (rewireable),
2. Cartridge fuse, and
3. High rupturing capacity (HRC) or high breaking capacity (HBC) fuses.

Figure 5.7 shows the types and classification of LV and HV fuses.

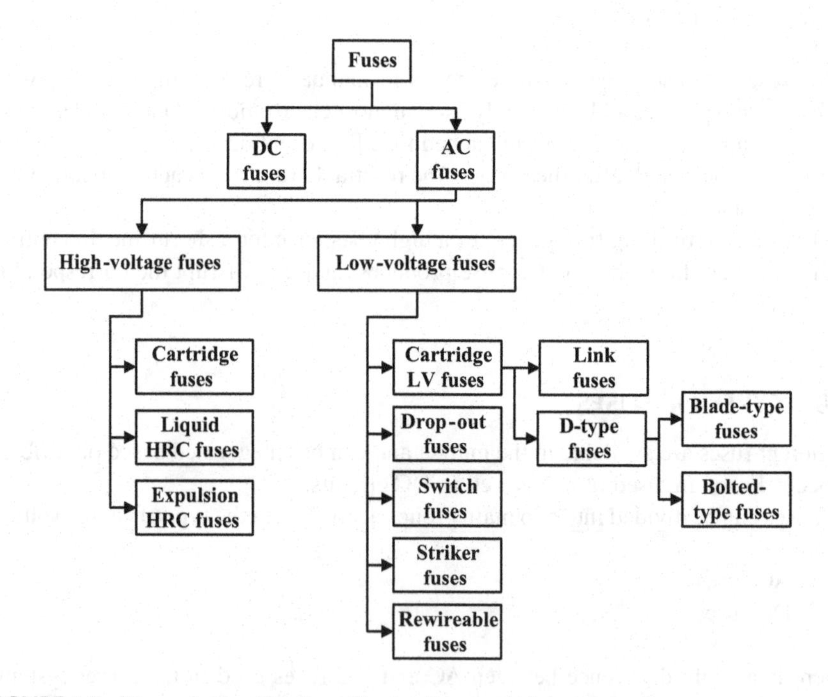

FIGURE 5.7 Types & classification of low-voltage and high-voltage fuses.

5.9 CARTRIDGE FUSES

Cartridge fuses protect electrical appliances and are commonly used where HV and currents are needed in industrial, commercial, and home distribution panels such as pumps, motor air-conditioners, refrigerators, and the like. Up to 600 A and 600 V AC are available.

Two types of cartridge fuses exist:

i. A general-purpose fuse with no pause in time.
ii. Heavy-duty, time-delay cartridge.

Both types range from 250 to 600 V AC and can be rated on the knife's end cap or blade. Cartridge fuses are connected to a base and can be split into more cartridge attachment fuses and cartridge fuses of type D.

5.10 D–TYPE CARTRIDGE FUSE

An adapter ring, foundation, cap, and cartridge are included in the D-types fuse. The fuse base is connected to the fuse cap within the fuse cap, where the cartridge is mounted. When the tip of the cartridge hits the fuse connection conductor, the circuit is completed (Figure 5.8).

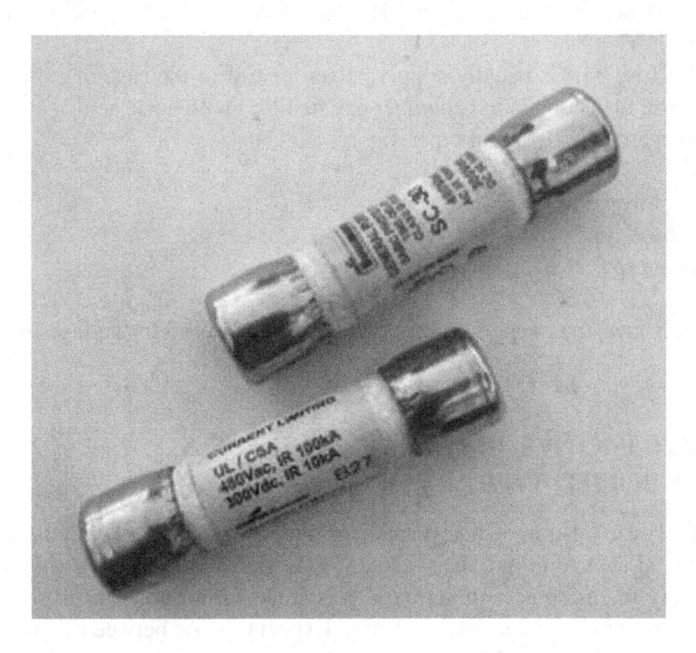

FIGURE 5.8 Cartridge fuses.

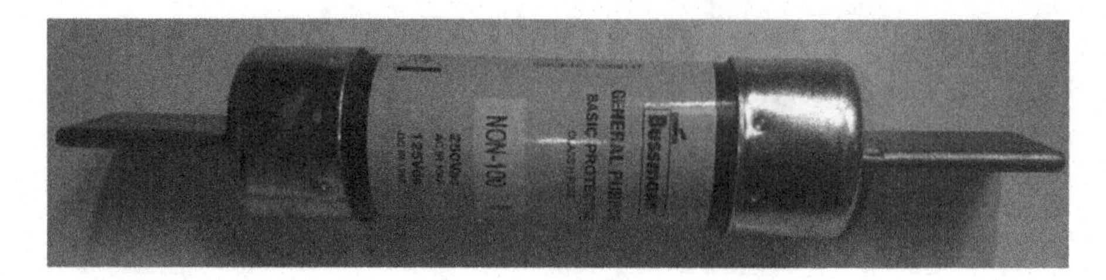

FIGURE 5.9 HRC fuse.

5.11 HRC (HIGH RUPTURING CAPACITY) FUSE OR LINK-TYPE CARTRIDGE FUSE

We have discussed fuse design, operation, and their applications in detail regarding High Rupturing Potential (HRC). It also includes various HRC fuses, such as DIN-type, NH-type, blade-type, HRC liquid-type, and HV-type expulsion fuse, and their benefits and drawbacks (Figure 5.9).

5.12 HV FUSES

In the power system, the high-voltage (HV) fuses are used to secure the power transformer, distribution transformers, instrument transformer, and the like, where circuit breakers may not protect the system. HV fuses are classified for more than 1500 V and up to 13 kV.

The HV fuse part is usually copper, silver, or tin. In the case of expulsion-style HV fuses, the fuse connection chamber can be filled with boric acid.

HV fuses are of three main types:

i. Open (drop out) type.
ii. Enclosed type.
iii. Spring type.

Figure 5.10 shows the open (drop out) type, commonly used for voltages up to 33 kV.

5.13 AUTOMOTIVE, BLADE TYPE, AND FUSES OF BOLTED TYPE

This type of fuse (also known as a spade or plug-in fuse) comes with a plastic body and two metal caps to fit into the socket. Mostly, they are used for wiring and short-circuit safety in cars. Fuse limiters and glass tube (also known as Bosch fuse) are commonly used in the automotive industry. Expect this. As between 12 and 42 V, the rating of car fuses is poor (Figure 5.11).

FIGURE 5.10 The open (drop out) type (https://electrical-engineering-portal.com/Power-fuses-in-high-voltage).

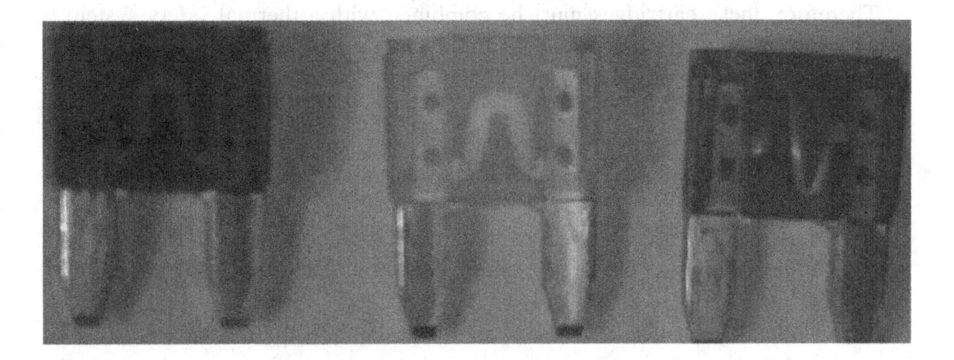

FIGURE 5.11 Blade-type fuses: Used in automobiles.

5.14 SMD FUSES (SURFACE MOUNT FUSE), CHIP, RADIAL, AND LEAD FUSES

SMD fuses (surface mount system and the name derived from SMT = Surface Mount Technology) are chip fuse types (often known as an electronic fuse) used in DC power applications such as hard drive, DVD players, cameras, cell phones, and so forth, where space plays an important role because SMD fuses are very limited in size and often difficult to replace.

Some additional types of SMD fuses and lead fuses are described as follows:

i. Very fast-acting chip fuses.
ii. Slow-blow chip fuses.
iii. High current-rated chip fuses.
iv. Radial fuse.
v. Fast-acting chip fuses.
vi. Telecom fuses.
vii. Pulse-tolerant chip fuses.
viii. Lead fuse.
ix. Through-hole styles fuse.
x. Axial fuse.

5.15 FUSE CHARACTERISTICS

5.15.1 Fuse Type

The fuses, according to their category of applications, are split into two letters. The gG and aM fuses are primarily used in LV installations.

The gG cartridges are for general use and protect the circuits from short and low overloads. The GG cartridges are labeled in black.

The AM cartridges are used on the circuits of motors and protect from high overloads and short circuits. They are designed to withstand such temporary loads (starting the engine).

Therefore, these cartridges must be combined with a thermal safety system to avoid low overloads. The aM cartridges will be labeled in green.

The first letter reveals the principal operation:

- *A (connected)*: An additional protective system must be associated with the fuse, as faults under the specified amount cannot break. It ensures the safety of the short circuit.
- *g (general)*: All defects are broken by the lowest current (even if they are) and the breaking capacity. It provides short-circuit and overload protection.

The second letter indicates the category of equipment to be protected:

Letter	Description
G	Conductors and cables protection
M	Motor circuits protection
R	Semiconductors protection
S	Semiconductors protection
Tr	Transformers protection
N	Conductors protection (North American standards)
D	Time-delay fuse for protecting motor circuits (North American standards)

5.15.2 Rated Currents and Voltages

The rated current will indefinitely cross a fuse without fusing or causing excessive temperature increase. The rated voltage is the voltage at which it is possible to use this fuse.

5.15.3 Conventional Non-Fusing and Fusing Currents

There is a difference between the two classic currents in a fuse: non-fusing and fusing.

- *Conventional non-fusing current (Inf)*: the value of by-pass current that a fuse cartridge can withstand for a conventional period without melting.
- *Conventional fuse current (If)*: a value of by-pass current causes the fuse cartridge to fuse before the conventional time has elapsed (Table 5.1).

5.15.4 Operating Zone

The standards define the fuse's operating zone to determine its operating time according to the current crossing. To calculate the various protective devices' discrimination, it is important to know the fuse's operating characteristics in the series.

5.15.5 Breaking Capacity

The breaking capacity should equal the possible short-circuit current when installing the fuse. The higher the braking power, the better the fuse can protect the installation against short circuits of high intensity. The HBC fuses restrict short circuits that can reach over 100 kA.

TABLE 5.1
Non-Fusing and Fusing Currents

Ratings Current (A)	(Inf) Non-Fusing Current	(If) Fusing Current	(*t*) Conventional Time (hours)
$In \leq 4$	1.5 In	2.1 In	1
$4 < In \leq 10$	1.5 In	1.9 In	1
$10 < In \leq 25$	1.4 In	1.75 In	1
$25 < In \leq 63$	1.3 In	1.6 In	1
$63 < In \leq 100$	1.3 In	1.6 In	2
$100 < In \leq 160$	1.2 In	1.6 In	2
$160 < In \leq 400$	1.2 In	1.6 In	3
$400 < In$	1.2 In	1.6 In	4

5.15.6 Selectivity

A current generally crosses several protected devices in series. These devices are distributed according to the various circuits to be protected. The protective device must select the appropriate area to interrupt if a fault occurs.

5.16 REWIREABLE FUSES

The most popular kit-kat fuse (re-plug fuse) is mostly used in home electrical wiring for small low-voltage (LV) systems and industrial applications.

A reconnected fuse contains two main parts. The internal fuse element is a holder made of tin-plated copper, aluminum, lead, and the like. The porcelain base has IN and OUT terminals that used to be in series with the circuit for protection.

The main advantage of a rewireable fuse is that it can be easily reconnected if it explodes due to a short circuit or overcurrent that causes the fuse elements to melt by placing another wire from the fuse elements of the same previous rating.

The fuse is so designed that the carrier can be safely withdrawn without the danger of touching live parts, and the fuse element is so enclosed that molten metal is safely contained and arcing effectively extinguished.

The simple wire fuse (Figure 5.12) is connected between two terminals in a porcelain carrier and is usually threaded through an asbestos tube.

5.17 THERMAL FUSES

Thermal fuses are the only fuses used only once and cannot be used again as they are temperature-sensitive fuses. The fuse element is made of a temperature-sensitive alloy. They are known as thermal breakers (TCO) or thermal bonds.

The fuse element maintains a mechanical spring connection normally closed in a thermal fuse. When high currents due to overcurrent or short-circuit flow through the fuse elements, the fuse elements melt, freeing the spring mechanism, preventing arc and fire, and protecting the connected circuit.

5.18 RESETTABLE FUSES

Resettable fuses can be used multiple times without replacing them. This fuse opens the circuit when an over event occurs, and after a specified time, they connect the

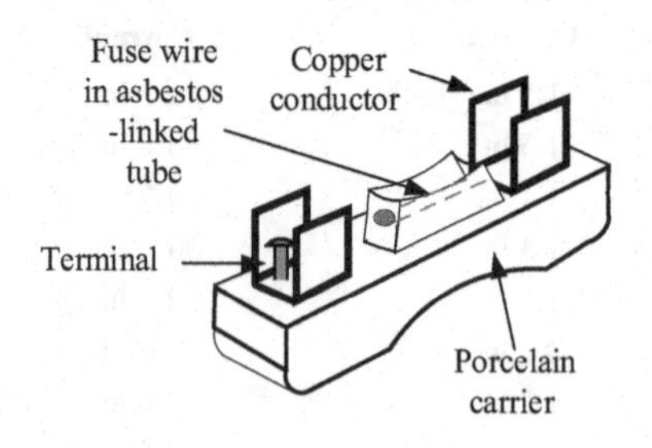

FIGURE 5.12 Rewireable fuses.

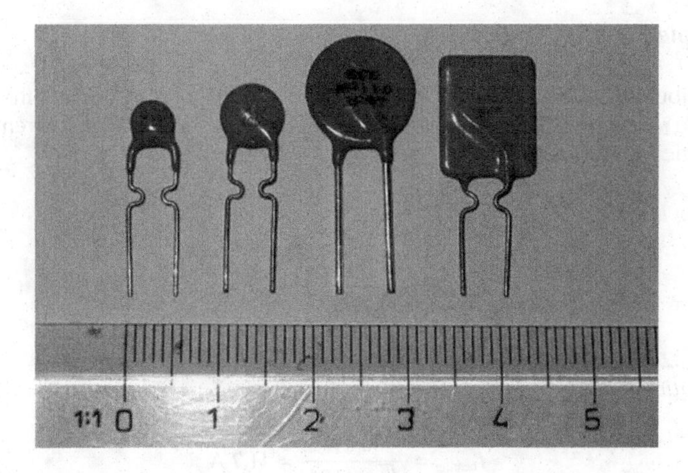

FIGURE 5.13 Resettable fuses (https://en.wikipedia.org/wiki/Resettable_fuse).

circuit again. A polymeric positive temperature coefficient device (PPTC, commonly known as a resettable fuse, multi-switch, or multi-fuse) is a passive electronic component that protects against short-current faults in electronic circuits (Figure 5.13).

The application of resettable fuses is overcome where manual fuse replacement is difficult or nearly impossible; this type is used in specific applications such as fuses in the nuclear system or the space system.

5.19 USES AND APPLICATIONS OF FUSES

Different types of electrical and electronic fuses can be used in all types of electrical and electronic systems and applications, including:

i. Cell phones.
ii. Game systems.
iii. DVD players.
iv. LCD monitors.
v. Scanners.
vi. Hard disk drives.
vii. Digital cameras.
viii. Motors and transformers.
ix. Home distribution boards.
x. Laptops.
xi. Air-conditioners.
xii. General electrical appliances and devices.
xiii. Printers.
xiv. Battery packs.
xv. Portable electronics.
xvi. Power converters.

Example 5.4

A distribution system operating at 415 V is shown in Figure 5.14. Select the fuse's suitable rating for each load and incoming circuit using the fuse time–current characteristic in Figure 5.15.

Lighting load:

$$I_{\text{lighting}} = \frac{20\text{ kW}}{\sqrt{3}\times 415\text{ V}} = 27.8\text{ A}$$

Select 32 A fuse

Heating load:

$$I_{\text{heating}} = \frac{30\text{ kW}}{\sqrt{3}\times 415\text{ V}} = 41.7\text{ A}$$

Select 50 A fuse

Motor load:

$$P_{\text{in}} = \frac{P_{\text{out}}}{\eta} = \frac{30\text{ kW}}{0.92} = 32.6\text{ kW}$$

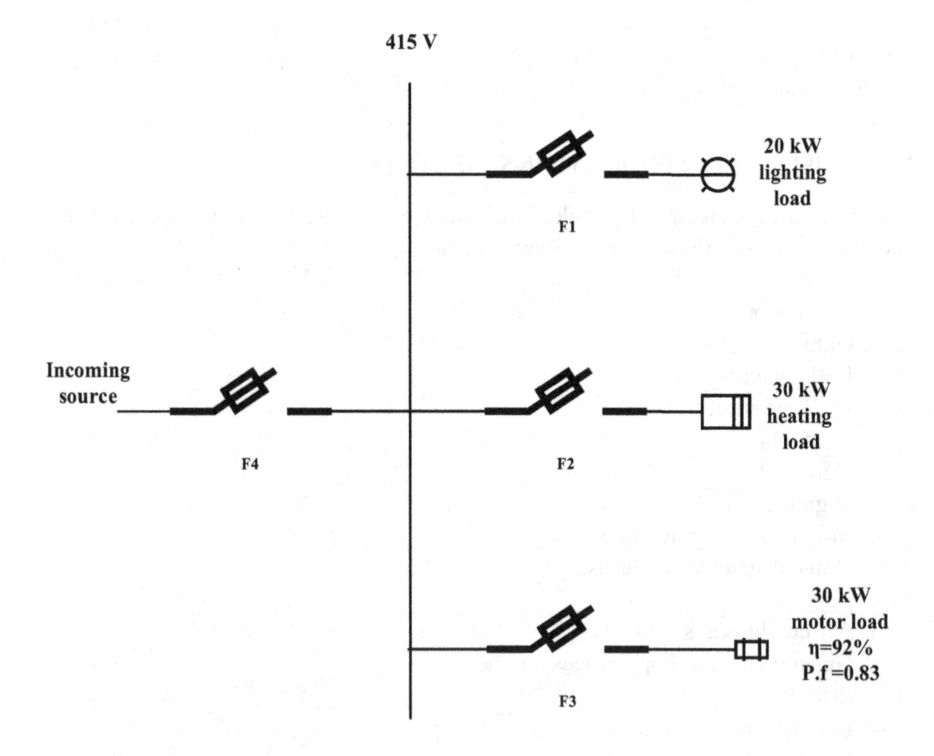

FIGURE 5.14 Circuit diagram for Example 5.4.

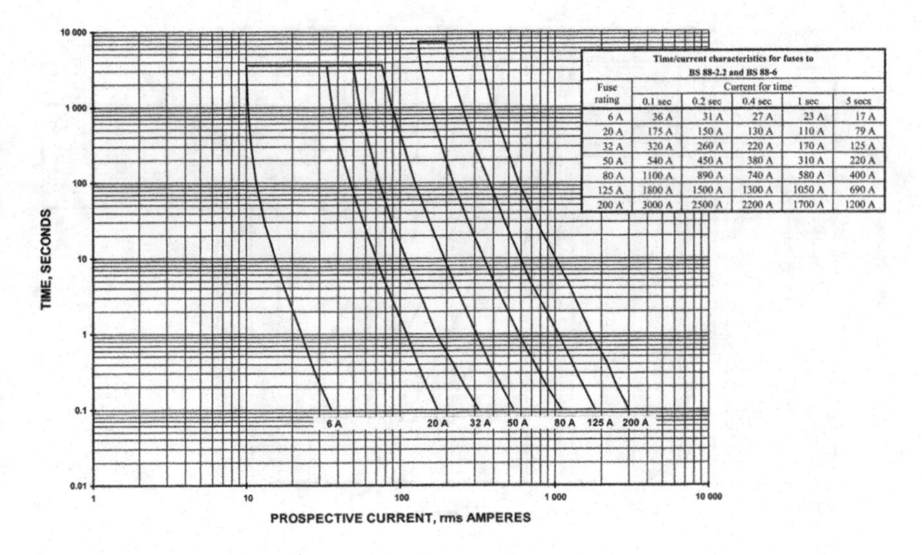

Time/current characteristics for fuses to BS 88-2.2 and BS 88-6

Fuse rating	Current for time				
	0.1 sec	0.2 sec	0.4 sec	1 sec	5 secs
6 A	36 A	31 A	27 A	23 A	17 A
20 A	175 A	150 A	130 A	110 A	79 A
32 A	320 A	260 A	220 A	170 A	125 A
50 A	540 A	450 A	380 A	310 A	220 A
80 A	1100 A	890 A	740 A	580 A	400 A
125 A	1800 A	1500 A	1300 A	1050 A	690 A
200 A	3000 A	2500 A	2200 A	1700 A	1200 A

FIGURE 5.15 Fuse time–current characteristics.

The motor full load current is

$$I_{\text{motor}} = \frac{32.6\ \text{kW}}{\sqrt{3} \times 415\ \text{V} \times 0.83} = 54.7\ \text{A}$$

The starting current for ten seconds is seven-time than the full load current, therefore

$$I = 7 \times 54.7 = 383\ \text{A}$$

From the time–current curve shown in Figure 5.16, an 80 A fuse would withstand 383 A for only six seconds.

Therefore, a 125 A fuse, which would withstand 383 A for longer than ten seconds, would be necessary.

To provide discrimination, the fuse at the incoming circuit must meet the following requirements:

It must carry the normal load current:

$$I = 27.8 + 41.7 + 54.7 = 124.2\ \text{A}$$

It must carry the load + the starting current of the motor

$$I = 27.8 + 41.7 + 383 = 452.5\ \text{A} \quad \text{for} \quad 10\ \text{seconds}$$

From the time–current curve, a 125 A fuse would withstand 452.5 A for more than ten seconds.

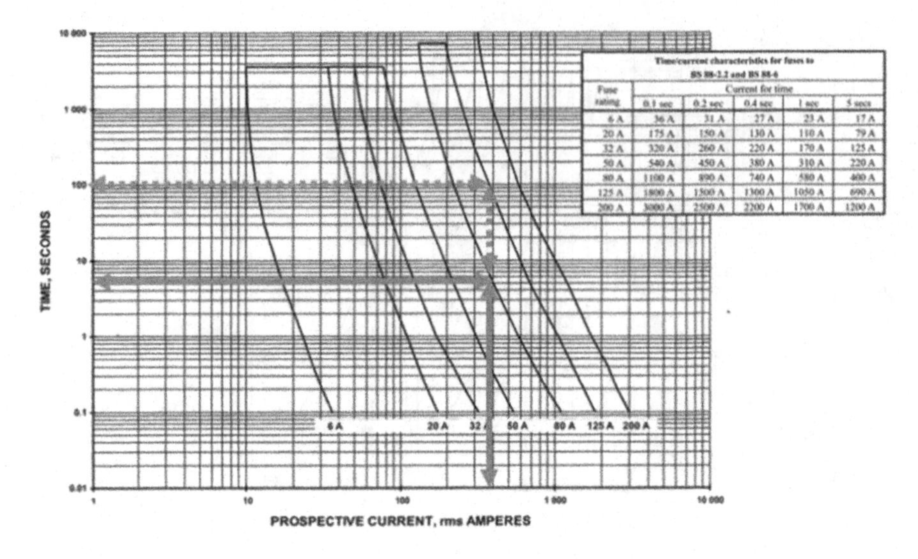

Fuse rating	Current for time				
	0.1 sec	0.2 sec	0.4 sec	1 sec	5 secs
6 A	36 A	31 A	27 A	23 A	17 A
20 A	175 A	150 A	130 A	110 A	79 A
32 A	320 A	260 A	220 A	170 A	125 A
50 A	540 A	450 A	380 A	310 A	220 A
80 A	1100 A	890 A	740 A	580 A	400 A
125 A	1800 A	1500 A	1300 A	1050 A	690 A
200 A	3000 A	2500 A	2200 A	1700 A	1200 A

FIGURE 5.16 Fuse time–current characteristic of Example 5.4.

Example 5.5

A 400 V distribution system is shown in Figure 5.17. Select the fuse's suitable rating for each load and incoming circuit using the fuse time–current characteristic in Figure 5.15.

Solution

Fuse selection for each circuit

Lighting load:

$$I_{\text{lighting}} = \frac{30\ \text{kW}}{\sqrt{3} \times 400\ \text{V}} = 43.3\ \text{A}$$

Select 50 A fuse

Heating load:

$$I_{\text{heating}} = \frac{30\ \text{kW}}{\sqrt{3} \times 400\ \text{V}} = 43.3\ \text{A}$$

Select 50 A fuse

Motor load:

$$P_{\text{in}} = \frac{P_{\text{out}}}{\eta} = \frac{20\ \text{kW}}{90\%} = 22.2\ \text{kW}$$

The motor full load current is

$$I_{\text{motor}} = \frac{22.2\ \text{kW}}{\sqrt{3} \times 400\ \text{V} \times 0.8} = 40\ \text{A}$$

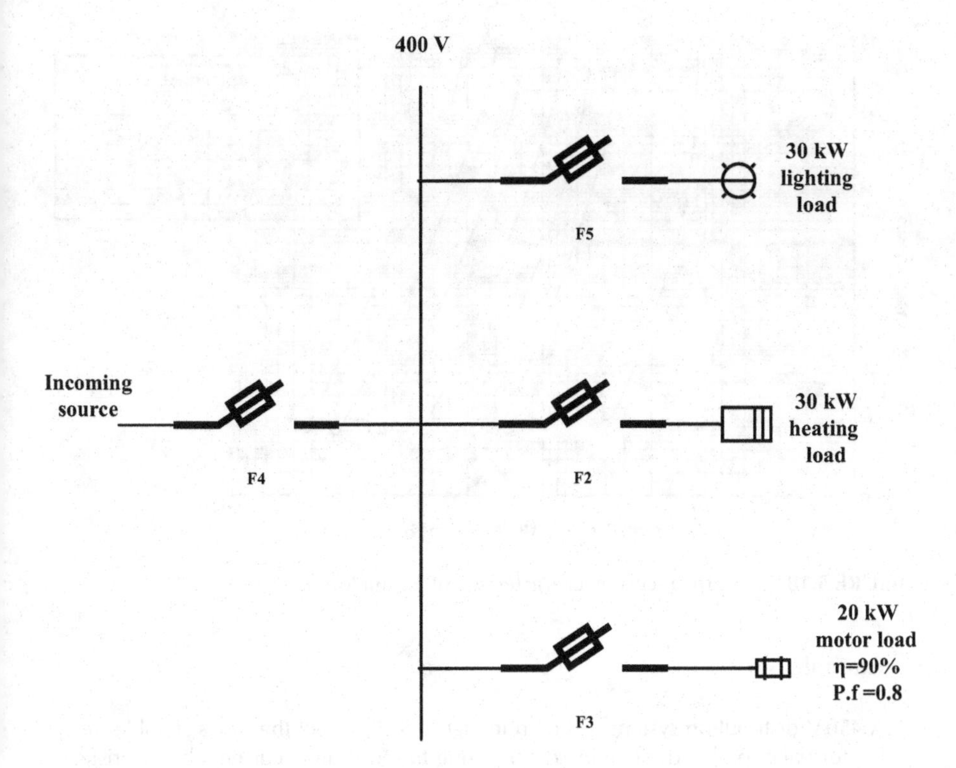

FIGURE 5.17 Circuit diagram of Example 5.5.

The starting current for ten seconds is seven-time than the full load current, therefore

$$I = 7 \times 40 = 280 \text{ A}$$

From the time–current curve shown in Figure 5.18, an 80 A fuse would withstand 280 A for 20 seconds. More than ten seconds.

To provide discrimination, the fuse at the incoming circuit must meet the following requirements:

It must carry the normal load current:

$$I = 34.3 + 34.3 + 40 = 108.6 \text{ A}$$

It must carry the load + the starting current of the motor

$$I = 34.3 + 34.3 + 280 = 348.6 \text{ A for 10 seconds.}$$

From the time–current curve, a 125 A fuse would withstand 348.6 A for more than ten seconds.

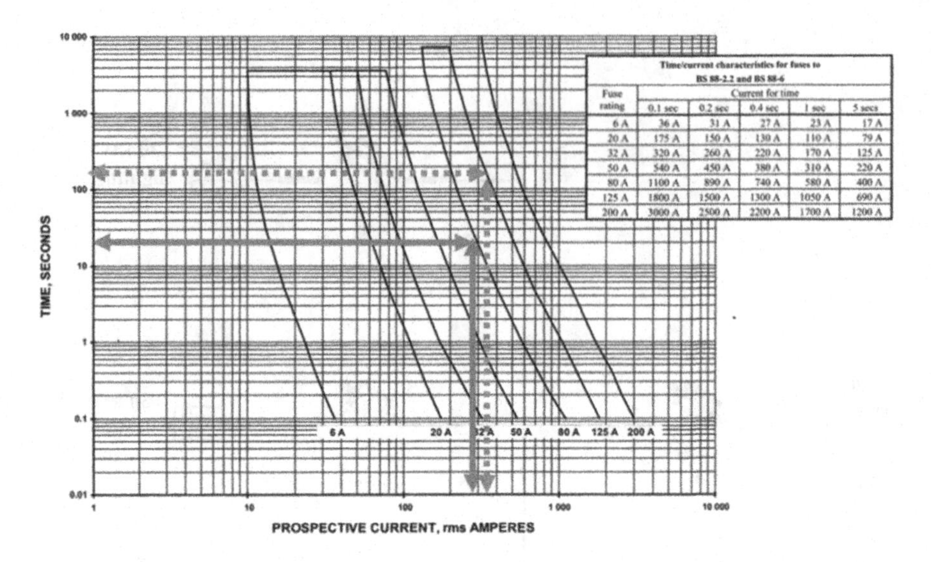

Time/current characteristics for fuses to BS 88-2.2 and BS 88-6					
Fuse rating	Current for time				
	0.1 sec	0.2 sec	0.4 sec	1 sec	5 secs
6 A	36 A	31 A	27 A	23 A	17 A
20 A	175 A	150 A	130 A	110 A	79 A
32 A	320 A	260 A	220 A	170 A	125 A
50 A	540 A	450 A	380 A	310 A	220 A
80 A	1100 A	890 A	740 A	580 A	400 A
125 A	1800 A	1500 A	1300 A	1050 A	690 A
200 A	3000 A	2500 A	2200 A	1700 A	1200 A

FIGURE 5.18 Fuse time–current characteristic of Example 5.5.

Example 5.6

A 450 V distribution system is shown in Figure 5.19. Select the fuse's suitable rating for each load and incoming circuit using the fuse time–current characteristic in Figure 5.15.

Solution

The fuse selection for each circuit:

Heating load:

$$I_{\text{heating}} = \frac{50\ \text{kW}}{\sqrt{3} \times 450\ \text{V}} = 64.15\ \text{A}$$

Select 50 A fuse

Lighting load:

$$I_{\text{lighting}} = \frac{30\ \text{kW}}{\sqrt{3} \times 450\ \text{V}} = 38.5\ \text{A}$$

Select 32 A fuse

Motor load:

$$P_{\text{in}} = \frac{P_{\text{out}}}{\eta} = \frac{30\ \text{kW}}{0.92} = 32.6\ \text{kW}$$

The motor full load current is

$$I_{\text{motor}} = \frac{32.6\ \text{kW}}{\sqrt{3} \times 450\ \text{V} \times 0.85} = 49.2\ \text{A}$$

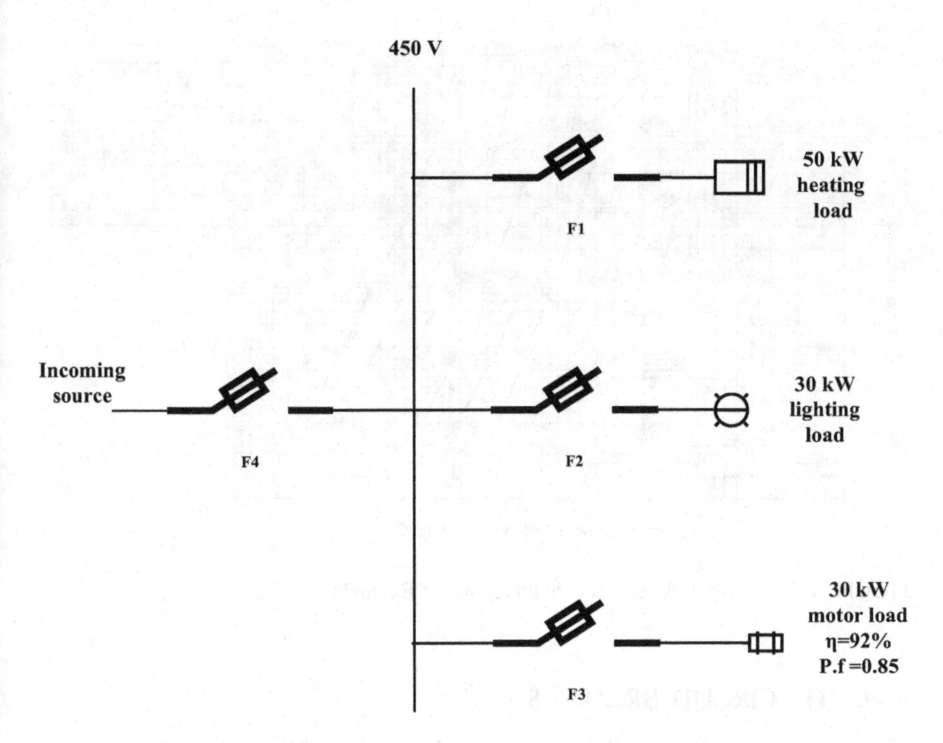

FIGURE 5.19 Circuit diagram of Example 5.6.

The starting current for ten seconds is seven-time than the full load current, therefore

$$I = 7 \times 49.2 = 344.44 \text{ A}$$

From the time–current curve shown in Figure 5.20, an 80 A fuse would withstand 344.44 A for only eight seconds.

Therefore, a 125 A fuse, which would withstand 344.44 A for longer than ten seconds, would be necessary.

To provide discrimination, the fuse at the incoming circuit must meet the following requirements:

It must carry the normal load current:

$$I = 64.15 + 38.5 + 49.2 = 151.85 \text{ A}$$

It must carry the load + the starting current of the motor

$$I = 64.15 + 38.5 + 344.44 = 447.1 \text{ A for } 10 \text{ seconds}$$

From the time–current curve, a 125 A fuse would withstand 447.1 A for more than ten seconds.

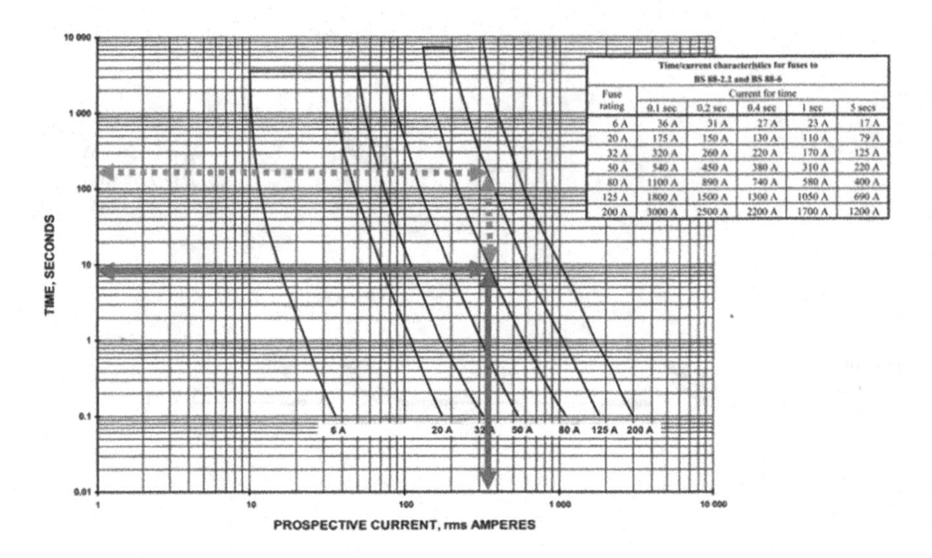

Fuse rating	Current for time				
	0.1 sec	0.2 sec	0.4 sec	1 sec	5 secs
6 A	36 A	31 A	27 A	23 A	17 A
20 A	175 A	150 A	130 A	110 A	79 A
32 A	320 A	260 A	220 A	170 A	125 A
50 A	540 A	450 A	380 A	310 A	220 A
80 A	1100 A	890 A	740 A	580 A	400 A
125 A	1800 A	1500 A	1300 A	1050 A	690 A
200 A	3000 A	2500 A	2200 A	1700 A	1200 A

FIGURE 5.20 Fuse time–current characteristic of Example 5.6.

5.20 HV CIRCUIT BREAKERS

The most important types of HV circuit breakers are the following:

i. Oil circuit breakers (OCB).
ii. SF6 circuit breakers.
iii. Vacuum circuit break.
iv. Air-blast circuit breaker.

The triggering action that causes a circuit breaker to open is usually produced through an overload relay that can detect abnormal line conditions. For example, the relay coil in Figure 5.21 is connected to the secondary of a current transformer. The primary carries the line current of the phase that must be protected. If the line current exceeds a preset limit, the secondary current will cause relay contacts to close. As soon as they close, the tripping coil is energized by an auxiliary DC source. This causes the three mainline contacts to open, thus interrupting the circuit.

5.20.1 Oil Circuit Breakers

The oil circuit breaker's operation: When the current-carrying contacts in the oil are separated, an arc is established between the separated contacts. When the separation of contacts has just started, the distance between the current contacts is small. As a result, the voltage gradient between contacts becomes high. This HV gradient between the contacts ionized the oil and consequently initiated arcing between the contacts. This arc will produce a large amount of heat in the surrounding oil, vaporize the oil, and decompose the oil in to mostly hydrogen and a small amount

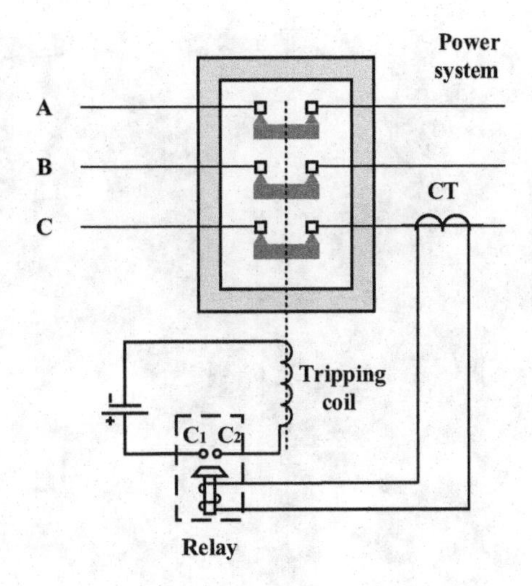

FIGURE 5.21 HV circuit breakers.

of methane, ethylene, and acetylene. The hydrogen gas cannot remain in molecular form and is broken into its atomic form, releasing a lot of heat. The arc temperature may reach up to 5000 K. Due to this high temperature, the gas is liberated and surrounds the arc rapidly, forming an excessively fast-growing gas bubble around the arc. It is found that the mixture of gases occupies a volume about 1000 times that of the oil decomposed. We can assume how fast the gas bubble around the arc will grow. Suppose this growing gas bubble around the arc is compressed by any means. In that case, the rate of the deionization process of ionized gaseous media in between the contacts will accelerate rapidly increasing the dielectric strength between the contacts.

Consequently, the arc will be quenched at zero crossings of the current cycle. This is the basic operation of the oil circuit breaker. The cooling effect of hydrogen gas surrounding the arc path also helps the quick arc quenching in the oil circuit breaker.

There are mainly two types of OCB available, discussed subsequently.

5.20.1.1 Bulk Oil Circuit Breaker (BOCB)

The bulk oil circuit breaker, or BOCB, is a circuit breaker where oil is used as the arc-quenching media and insulating media between current-carrying contacts and earthed parts of the breaker. The oil used here is the same as the transformer insulating oil.

BOCBs are steel tanks filled with insulating oil. In one version (Figure 5.22), three porcelain bushings channel the three-phase line currents to a set of fixed contacts. Three movable contacts, actuated simultaneously by an insulated rod, opened and closed the circuit. When the circuit breaker is closed, the line current for each phase penetrates the tank by way of one porcelain bushing, flows through the first fixed contact, the movable contact, the second fixed contact, and then on out a second bushing.

FIGURE 5.22 Bulk oil circuit breaker or BOCB (https://in.pinterest.com/eeetblog/).

If an overload occurs, the tripping coil releases a powerful spring that pulls on the insulated rod, causing the contacts to open. As soon as the contacts separate, a violent arc is created, which volatilizes the surrounding oil. The pressure of the hot gases creates turbulence around the contacts. This causes cool oil to swirl around the arc, thus extinguishing it.

5.20.1.2 Minimum Oil Circuit Breaker (MOCB)

As the volume of the oil in the BOCB is huge, the chances of fire hazards in the bulk oil system are more. To avoid unwanted fire hazards in the system, one important development in the design of an oil circuit breaker has been introduced where the use of oil in the circuit breaker is much less than that of the BOCB. It has been decided that the oil in the circuit breaker should be used only as arc-quenching media, not as an insulating media. Then the concept of *minimum oil circuit breaker* comes.

These circuit breakers contain a minimum quantity of oil. The three phases are separated into three chambers, as shown in Figure 5.23. Unlike a BOCB, the insulating oil is available only in the interrupting chamber.

FIGURE 5.23 Minimum oil circuit breaker (MOCB) (https://www.deepakkumaryadav.in/2020/03/minimum-oil-circuit-breaker.html).

5.20.2 SF6 Circuit Breakers

These enclosed circuit breakers, insulated with SF6 gas, are used whenever space is available. Several characteristics of SF6 circuit breakers can explain their success:

i. The simplicity of the interrupting chamber, which does not need an auxiliary braking chamber.
ii. Autonomy provided by the puffer technique.
iii. The possibility to obtain the highest performance, up to 63 kA, with a reduced number of interrupting chambers.
iv. Short break time of 2–2.5 cycles.
v. High electrical endurance, allowing at least 25 years of operation without reconditioning.
vi. Possible compact solutions when used for GIS or hybrid switchgear.
vii. Integrated closing resistors or synchronized operations to reduce switching overvoltages.
viii. Reliability and availability.
ix. Low noise levels.

Reducing the number of interrupting chambers per pole has simplified circuit breakers considerably and the number of parts and seals required. As a direct consequence, the reliability of circuit breakers improved. Figure 5.24 shows an SF6 circuit breaker.

FIGURE 5.24 SF6 circuit breakers (https://upload.wikimedia.org/wikipedia/commons/2/21/Circuit_Breaker_115_kV.jpg).

5.20.2.1 Disadvantages

The main disadvantages of SF6 circuit breakers are:

1. SF6 breakers are costly due to the high cost of SF6.
2. Since SF6 gas must be reconditioned after every breaker operation, additional equipment is required.

5.20.2.2 Applications

A typical SF6 circuit breaker consists of interrupter units capable of dealing with currents up to 60 kA and voltages in the range of 50–80 kV. Several units are connected in series according to the system voltage. SF6 circuit breakers have been developed for voltages 115–230 kV, power ratings 10–20 MVA, and interrupting time of fewer than three cycles.

5.20.3 Vacuum Circuit Breakers

These circuit breakers operate on a different principle from other breakers because there is no gas to ionize when the contacts open. They are hermetically sealed; consequently, they are silent and never become polluted (Figure 5.25). Their interrupting capacity is limited to about 30 kV. For higher voltages, several circuit breakers are connected in series.

FIGURE 5.25 VCB circuit breaker (http://www.marineeto.we.bs/2020/06/10/vcb-facts-you-must-know/).

5.20.3.1 VCB Circuit Breaker Components

i. Upper connection
ii. Vacuum interrupter
iii. Lower connection
iv. Roller contact (swivel contact for 630 A)
v. Contact pressure spring
vi. Insulated coupling rod
vii. Opening spring
viii. Shift lever
ix. Mechanism housing with spring operating mechanism
x. Driveshaft
xi. Pole tube
xii. Release mechanism

5.20.4 Air-Blast Circuit Breakers

These circuit breakers interrupt the circuit by blowing compressed air at supersonic speed across the opening contacts. Compressed air is stored in reservoirs at a pressure of about 3 MPa (435 psi) and replenished by a compressor in the substation.

FIGURE 5.26 A typical three-phase contact module of an air-blast circuit breaker (http://zhwjpe.com/air-blast-circuit-breakers/).

The most powerful circuit breakers can typically open short-circuit currents of 40 kA at a line voltage of 750 kV in three to six cycles on a 50 Hz line. The air-blast noise is so loud that noise-suppression methods must be used when the circuit breakers are installed near residential areas. Figure 5.26 shows a typical single-phase contact module of an air-blast circuit breaker.

5.20.4.1 Types of Air-Blast Circuit Breakers

Figure 5.27 shows a typical three-phase air-blast circuit breaker. Each phase is composed of three contact modules connected in series.

Air-blast circuit breakers are manufactured with the help of three technologies:

1. *Axial blast type:* The air blast is directed along the arc path, as shown in Figure 5.27a.
2. *Cross-blast type:* The air blast is directed at right angles to the arc path, as shown in Figure 5.27b.
3. *Radial blast type:* The air blast is directed radially, as shown in Figure 5.27c.

The advantages of air-blast circuit breaker are

i. Cheapness, chemically stable, and inertness of air.
ii. Short and consistent arcing time.

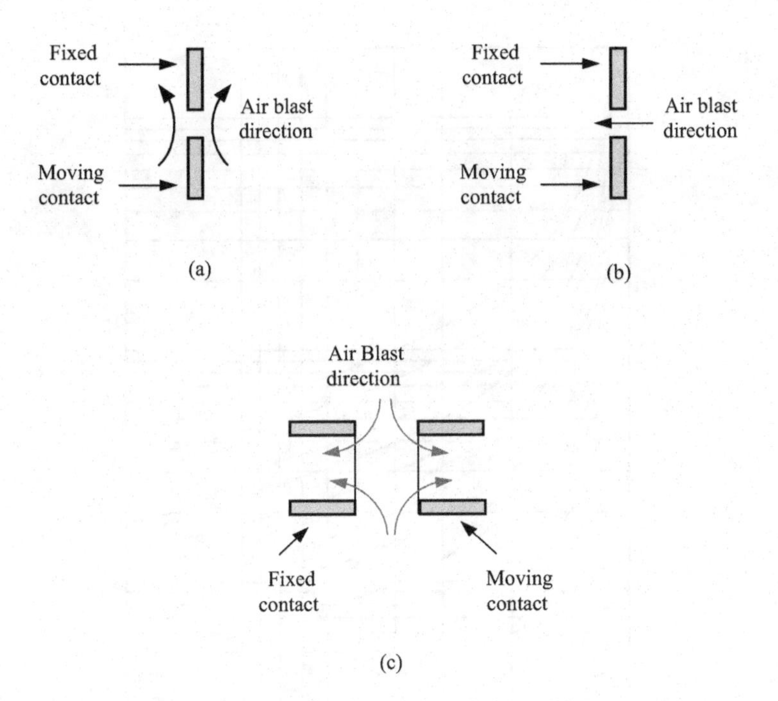

FIGURE 5.27 A typical three-phase air-blast circuit breaker. Each phase is composed of three contact modules connected in series. (a) Axial blast type, (b) cross-blast type, and (c) radial blast type.

iii. No fire hazard.
iv. High-speed operation.
v. Less maintenance and easy operation.

5.21 DIRECTIONAL OVERCURRENT TIME PROTECTION

This section deals with directional overcurrent protection by studying its theoretical principles in practical experiments. Determination of inherent system times provides insight into the chronological sequences involved in protection technology. Subsequently, advanced applications of overcurrent protection are revealed by recording a dependent overcurrent characteristic and testing recognition of energy flow direction.

Directional, maximum-overcurrent time protection follows a continuous time function $t(I)$ in dependence on the fault current (Figure 5.28). Every overcurrent value is associated with a fixed delay time. The relay stores several characteristics, which can be set via the delay time $t_{I>}$ serving as a multiplier in the case of directional, maximum-overcurrent time protection's trip characteristic. Apart from different time ranges, it is possible to choose between three characteristics (normal, strong, and extremely inverse). Areas of application mainly comprise overload protection for motors and transformers. Figure 5.29 shows directional overcurrent characteristics.

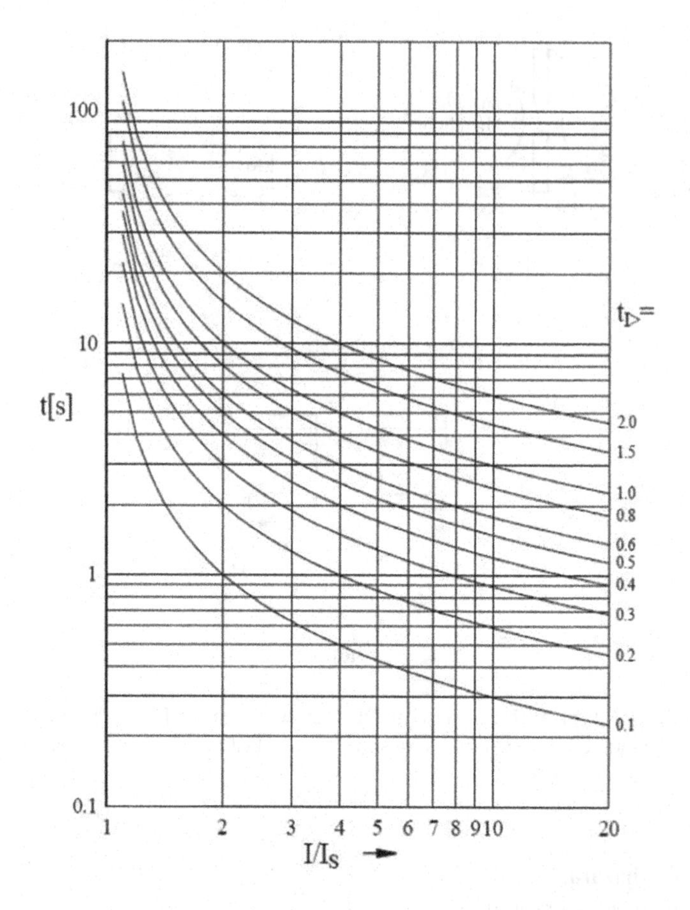

FIGURE 5.28 Lucas-Nülle GmbH normal inverse trip characteristic of directional, maximum-overcurrent time protection.

Record the characteristics $T_A(I)$ for the three different sensitivity stages of $t_>$ in a diagram given in Figure 5.29.

Current I (A)	0.3	0.4	0.5	0.6	0.7	0.8	0.9	1.0
$tI_>=1$	17.7	10.1	7.5	6.2	5.5	4.9	4.6	4.2
$tI_>=0.5$	8.9	5.1	3.8	3.1	2.8	2.4	2.3	2.1
$tI_>=0.1$	1.8	1.1	0.8	0.6	0.5	0.5	0.4	0.4

$t(I)$ = Tripping time = $tI_> \times (0.14/((I/I_s)^{0.02} - 1))$
$tI_>$ = Time multiplier = 1/0.5/0.1
I = Fault current = 0.4–1.3 A
I_s = Pick-up value = 0.3 A

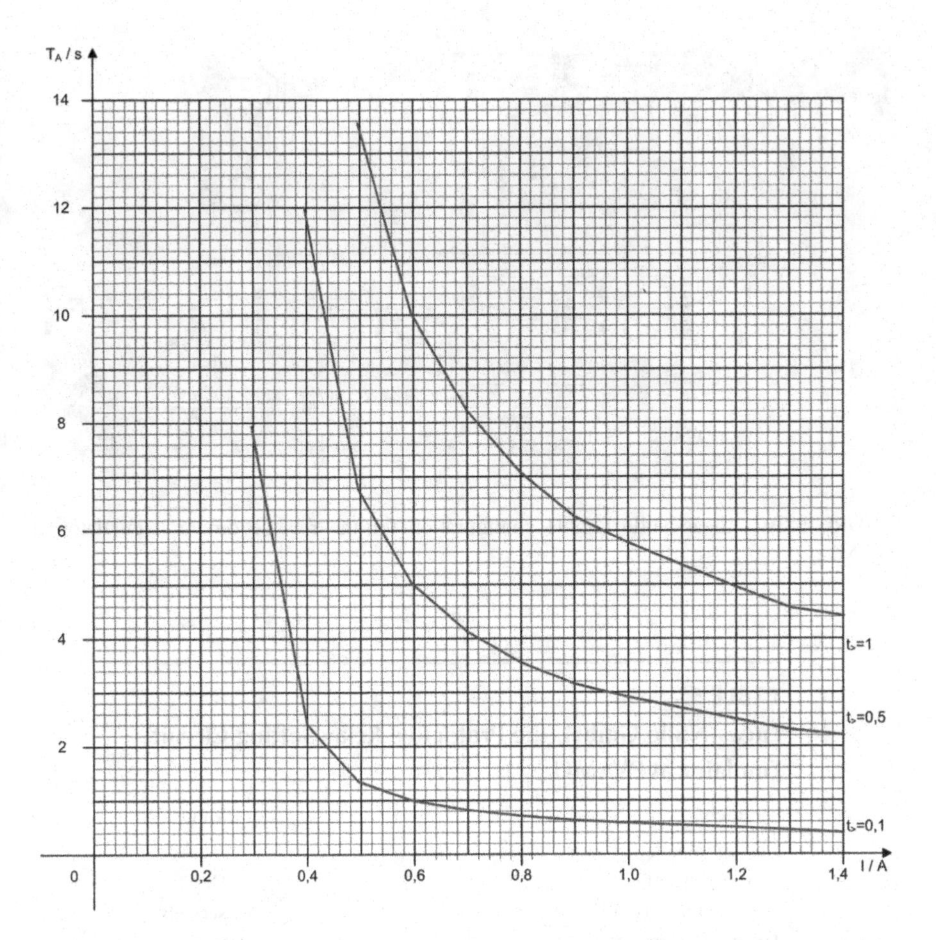

FIGURE 5.29 Directional overcurrent characteristic by Lucas-Nülle GmbH at PVAMU.

5.22 TESTING DIRECTION RECOGNITION

Reverse the direction of energy flow by interchanging the plug connections on the protective relay, as shown in Figure 5.30.

The protective relay is described in Table 5.2.

The line-to-line voltage is set to 220 V and the resistive load is set at maximum value. A current of 0.24 A should then flow. When the protective relay is energized, **TRIP** is displayed, and the LED of the energized phase flashes. The direction LEDs (**green** and **red arrows**) indicate the direction of energy flow. The **green LED** flashes to indicate the forward direction; the **red LED** flashes to indicate the reverse direction

The transmission line is connected by closing power switch 1 (**ON** button). A current of 0.24 A should flow again. Produce a short circuit between phases L_1 and L_2 by actuating power switch 2 (**ON** button).

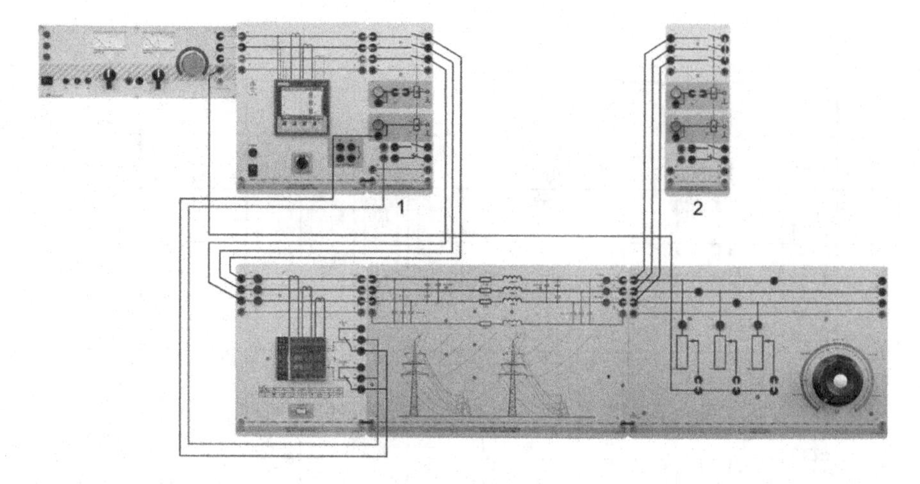

FIGURE 5.30 Lucas-Nülle GmbH configuration for the reverse direction emulator at PVAMU.

TABLE 5.2
Lucas-Nülle GmbH, the Protective Relay Setting Circuit Emulator at PVAMU

Parameter	Value
Pick-up value $I_>$	0.3 A
Trip characteristic	DEFT
Trip delay $tI_>$ forward	1 seconds
Trip delay $tI_>$ reverse	10 seconds
Characteristic angle	49
Frequency	60
Serial interface's address	1

PROBLEMS

5.1. It is proposed to use a No.30 AWG copper wire as a fuse element. If its initial temperature is 60°C, calculate the following:
  a. The I^2t needed to melt the wire (copper melt at 1000°C).
  b. The time needed to melt the wire if the short-circuit current is 40 A.

5.2. It is proposed to use a No.30 AWG aluminum wire as a fuse element. If its initial temperature is 55°C, calculate the following:
  a. The I^2t needed to melt the wire (copper melt at 900°C).
  b. The time needed to melt the wire if the short-circuit current is 45 A.

5.3. It is proposed to use a No.30 AWG copper wire as a fuse element. If its initial temperature is 50°C, calculate the following:
   a. The I^2t needed to melt the wire (copper melt at 1083°C).
   b. The time needed to melt the wire if the short-circuit current is 30 A.
5.4. A distribution system operating at 415 V is shown in Figure 5.31. Select the fuse's suitable rating for each load and incoming circuit using the fuse time–current characteristic in Figure 5.32.

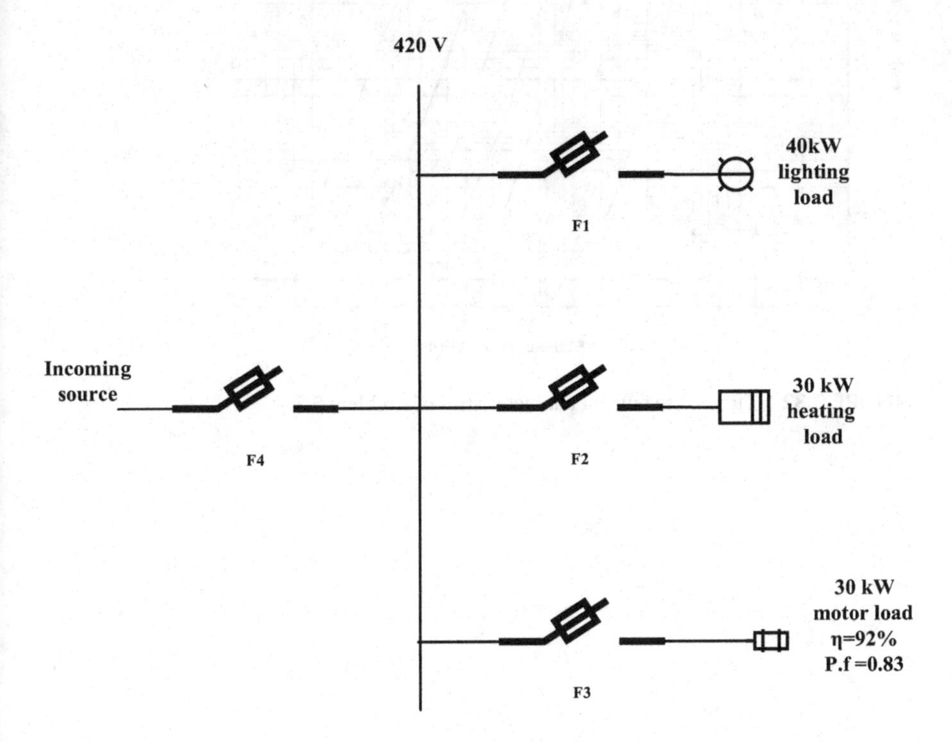

FIGURE 5.31 Circuit diagram of Problem 5.4.

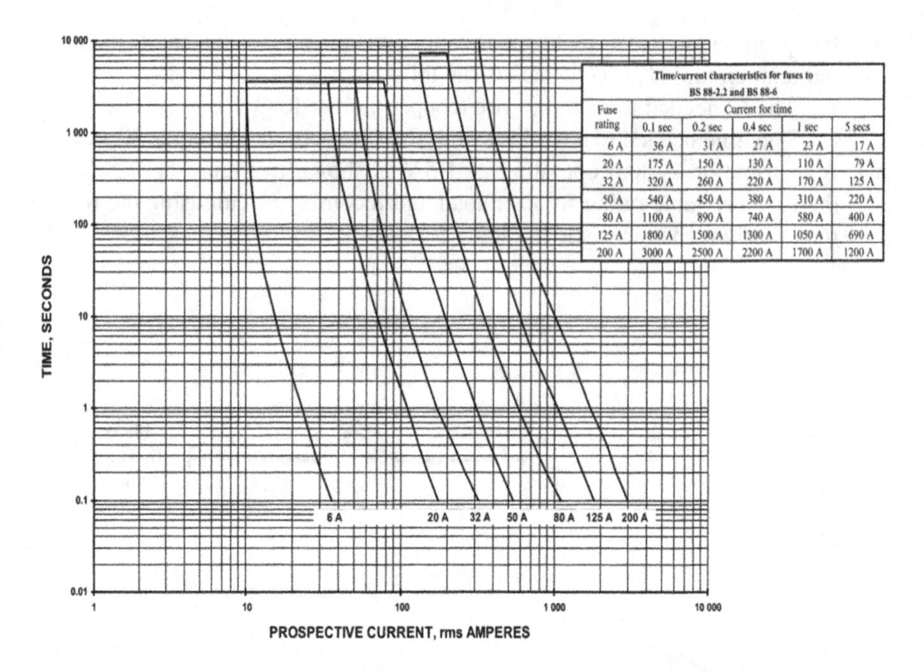

Time/current characteristics for fuses to BS 88-2.2 and BS 88-6

Fuse rating	Current for time				
	0.1 sec	0.2 sec	0.4 sec	1 sec	5 secs
6 A	36 A	31 A	27 A	23 A	17 A
20 A	175 A	150 A	130 A	110 A	79 A
32 A	320 A	260 A	220 A	170 A	125 A
50 A	540 A	450 A	380 A	310 A	220 A
80 A	1100 A	890 A	740 A	580 A	400 A
125 A	1800 A	1500 A	1300 A	1050 A	690 A
200 A	3000 A	2500 A	2200 A	1700 A	1200 A

FIGURE 5.32 Time–fuse current characteristics of Problem 5.5.

6 Overcurrent Relay

6.1 INTRODUCTION

When a fault occurs in the power system, the fault current is almost always greater than the pre-fault current in the power system elements. Overcurrent relays protect power system elements such as transmission lines, transformers, generators, motors, and the like. Electromechanical overcurrent relays are generally of the induction disk type.

An overcurrent relay is a sensor relay that operates when the current exceeds a preset value. The overcurrent relay in the transmission network also works as a backup relay. It is the basic protection in the distribution network, which mainly protects the feeders.

This chapter presents the overcurrent relay, PSM, time grading, and relay coordination method; the chapter also discusses requirements for proper relay coordination and hardware and software for overcurrent relays. Overvoltage and undervoltage protection using Lucas-Nülle GmbH power system/SCADA network devices are explained.

6.2 OVERCURRENT RELAY

Depending upon the time of operation, the relay is classified (Figure 6.1) as follows:

i. Constant time relay.
ii. Instantaneous relay (as attracted armature, moving iron type, permanent magnet, moving coil type, and static relay).
iii. Inverse relay (electromagnetic induction type, the permanent magnet moving coil type, and static relay) can be of
   a. Inverse time-–current relay.
   b. Inverse definite minimum time (IDMT) overcurrent relay.
   c. Very inverse relay.
   d. Extremely inverse relay.
iv. Directional overcurrent relay

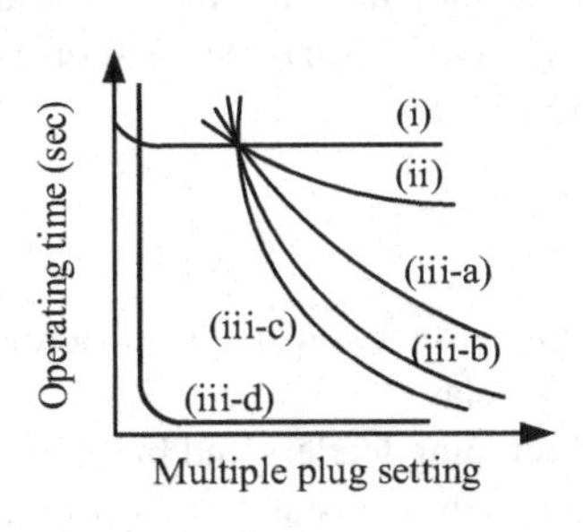

FIGURE 6.1 Operating time multiplier plug setting characteristics.

DOI: 10.1201/9781003394389-6

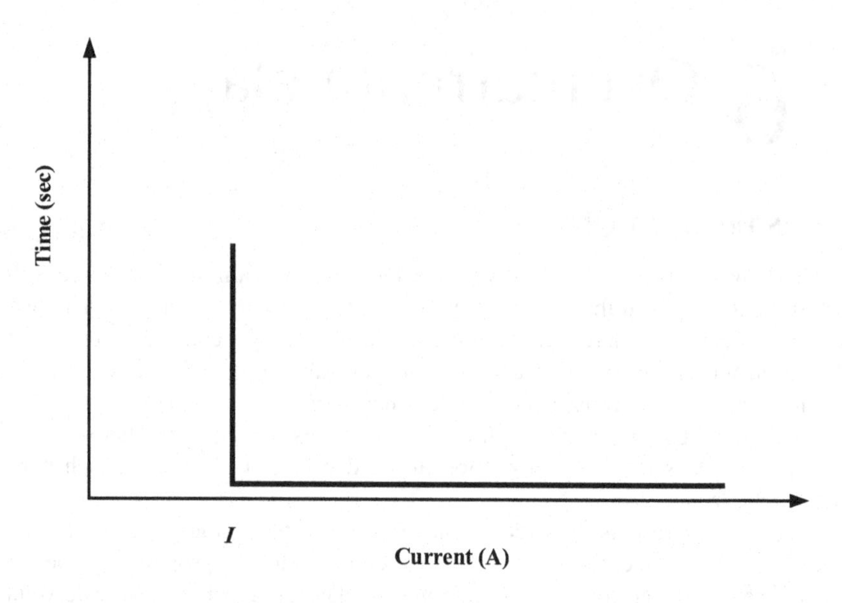

FIGURE 6.2 Instantaneous overcurrent relay characteristics.

6.2.1 Instantaneous Overcurrent Relay

The relay operates at a specified time when the current exceeds the pick-up value. The relay's operation depends only on the current magnitude since the operating time is constant. In this type, there is no time delay. The principle of coordination of the instantaneous overcurrent relay is that the fault current varies with the fault's location due to the difference in impedance between the fault and the source. If the relay operating current increases gradually relative to other relays when moving toward the source while the relay located further from the source is running to a lower current value (Figure 6.2).

6.2.2 Definite Time Overcurrent Relay

The working principle of a specific time overcurrent relay is that the current should exceed the preset value, and the fault must be continuous at least at a time equal to the time setting of the relay. Its operation is independent of the current magnitude above the pick-up value (Figure 6.3).

6.2.2.1 Application

i. It also acts as a backup protection to the differential relay of the power transformer with a time relay.
ii. For the protection of outgoing feeders and bus couplers.
iii. It acts as backup protection to distance relays in a transmission line with time delay.

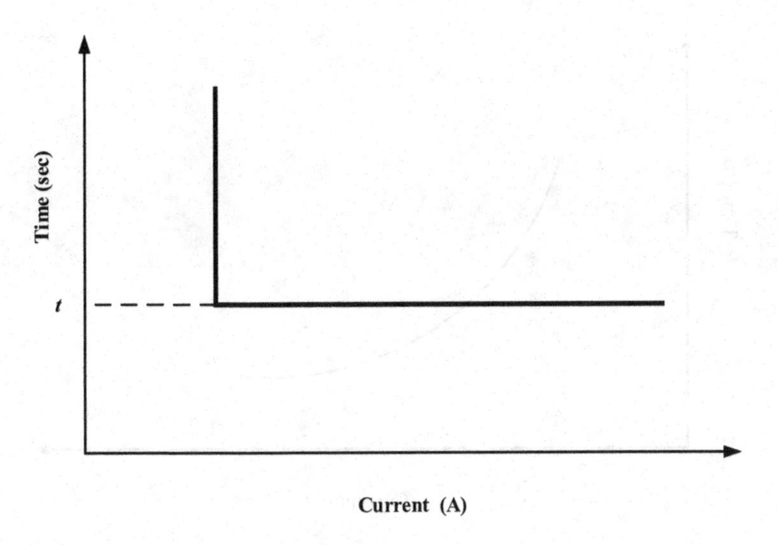

FIGURE 6.3 Definite time overcurrent relay characteristics.

6.2.2.2 The Drawback of the Relay

In the event of a fault, the supply continuity cannot be maintained at the end of the load.

Time lags are provided, which is not desirable in short circuits. It is difficult to coordinate and requires changes as the load is added.

Not suitable for long-distance transmission lines where fast fault clearance is essential for stability.

The poor distinction as the relay has difficulty distinguishing fault currents at one point or another when the defect resistance between these points is small.

6.2.3 Inverse Time Overcurrent Relay

The inverse time overcurrent relay is when the relay operating time decreases as the fault current increases. The operating time of the relay will be less as the fault current increases, and vice versa. The relay will take unlimited time to operate if the fault current is equal to the pick-up value (Figure 6.4).

6.2.4 IDMT Relay

In this type of relay, the operating time is inversely proportional to the fault current. The operating time of the relay can be reduced by adjusting the time dial setting. The relay works when the current exceeds the pick-up value, and the operating time depends on the current value. In the lower values of the fault current, the relay gives the reverse time's current characteristics. In the case of obtaining higher values of the fault current, it gives specific time characteristics. This type is used to protect distribution lines (Figure 6.5).

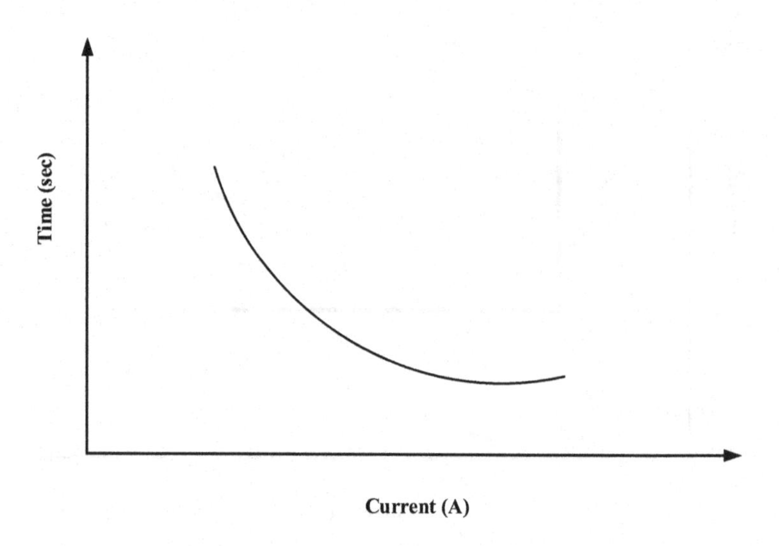

FIGURE 6.4 Inverse time overcurrent relay characteristics.

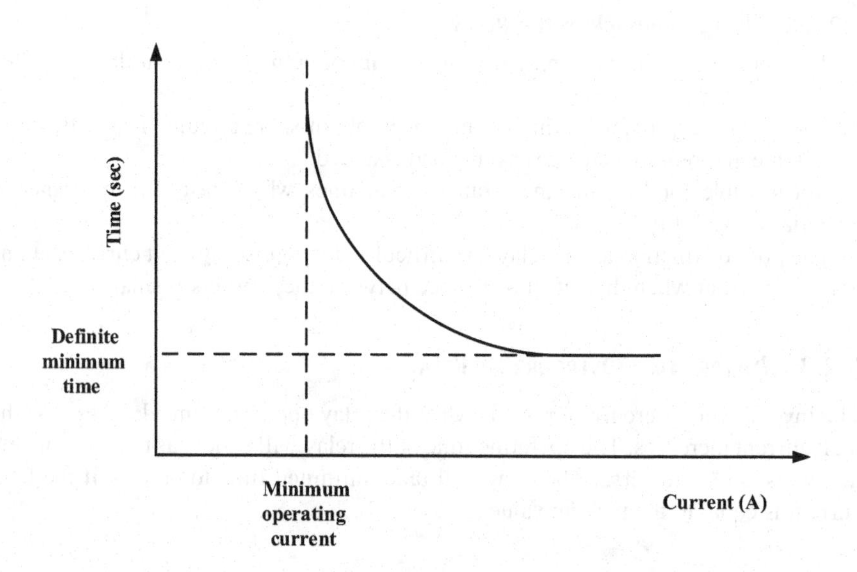

FIGURE 6.5 Inverse definite minimum time relay characteristics.

6.2.5 Very Inverse Relay

In this type of relay, the operating time range is inversely proportional to the fault current over a wide range. It is effective for protecting against the earth's faults, protecting feeders, and long transmission lines.

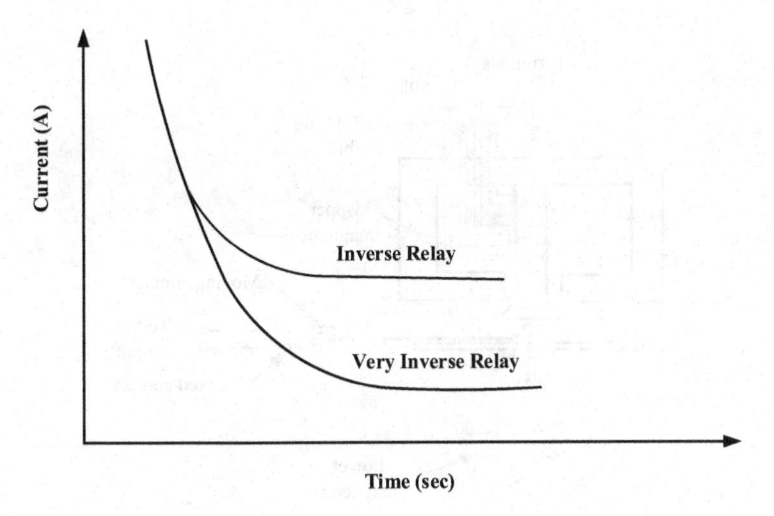

FIGURE 6.6 Very inverse overcurrent relay characteristics.

6.2.5.1 Application of the Very Inverse Relay

i. It is used when the fault current is dependent on the fault location.
ii. It is used when the fault current is independent of normal changes in generating capacity.
iii. Suitable for the application if there is a reduction in fault current as the power source's fault distance increases (Figure 6.6).

6.2.6 Extremely Inverse Relay

The operating time of this relay is inversely proportional to the square of the current. It gives more inverse characteristics than that IDMT and a very inverse overcurrent relay (Figure 6.7). Figure 6.8 shows a trigger and output signal discrimination time relay.

6.2.7 Directional Overcurrent

Directional overcurrent protection is achieved through a relay that utilizes a dual-actuating quantity induction mechanism.

When fault current can follow in both directions through the relay location, it is necessary to respond to the relay direction by adding directional control elements. Directional control is a design feature that is highly desirable for this type of relay. With this feature, an overcurrent unit is inoperative, no matter how large the current may be unless the contacts of the directional unit are closed. This is accomplished by connecting the directional-unit contacts in series with the shading-coil circuit or with one of the two flux-producing circuits of the overcurrent unit. When this circuit is open, no operating torque is developed in the overcurrent unit. The contacts of the overcurrent unit alone are in the trip circuit.

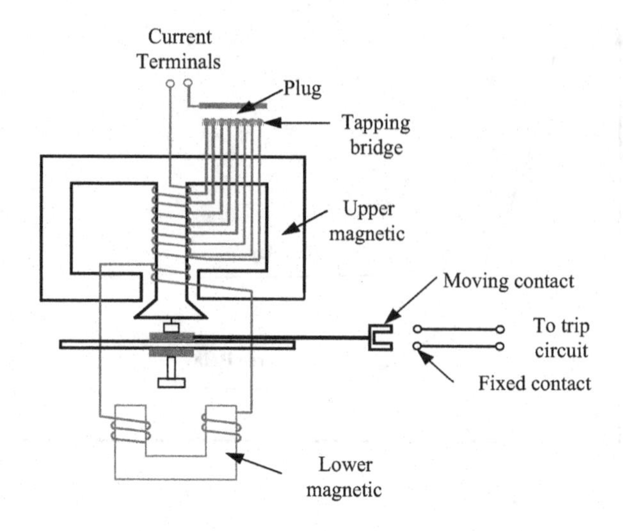

FIGURE 6.7 Sample induction disk overcurrent relay.

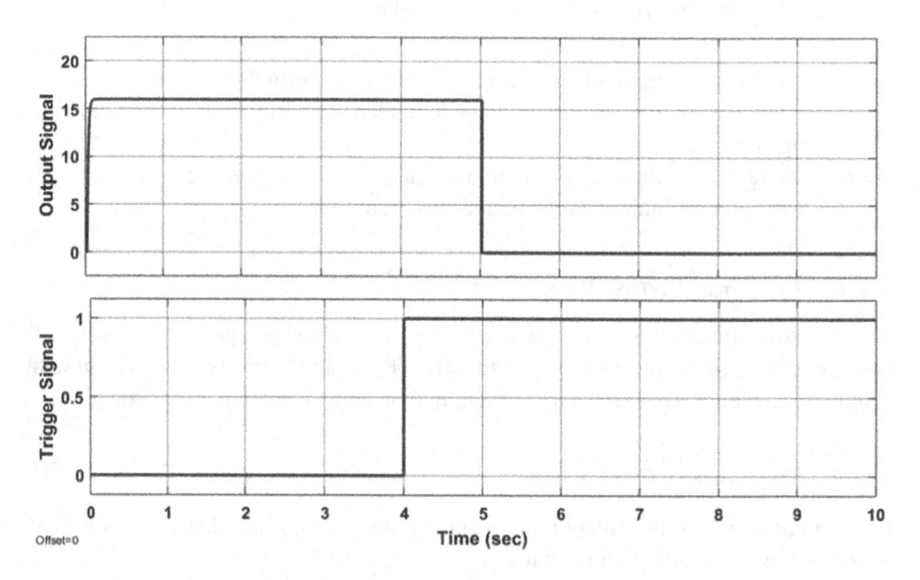

FIGURE 6.8 Discrimination time relay circuit.

Directional relays are usually used in conjunction with other forms of relays, usually over the current type; when used as an overcurrent relay, the combination uses to select to respond to the fault current only in the protective zone's direction (Figures 6.9 and 6.10).

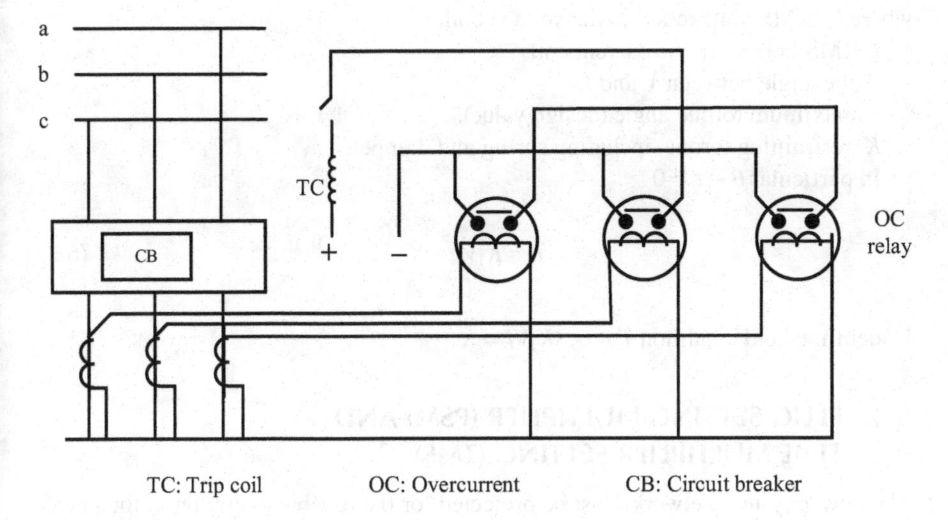

FIGURE 6.9 OC protection with 3 OC relays.

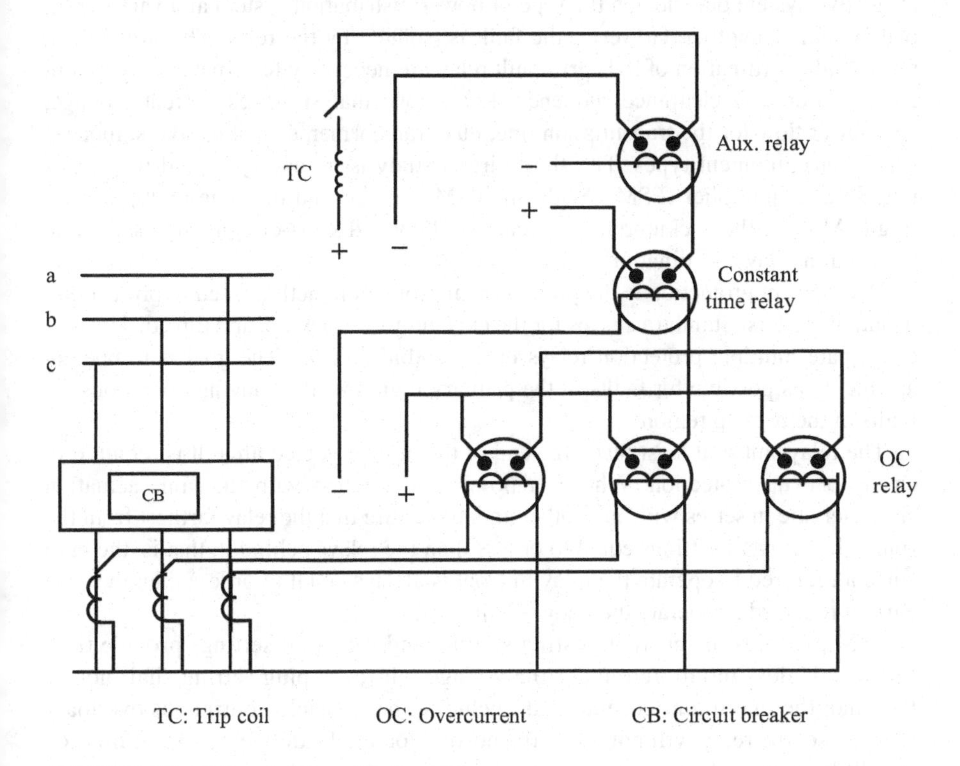

FIGURE 6.10 OC protection with time delay.

The torque developed by the directional unit is

$$T = VI\cos(\theta - \tau) - K \tag{6.1}$$

where V: RMS voltage fed to the voltage coil.
I: RMS current in the current coil.
θ: the angle between V and I.
τ: maximum torque angle (design value).
K: restraining torque, including spring and damper.
In particular $\theta - \tau = 0$

$$T = K_1 VI - K \tag{6.2}$$

Under threshold condition $T = 0,\ K_1 VI = K$.

6.3 PLUG SETTING MULTIPLIER (PSM) AND TIME MULTIPLIER SETTING (TMS)

The power system network must be protected for the reliable operation of the power sources; relays and circuit breakers achieve this protection. The size and design of protective system depends on the type of power distribution system and varies from one system to another. However, the fault is isolated by the relay. The overcurrent phase and coordination of the earth fault relay are necessary to achieve correct fault identification and clearance sequence. Load flow analysis gives current, voltage, and power flow for the transmission line, bus, transformer, circuit breakers, motors, and other equipment types. The short-circuit study is necessary to find the relay's plug setting multiplier (PSM). With this PSM, we can find the time multiplier setting (TMS) for the backup relay. Hence, load flow and short-circuit study should be required in relay coordination.

Overcurrent protection is the predominant protection method used to protect distribution feeders. Standard curves for the current–time, pick-up, and time disk values coordinate multiple protection relays on the radial feeders. The goal is to operate as quickly as possible for faults in the primary region while delaying operations for faults in the backup region.

The relays must at least reach the end of the next protected area. It is required to ensure backup protection. Whenever possible, use relays with the same actuation characteristic in series with each other and make sure that the relay farthest from the source has current settings equal to or less than the relays behind it, that is, the base current required to operate the relay in front is always equal to or less than the base current required to operate the relay is behind it.

PSM provides the current posting setting, and time dial settings provide time settings. Unless monitored under the voltage relay, the plug setting shall not be less than the maximum normal load, including permissible continuous overload. Otherwise, the relay will not allow the normal load to be delivered. An allowance should be made when estimating the set of components because the relay pick-up varies from 1.05 to 1.3 times that of the clay settings, depending on the standards.

The feeder overcurrent relay's coordination curve should fall below the feeder overload curve and the feeder short-circuit damage curves on the time–current characteristics graph. An overcurrent relay (OCR) is a protection relay that operates when the load current exceeds a preset value. The coordination curve of the feeder overcurrent relay should fall above the capacitance curve of the feeder. Overcurrent relays generally contain multiples of the current setting of 50%–200% in steps of 25%, which are referred to as plug setting (PS).

Two parameters determine each relay's PS: maximum load current and minimum fault current.

The protection relay coordination setting is made during the system design process based on the fault current calculation. In coordinating overcurrent relays, the goal is to define the time multiplier setting (TSM) and PSM for each relay. The total operating time of the primary relays is appropriately reduced.

The overcurrent relay has a minimum operating current, known as the relay current setting. The current setting should be chosen. The relay does not operate at the maximum load current in the shielded circuit and operates at a current equal to or greater than the expected minimum fault current. The relay's current setting closest to the source should always be higher than the previous setting of the relay. The relays should have higher current settings than any current that could flow through the relays under normal conditions: 110% of the rated current. Electronic relays and microprocessors have current adjusting steps of 5%.

It is necessary to modify the timescale of the current–time characteristic to implement the power system's relay. The TMS should be chosen to relay the minimum possible time at the end of the radial feeder. A time multiplier should be selected in the preceding sections toward the source and given a selected interval of the downstream relay at the maximum error conditions. The TMS should allow for the cutter time and bypass the relay and the time errors allowed in the successive relays' running time.

6.4 STANDARD FORMULA FOR OVERCURRENT RELAY

By using the general equation of the IEC (International Electro-Technical Commission) standard (Figure 6.11 and Table 6.1):

$$T_p = \frac{C}{\left(I/I_p\right)^{\alpha} - 1} \times \text{TMS} \tag{6.3}$$

T_p = Operating time in seconds.

$\left(I/I_p\right)^{\alpha}$ = Applied multiples of set current value.

C and α = constant of relay

Constants for IEC Standard Time Overcurrent Characteristics IEC Standard.

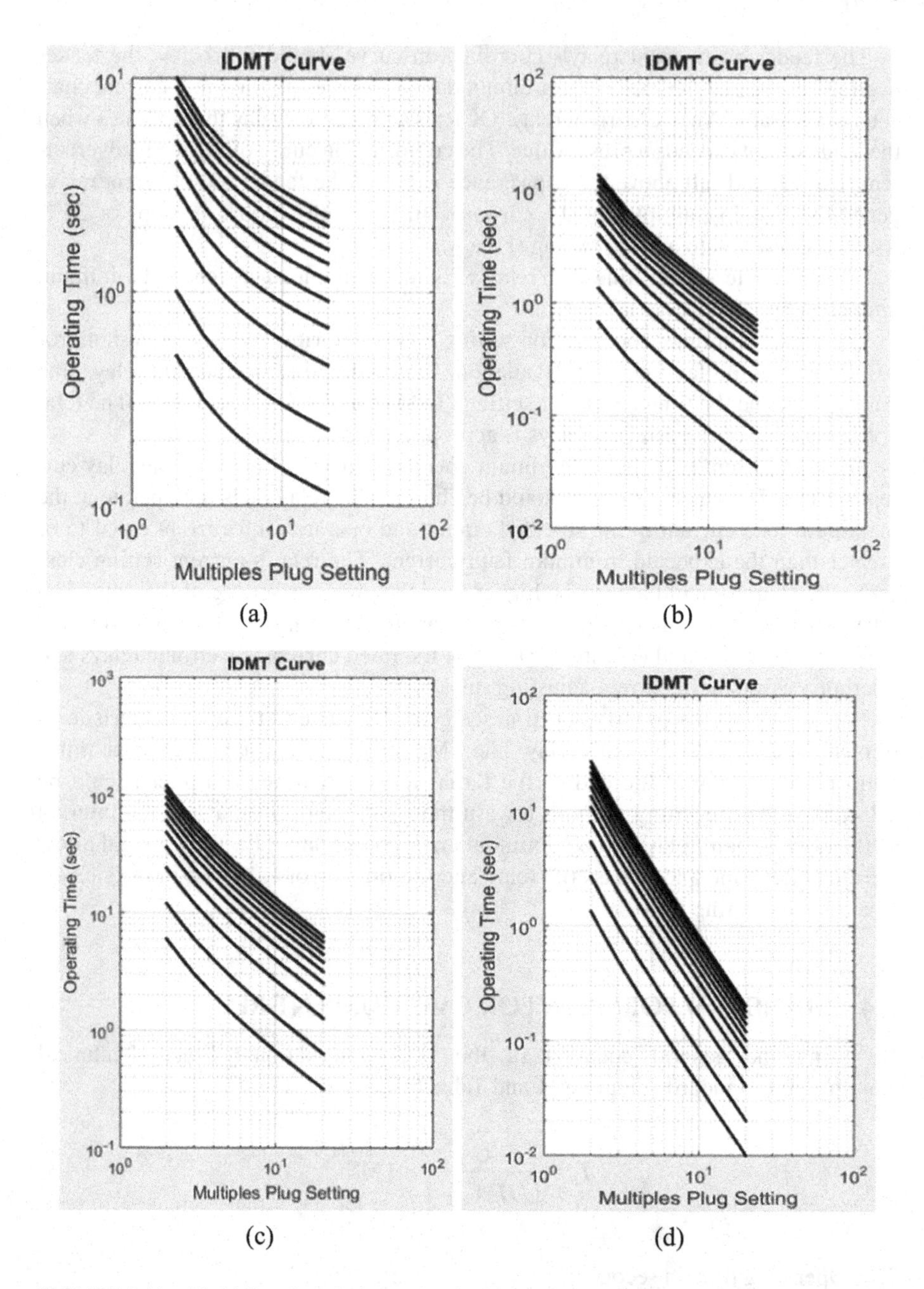

FIGURE 6.11 Multiples plug setting operating time characteristics. (a) IEC normal inverse, (b) IEC very inverse, (c) IEC long inverse, (d) IEC extremely inverse, (e) IEC ultrainverse, (f) IEEE moderately inverse, (g) IEEE very inverse, and (h) IEEE extremely inverse.

(Continued)

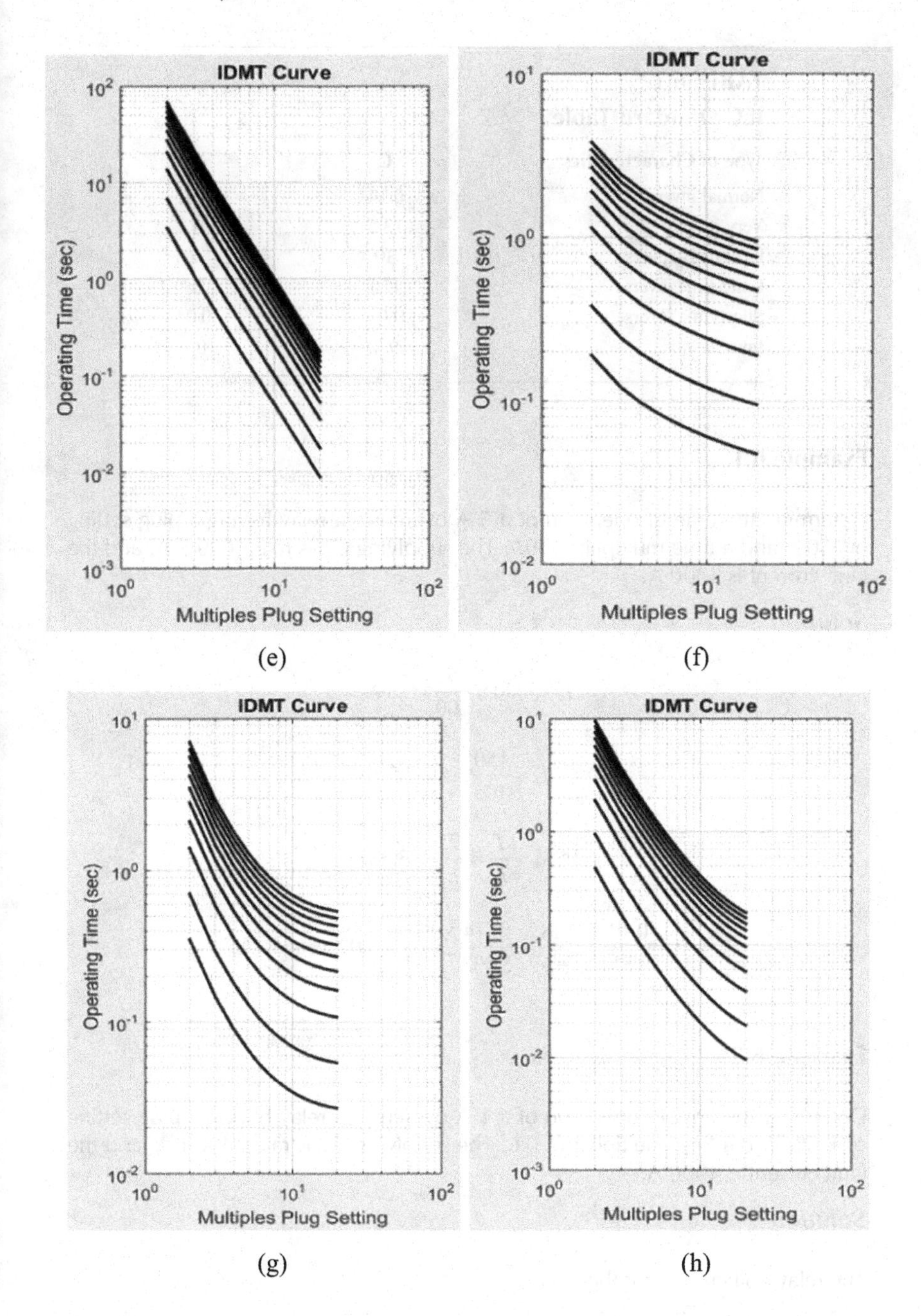

FIGURE 6.11 (*Continued*) Multiples plug setting operating time characteristics. (a) IEC normal inverse, (b) IEC very inverse, (c) IEC long inverse, (d) IEC extremely inverse, (e) IEC ultrainverse, (f) IEEE moderately inverse, (g) IEEE very inverse, and (h) IEEE extremely inverse.

TABLE 6.1
IEC Standard Table

Type of Characteristics	C	α
Normal inverse	0.14	0.02
Very inverse	13.5	1
Extremely inverse	80	2
Long-time inverse	120	1
Short time inverse	0.05	0.04
Inverse	9.4	0.7

Example 6.1

Determine the time of operation of a 5 A overcurrent relay having a plug setting of 150% and a time multiplier of 0.7. The supplying CT is rated 600:5 A, and the fault current is 3000 A.

Solution

$$I = 3000 \times \frac{5}{600} = 25\,\text{A}$$

$$I_s = \frac{150}{100} \times 5 = 7.5$$

$$\text{PSM} = \frac{I}{I_s} = \frac{25}{7.5} = 3.333$$

$$t = \frac{0.14 \times \text{TMS}}{\text{PSM}^{0.02} - 1} = \frac{0.14 \times 0.7}{3.333^{0.02} - 1} = 4.02\,\text{seconds}.$$

Example 6.2

Determine the time of operation of a 1 A overcurrent relay having a plug setting of 125% and a time multiplier of 0.6. The supplying CT is rated 400:1 A, and the fault current is 4000 A.

Solution

The relay coil current for the fault

$$I = \frac{1}{400} \times 4000 = 10\,\text{A}$$

The nominal relay coil current

$$I_s = \frac{125}{100} \times 1\,\text{A} = 1.25\,\text{A}$$

$$\text{PSM} = \frac{I}{I_s}$$

The PSM

$$\text{PSM} = \frac{10}{1.25} = 8$$

The time of operation

$$t = \frac{0.14 \times \text{TMS}}{\text{PSM}^{0.02} - 1} = \frac{0.14 \times 0.6}{8^{0.02} - 1} = 1.98\,\text{seconds}$$

Example 6.3

Determine the operation time of a relay of rating 5 A and having a plug setting of 125% and a time multiplier of 0.5. The supplying CT is rated 600:5 A, and the fault current is 4000 A.

$$\text{PSM} = \frac{I_f}{I_{\text{CT secondary}} \times \text{PS} \times \text{CT}}$$

$$\text{PSM} = \frac{4000}{5 \times 125\% \times 600/5} = 5.3$$

From the plug setting multiplier time log–log data given in Figure 6.12, for PSM = 5.3

$$t = 4.2\,\text{seconds}$$

$$t_{\text{op}} = t \times \text{TMS} = 4.2 \times 0.5 = 2.1\,\text{seconds}.$$

Example 6.4

An IDMT overcurrent relay has a current setting of 150% and a TMS of 40%. The time of operation of a relay rating is 5 A. The relay is connected in the circuit through a CT ratio of 500/5. Calculate the time of operation of the relay if the circuit carries a fault current of 6000 A.

$$\text{PSM} = \frac{I_f}{I_{\text{CT secondary}} \times \text{PS} \times \text{CT}}$$

$$\text{PSM} = \frac{6000}{5 \times 150\% \times 500/5} = 8$$

From the plug setting multiplier time log–log data given in Figure 6.12, for PSM = 8

$$t = 3.3\,\text{seconds}$$

$$t_{\text{op}} = t \times \text{TMS} = 3.3 \times 0.4 = 1.32\,\text{seconds}.$$

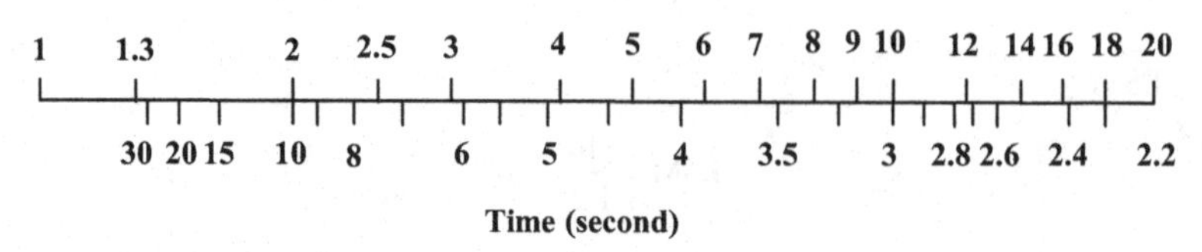

FIGURE 6.12 PSM–time characteristics.

Example 6.5

For the relay R in the system shown, determine the current tap setting (CTS). If the maximum three-phase fault current is 2400 A and the TDS = 2.0 finds the operating time, the relay type is CO-8 (inverse type) (Figure 6.13).

Solution

The load current

$$I_L = \frac{S}{\sqrt{3}\ V} = \frac{4.5 \times 10^6}{\sqrt{3} \times 13.2 \times 10^3} = 196.82\ \text{A}$$

The relay current

$$I_R = 196.82 \times \frac{5}{300} = 3.28\ \text{A}$$

Since the CTS of CO-8 relay available are 4, 5, 6, 7, 8, 10, and 12

$$\text{Hence we choose CTS} = 4$$

$$\text{Fault current} = 2400\ \text{A}$$

$$\text{Relay side fault current} = 2400 \times \frac{5}{300} = 40\ \text{A}$$

$$\text{Multiple selected CTS} = \frac{40}{4} = 10$$

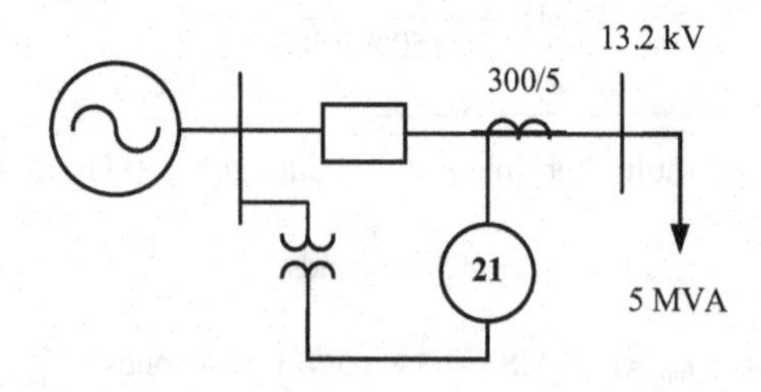

FIGURE 6.13 Connection of Example 6.5.

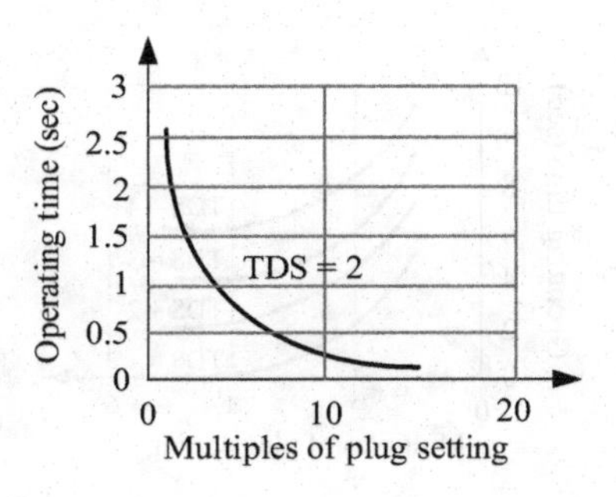

FIGURE 6.14 Operating time multiple of plug setting characteristics at TDS = 2.

From the OC curve (Figure 6.14)

$$t_{op} = 0.31\,\text{seconds}.$$

Example 6.6

If the rated current (pick-up current) of a relay is 3 A, and the time dial setting is 1.

a. How long does it take the relay to trip if the supply CT is rated at 400:5 A, and the fault current is 480 A? The type of OC relay is CO-8 (Figure 6.15).
b. Solve using the standard curve equation and compare the results.

Solution

$$I_s = 480 \times \frac{5}{400} = 6\text{ A}$$

$$\text{Tap value of current} = 3\text{ A}$$

$$\therefore \text{Multiple tap value current} = \frac{I_s}{I_{tap}} = \frac{6}{3} = 2$$

From the CO-8 characteristic curves (see Figure 6.16):

$$\text{Operating time} = 2.1\,\text{seconds}.$$

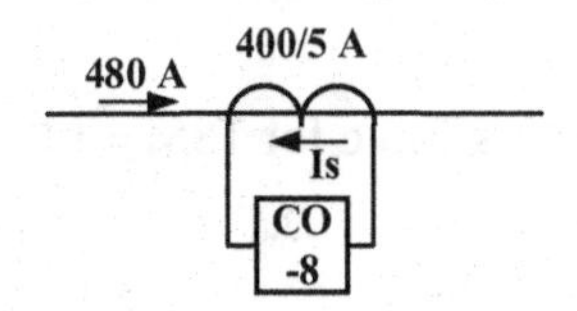

FIGURE 6.15 Connection of Example 6.6.

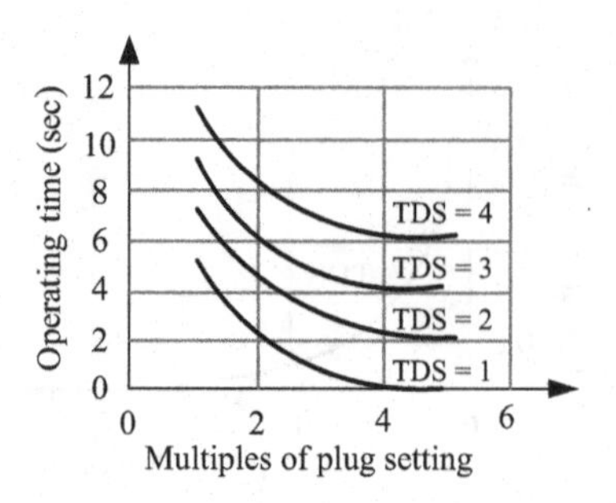

FIGURE 6.16 Operating time multiple of plug setting characteristics.

Example 6.7

The 60 MVA, 10/30 kV transformer shown in Figure 6.17 operates at 20% overload. The transformer's CB is equipped with a 1000/5 CT ratio, the feeder's CBs have a 500/5 CT ratio, the feeder relays are set at 125%, and TMS = 0.4. Use the time–PSM given in Table 6.2 and a discrimination margin of 0.5 seconds for a three-phase fault current of 3000 A at point F, find (Table 6.3)

i. The operating time of the feeder relay.
ii. The minimum setting of the transformer relay.
iii. The TMS of the transformer.

Solution

i.

$$\text{PSM} = \frac{3000 \times 5/500}{5 \times 125\%} = 4.8$$

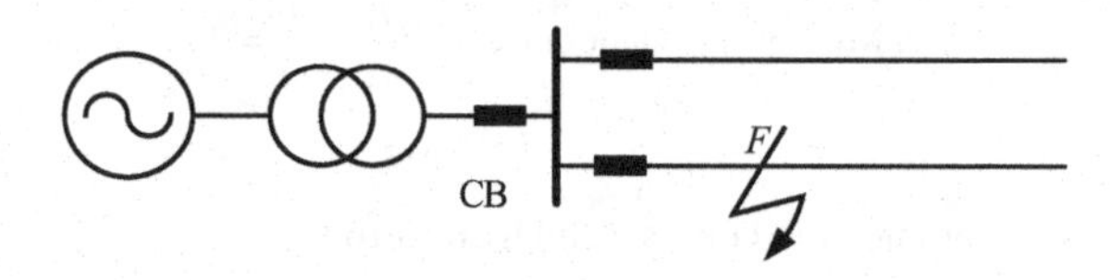

FIGURE 6.17 The power system configuration of Example 6.7.

TABLE 6.2
The Time–PSM Characteristics Table for TSM = 1.0

Time (seconds)	2.2	2.5	2.8	3.5	4.0	5.0	8.0	10.0	∞
PSM	20.0	15.0	10.0	6.4	5.0	4.0	3.2	2.0	1.0

TABLE 6.3
System Data

Generator	10 kV	40 MVA	$X=0.15$ p.u.
Transformer	10/30 kV	60 MVA	$X=0.2$ p.u.
Feeder (each)	30 kV	50 MVA	$R=0.05$ p.u., $X=0.25$ p.u.

$t=4.2$ seconds from time–PSM characteristics

$$t_{op} = \text{TMS} \times t$$
$$= 0.4 \times 4.2 = 1.92 \text{ seconds}$$

ii. The overload current $= 1.25 \dfrac{60 \times 10^3}{\sqrt{3} \times 30} = 1443.37$ A

$$\text{PSM} = \frac{1443.37 \times \frac{5}{1000}}{5 \times \text{PS}}$$
$$= \frac{1.443}{\text{PS}}$$

So, the result set of PS should be >1.443

Therefore, PS = 1.5 or 150%.

iii.

$$\text{PSM} = \frac{3000 \times 5/1000}{5 \times 150\%} = 2$$

$$t = 10 \text{ seconds}$$

The transformer operating time

$$t_{op} = 0.5 + 1.92 = 2.42 \text{ seconds}$$

$$\text{TMS} = \frac{t_{op}}{t} = \frac{2.42}{10} = 0.242$$

6.5 RELAY COORDINATION

The protective relay coordination is done during system design based on the short-circuit current level. It determines the sequence of relay operations for various faults in the power system so that the faulted section is cleared minimally. For proper relay coordination, it is necessary to determine an appropriate TSM and PSM for each relay to minimize the relay's operating time. Besides the TSM and PSM, the type of network, either radial or interconnected, plays a vital role in the optimum relay coordination.

6.5.1 Primary and Backup Protection

The first line of protection providing a quick and selective clearing of faults in the system is called primary protection. The protection given to the system when the main protection fails is called backup protection.

Failure of the main protection may be due to any of the following reasons

i. DC supply to the tripping circuit fails
ii. The current or voltage supply to the relay fails
iii. The tripping mechanism of the circuit breaker fails
iv. The circuit breaker fails to operate
v. The main protective relay fails

6.5.2 Method of Relay Coordination

There are three types of discrimination:

i. Discrimination by time,
ii. Discrimination by current, and
iii. Discrimination by both current and time.

The three methods used for correct relay coordination. Though the methods are different, they follow the same goal of isolating only the system's faulty section and leaving the rest of the system undisturbed.

6.5.2.1 Discrimination by Time

In this method, an appropriate time setting keeping the same fault current level is given to each relay controlling the circuit breakers in the power system to ensure that the relay nearest to the fault operates first. The relay near the source will have the maximum time compared to the relay from the source's far end (Figure 6.18).

Overcurrent protection is provided at A, B, C, and D at the infeed end of each power system section. If fault F occurs, relay C will have the least operating time compared to other relays. If relay C can clear the fault, there is no need for other relay operations, but if it fails to clear the fault in a given time, then relay B will act as the backup relay.

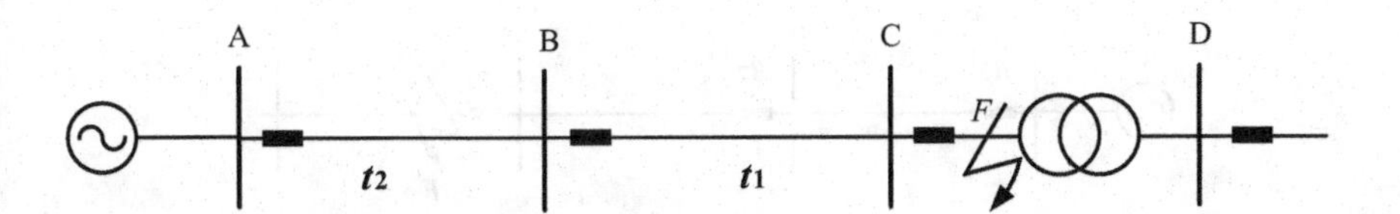

FIGURE 6.18 Discrimination by time.

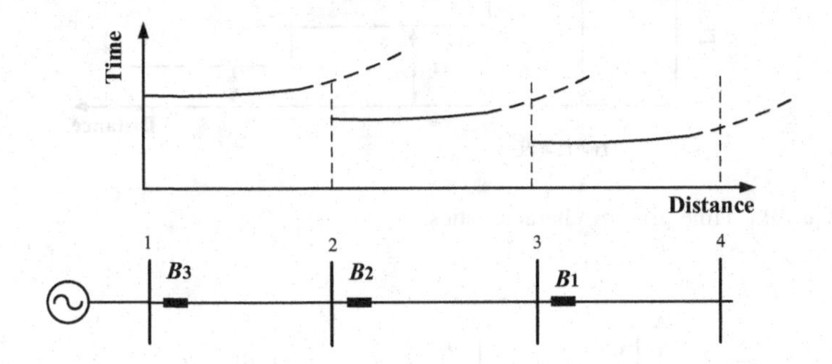

FIGURE 6.19 Coordinate delay time (CDT) (or time step delay) (or discrimination margin).

The relay's operating time from the source should be minimum and increase as we go toward the source (Figure 6.19).

The increase is by a discrimination margin of 0.4 or 0.5 seconds, which lates for the first relay to operate time plus to overtravel the next relay plus safety time factor.

$$\text{Overtravel time} = 0.1 \text{ seconds}$$

$$\text{Safety factor} = 0.1 - 0.2\,\text{seconds}$$

$$\text{CB time} = 0.2\,\text{seconds}$$

The main application is when the fault levels at various locations do not vary greatly (time-delay step) (Figure 6.20).

6.5.2.2 Discrimination by Current

An alternative to time-graded system, or time grading, current grading can be applied when the impedance between two substations is sufficient. The current-graded system normally employs a high-speed, high-set overcurrent relay (instantaneous). Discrimination by current relies on the fact that the fault current varies with the fault position because of the difference in impedance values between the source and the fault. Therefore, the relays controlling the various circuit breakers are set to operate at suitably tapered current values such that only the relay nearest to the fault trips its breaker (Figures 6.21 and 6.22).

The drawback of this type is if a fault is very near to station B in section BC, relay A may feel that it is in section AB (no discrimination at section end). To obtain discrimination, only about 80% of the line length is protected by a relay at one station.

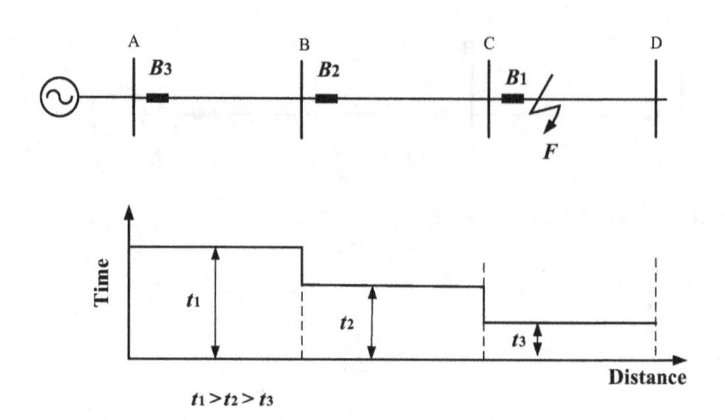

FIGURE 6.20 Time-grading characteristics.

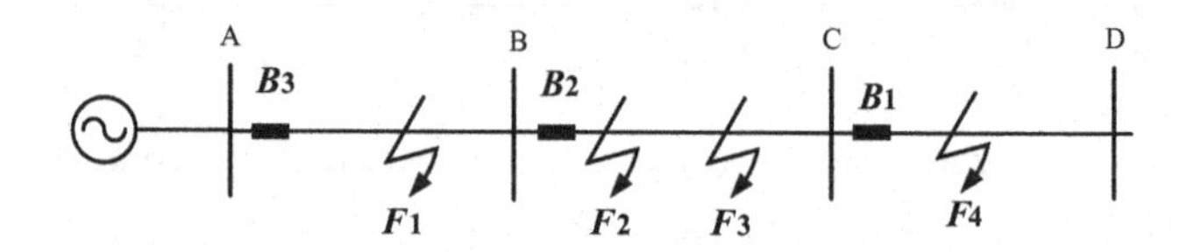

FIGURE 6.21 Discrimination by the current.

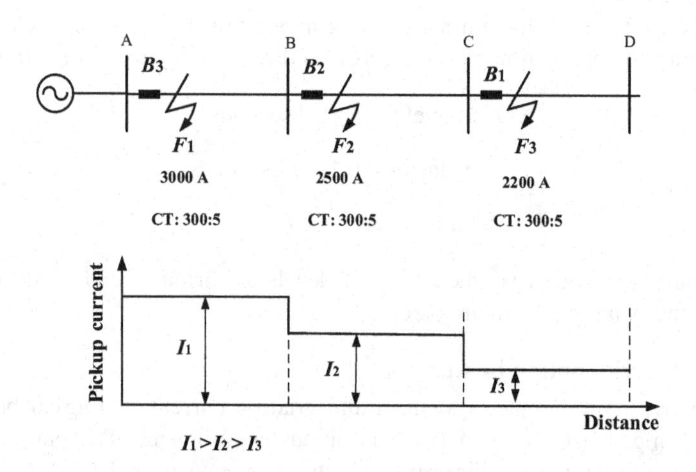

FIGURE 6.22 Current grading characteristics.

6.5.2.3 Discrimination by Both Time and Current

This type of grading is achieved with an inverse time overcurrent relay, and the most widely used is the IDMT relay. If the IDMT relay is slow at low values of overcurrent, extremely inverse is used, and if the fault current reduces substantially as the fault section moved away from the source, very inversely is used.

For proper coordination on a radial system, the pick-up of a relay should be such that it will operate for all faults in its zone and provide backup for faults in the adjoined zones.

The backup setting should equal the fault current value when the fault is at the far end. The next zone with minimum generation is connected to the system

With this characteristic, the operation time is inversely proportional to the current level, and the actual characteristics are a function of both time and current settings. For a large variation in fault current between the two ends of the feeder, faster operating times can be achieved by the relays nearest to the source, where the fault level is the highest.

Overcurrent relay characteristics generally start with selecting the correct characteristics for each relay, followed by choosing the relay current settings. Finally, the grading margins and hence time settings of the relays are determined.

IDMT relays are used in this feeder protection for proper coordination among various relays. A relay's pick-up should be used to operate for all short circuits in its line section and provide backup for short circuits in the immediately adjoining section. On a radial system, the current setting of the relay farthest from the source should be minimum and increase as it goes toward the source.

- *Fault at P,* Relay C operates at t_1
  Relay B will not operate ($t_2 > t_1$)
  Relay A will not pick up.
  Relay B will clear the fault if relay C fails in time t_2.

- *Fault at Q,* Relay B operates at t_3.
  Relay A will not operate ($t_4 > t_3$)
  If relay B fails, Relay A will clear the fault in time t_4.

- *Fault at R,* Relay A operates to clear the fault at t_5.
  The generator at bus one can be an equivalent representation of one or more transformers feeding bus one from a higher voltage supply point (Figure 6.23).

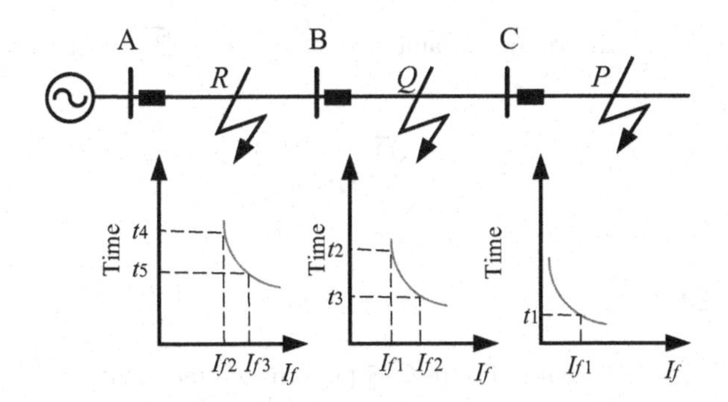

FIGURE 6.23 Time–current characteristics.

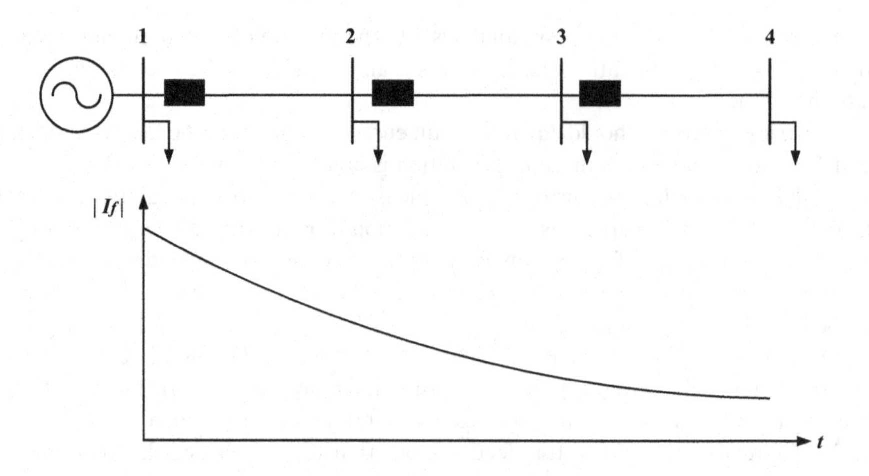

FIGURE 6.24 Fault current is inversely proportional to distance.

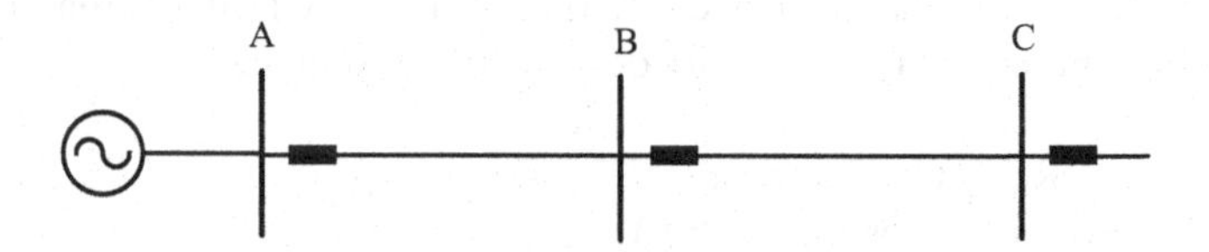

FIGURE 6.25 Radial system of Example 6.8.

Since the power source is only to the left of each line section, it is sufficient to provide only one circuit breaker for each section at the source end (Figure 6.24).

Fault current is inversely proportional to distance, as shown in Figure 6.24.

The principle of backup protection with an overcurrent relay for any relay X backing up the next downstream relay Y is that relay X must pick up:

i. For one-third of the minimum current seen by Y.
ii. For the maximum current seen by Y, no sooner than coordinate delay time (CDT) after Y should have picked up for the current.

A line-to-line fault will produce a fault current equal to $\sqrt{3}/2$ times the three fault currents

$$I_{f\,(\text{L-L})} = \frac{\sqrt{3}}{2}\, I_{f\,(\text{three-phase})} \tag{6.4}$$

Example 6.8

The power system shown in Figure 6.25 performs current–time grading calculation, using Table 6.4 for data of the system and Table 6.5 as the time–PSM characteristics table for TSM = 1.0.

TABLE 6.4
Data of the System

Relay	CT Ratio	Fault Current (A)
c	200/5	4000
b	200/5	5000
a	400/5	6000

TABLE 6.5
The Time–PSM Characteristics Table for TSM = 1.0

Time (seconds)	2.2	2.5	2.8	3.5	4.0	5.0	8.0	10.0	∞
PSM	20.0	15.0	10.0	6.4	5.0	4.0	3.2	2.0	1.0

Solution

START WITH RELAY C

The secondary current with maximum fault current at bus C is $4000/(200/5) = 100$ A

For 100% relay setting $\text{PSM} = \frac{100}{5} = 20$

From the log–log characteristics, the operating time = 2.2 seconds, but these characteristics are used for TMS = 1.0.

For first coordination, use TMS = 0.1.

For this relay, the operating time will be

$$t_{op} = 0.1 \times 2.2 = 0.22 \text{ seconds}$$

To achieve discrimination between the relay at B and that at C when fault takes place just before C or just after C (i.e., no change in fault current), let discrimination time = 0.5 seconds.

The operating time of relay B when fault takes place near C will be $0.22 + 0.5 = 0.72$ seconds.

The secondary current in the relay at B when fault takes place near C will be 4000/40 = 100 A.

Assume relay B current setting = 125%

$$\therefore \text{PSM} = \frac{100}{125\% \times 5} = 16$$

So, the operating time from the characteristics is 2.5 seconds, but relay B operating time when graded w.r.t relay C is 0.72 seconds

$$\text{TMS for relay B} = \frac{0.72}{2.5} = 0.29$$

When the fault is near relay B, the PSM with an operating current is 6.25 A is

$$\text{PSM} = \frac{5000}{6.25 \times 40} = 20$$

From the characteristics, the operating time = 2.2 seconds.

$$\text{The actual operating time} = 2.2 \times 0.29 = 0.638 \text{ seconds}$$

Since the CT ratio at A is 400/5, which is high compared to the relay at B, the current discrimination is inherent. Let the relay setting of relay A be 125%.

So, PSM of relay A when fault takes place near B

$$\text{PSM} = \frac{5000}{5 \times 125\% \times 80} = 10$$

From the characteristics, relay A operation time is three seconds.

But, the operating time of relay A, when graded with relay, should be 0.638 + 0.5 = 1.138 seconds

$$\therefore \text{TMS of relay A} = \frac{1.138}{3} = 0.379$$

When a fault is near A, the $\text{PSM} = \frac{6000}{5 \times 125\% \times 80} = 12$

From the characteristics, the corresponding time = 2.6 seconds.

The actual operating time of relay A

$$= 2.6 \times 0.379 = 0.985 \text{ seconds}$$

6.6 REQUIREMENTS FOR PROPER RELAY COORDINATION

- Relay current setting

  The relay current setting is the minimum current required for the relay to operate. Determining the current setting should be so that the relay does not operate for the maximum fault current level but does operate for a minimum fault current level. Suppose the current setting is set for the maximum fault current level in the power system. In that case, an overcurrent relay can provide a small degree of protection against overload and fault, but the main function of an overcurrent relay is to isolate primary system faults, not overload protection.

- Relay time-grading margin

  The time-grading margin is the minimum time interval between the primary and backup protective relays to achieve proper discrimination.

  If the grading margin is not provided, then more than one relay will operate for the same fault, leading to a failure to determine the fault location and a blackout in the power system.

- Time multiplier setting

  The operating time of an electrical relay mainly depends upon two factors

  i. How long distance to be traveled by the moving parts of the relay for closing relay contacts and

  ii. How fast do the moving parts of the relay cover this distance?

The adjustment of the traveling distance of a relay is commonly known as a time setting. This adjustment is commonly known as *the TSM of the relay.*

The following algorithm gives the theoretical calculation for overcurrent relay setting:

$$\text{Assuming Plug Setting (PS) of relay } 1 = 100\% \text{ and } \text{TMS} = 0.025$$

$$\text{PS of relay } 2 > \frac{1.3}{1.05} \times \text{PS of relay } 1 \tag{6.5}$$

$$\text{PS of relay } 3 > \frac{1.3}{1.05} \times \text{PS of relay } 2 \tag{6.6}$$

$$\text{PSM} = \frac{I_f}{\text{PS} \times \text{CT}_{\text{ratio}}} \tag{6.7}$$

$$T = \frac{0.14}{\text{PSM}^{0.02} - 1} \times \text{TMS} \tag{6.8}$$

Example 6.9

The power system is shown in Figure 6.26. The MVA range of the source is between 120 and 250 MVA. Perform the current grading.

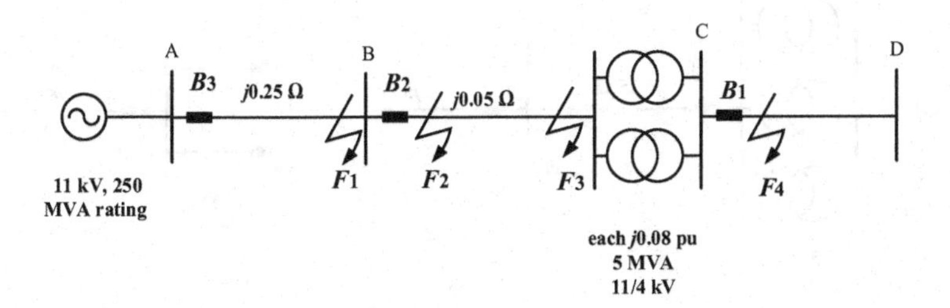

FIGURE 6.26 The power system configuration of Example 6.9.

Solution

For fault at F_1

$$I_{f1} = \frac{\frac{11,000}{\sqrt{3}}}{\frac{11^2}{250} + 0.25} = 8652.38 \text{ A}$$

Hence, the relay controlling the circuit breaker at A and set to operate at a fault current of 8652.38 A would protect the whole transmission line A–B.

For faults at F_1: It is not practical to distinguish between faults at F_1 and F_2 since the distance between these points can be only a few meters.

For fault at F_3

$$I_{f3} = \frac{\frac{11,000}{\sqrt{3}}}{\frac{11^2}{250} + 0.25 + 0.05} = 8100 \text{ A}$$

For fault at F_4

$$I_{f4} = \frac{\frac{11,000}{\sqrt{3}}}{\frac{11^2}{250} + 0.25 + 0.05 + \frac{0.08}{2} \times \frac{11^2}{5}} = 3625 \text{ A}$$

Hence a relay controlling the CB at B and set to operate at a current of 3625 A plus a safety margin would not operate for a fault at F4 and would thus discriminate with the relay at A.

Assume a safety margin of 30%; it is reasonable to choose a relay setting of 130% of 3625 or 4713 A.

Example 6.10

A portion of the 13 kV radial system is shown in Figure 6.27. Assume the reactances of transformers and transmission lines are given in ohm and referred to the generator side to calculate the minimum and maximum fault current as a fault at each bus.

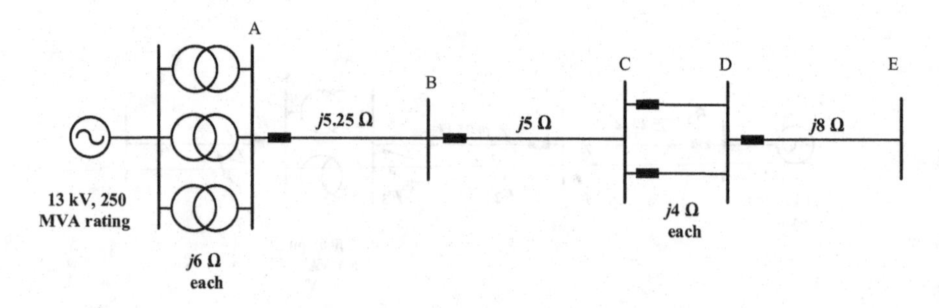

FIGURE 6.27 The power system configuration of Example 6.10.

Solution

Fault at bus E.

Maximum fault current (three-phase fault):

$$I_{f\,\text{three-phase}} = \frac{13{,}000/\sqrt{3}}{j\left(\frac{13^2}{250}+\frac{6}{3}+5.25+5+\frac{4}{2}+8\right)} = -j327.38\text{ A}$$

Minimum fault current (line-to-line fault):

$$I_{f\text{L-L}} = \frac{\sqrt{3}}{2}\, I_{f\,\text{three-phase}}$$

$$I_{f\text{L-L}} = \frac{\sqrt{3}}{2}(-j327.38) = -j283.52\text{ A}$$

Fault at bus D.

Maximum fault current (three-phase fault):

$$I_{f\,\text{three-phase}} = \frac{13{,}000/\sqrt{3}}{j\left(\frac{13^2}{250}+\frac{6}{3}+5.25+5+\frac{4}{2}\right)} = -j502\text{ A}$$

Minimum fault current (line-to-line fault):

$$I_{f\text{L-L}} = \frac{\sqrt{3}}{2}\, I_{f\,\text{three-phase}}$$

$$I_{f\text{L-L}} = \frac{\sqrt{3}}{2}(-j502) = -j435.5\text{ A}$$

Fault at bus C.

Maximum fault current (three-phase fault):

$$I_{f\,\text{three-phase}} = \frac{13{,}000/\sqrt{3}}{j\left(\frac{13^2}{250}+\frac{6}{3}+5.25+5\right)} = -j580.66\text{ A}$$

Minimum fault current (line-to-line fault):

$$I_{f\text{L-L}} = \frac{\sqrt{3}}{2}\, I_{f\,\text{three-phase}}$$

$$I_{f\text{L-L}} = \frac{\sqrt{3}}{2}(-j580.66) = -j502.86\text{ A}$$

Fault at bus B.

Maximum fault current (three-phase fault):

$$I_{f\,\text{three-phase}} = \frac{13{,}000/\sqrt{3}}{j\left(\frac{13^2}{250}+\frac{6}{3}+5.25\right)} = -j947\text{ A}$$

Minimum fault current (line-to-line fault):

$$I_{f\text{L-L}} = \frac{\sqrt{3}}{2} I_{f\,\text{three-phase}}$$

$$I_{f\text{L-L}} = \frac{\sqrt{3}}{2}(-j947) = -j820\text{ A}$$

Fault at bus A.

Maximum fault current (three-phase fault):

$$I_{f\,\text{three-phase}} = \frac{13{,}000/\sqrt{3}}{j\left(\frac{13^2}{250}+\frac{6}{3}\right)} = -j2804\text{ A}$$

Minimum fault current (line-to-line fault):

$$I_{f\text{L-L}} = \frac{\sqrt{3}}{2} I_{f\,\text{three-phase}}$$

$$I_{f\text{L-L}} = \frac{\sqrt{3}}{2}(-j2804) = -j2429\text{ A}$$

6.7 HARDWARE AND SOFTWARE FOR OVERCURRENT RELAYS

The general expression of the analytical relationship between time versus input current depends on inverses and is given by

$$T = \frac{K}{I^n - 1}$$

where K = TMS.

I = input current to overcurrent relay in multiples of PSs.

n = constant deciding the inferences.

For different values of n, the characteristics are

$$T = \frac{K}{I^0}(n = 0)\text{ DTOC relay}$$

6.8 OVERVOLTAGE AND UNDERVOLTAGE PROTECTION

6.8.1 Undervoltage Test

All devices use a power plug connected to the supply voltage, as shown in Figure 6.28. Initially both power switches are open (OFF buttons). The resistive load is set to its maximum value. The voltage relay's DIP switches were set as given in Table 6.6. The voltage is measured between the outer conductors (delta voltage) and has a value of 230 V.

6.8.2 Overvoltage Test

Figure 6.28 shows the Lucas-Nülle GmbH layout plan for the overvoltage test emulator at PVAMU.

All devices connect using a power plug to the supply voltage. The resistive load to its maximum value. The DIP switches should remain on the same settings as those in the test of **undervoltage protection**. The overvoltage excitation should lie between 110% and 115% at a voltage of 110 V.

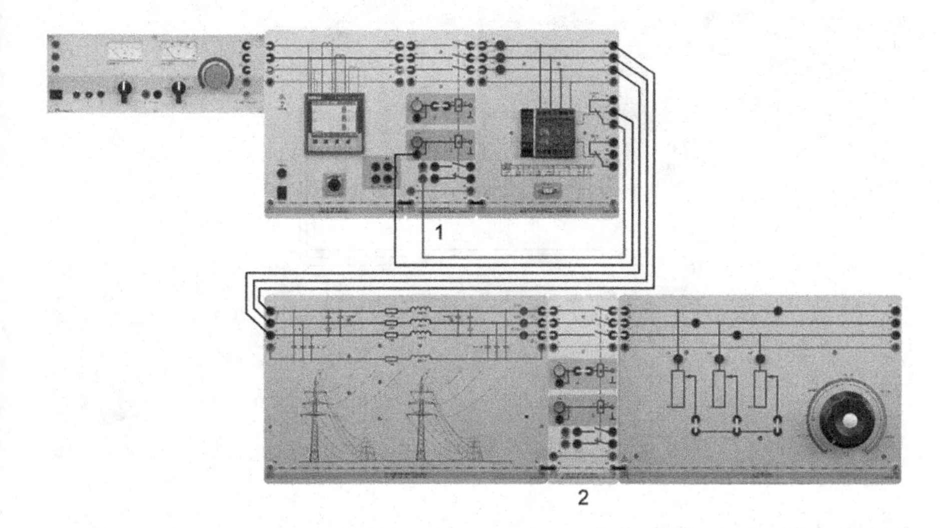

FIGURE 6.28 Lucas-Nülle GmbH layout plan for overvoltage test emulator at PVAMU.

TABLE 6.6
The Voltage Relay's DIP Switches Using the Lucas-Nülle GmbH Test Emulator at PVAMU

Function	Un	Un	Un	ps Supv.		Hyst.	Hyst.	tU<tU>
DIP Switch	**1 (V)**	**2 (V)**	**3 (V)**	**4**	**5**	**6 (%)**	**7 (%)**	**8**
ON	110	230	400	Active	Δ	6	10	×1
OFF	100	100	100	Inactive	Y	3	3	×0.1

Once all these settings have been made, connect the transmission line and the load by closing the power switches (ON buttons).

The operational state should be assumed (both red LEDs on the voltage relay are OFF). If a relay is still energized (red LED on), turn the power switch off again and check your settings.

6.8.3 Hysteresis Test

To test various hysteresis levels, which can be set for a protective relay. The set hysteresis always applies jointly to the overvoltage and undervoltage.

The setup from the previous system remains unchanged here. Just disconnect the feed line to the power switch so that it is not inadvertently tripped during the next round of measurements (Figure 6.29).

The resistive load is set to its maximum value. The voltage relay's DIP switches were set as indicated in Table 6.7. The (delta) voltage is measured between the outer conductors and equals 110 V.

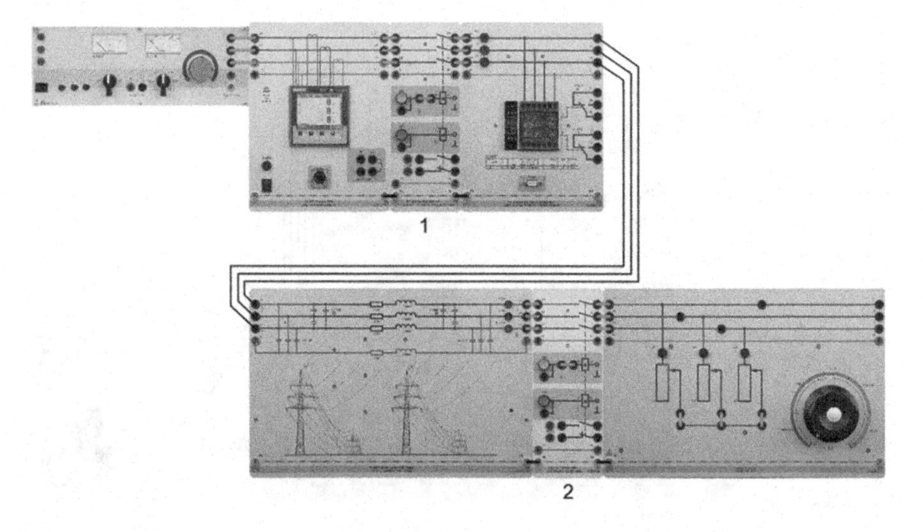

FIGURE 6.29 Lucas-Nülle GmbH layout plan for hysteresis test emulator at PVAMU.

TABLE 6.7
The Voltage Relay's DIP Switches Using the Lucas-Nülle GmbH Test Emulator at PVAMU

Function	Un	Un	Un	ps Supv.		Hyst.	Hyst.	tU<tU>
DIP Switch	1 (V)	2 (V)	3 (V)	4	5	6 (%)	7 (%)	8
ON	110	230	400	Active	Δ	6	10	×1
OFF	100	100	100	Inactive	Y	3	3	×0.1

TABLE 6.8
The Voltage Relay's DIP Switches Using the Lucas-Nülle GmbH Test Emulator at PVAMU

$U<$ (%)	$U>$ (%)	R_{PU} (V)	R_{Re} (V)	$R=R_{Re}/R_{PU}$
75	105	115	111	0.96
75	110	120	116	0.97
75	115	126	120	0.95
75	120	131	126	0.96

TABLE 6.9
Set the Hysteresis to 10% (DIP Switch 7) Using the Lucas-Nülle GmbH Test Emulator at PVAMU

Function	Un	Un	Un	ps Supv.		Hyst.	Hyst.	tU<tU>
DIP Switch	1 (V)	2 (V)	3 (V)	4	5	6 (%)	7 (%)	8
ON	110	230	400	Active	Δ	6	10	×1
OFF	100	100	100	Inactive	Y	3	3	×0.1

The adjustable three-phase power supply set the voltage to 110 V. The trip delay is zero seconds in both cases, and the potentiometer to the first overvoltage value is indicated in Table 6.8. Slowly increase the voltage on the three-phase power supply until the red LED "U>" comes on, and note the pick-up value under R_{PU}. Gradually decrease the voltage and record the release (or reset) value R_{Re} in Table 6.8.

- Repeat these steps on the remaining settings.
- Calculate the corresponding reset ratios and determine their average.

Calculate the average hysteresis with the help of the reset ratios.

$$H = 4.3\%$$

Set the hysteresis to 10% (DIP switch 7), as in Table 6.9.

Proceed as in the previous experiment phase and note the measured values until the red LED "U<" comes on.

The calculation of the reset ratios and determining their average is given in Table 6.10 using the Lucas-Nülle GmbH test emulator at PVAMU.

Calculate the average hysteresis with the help of the reset ratios.

$$H = 8.9\%$$

TABLE 6.10
Set the Hysteresis to 10% (DIP Switch 7) Using the Lucas-Nülle GmbH Test Emulator at PVAMU

U< (%)	*U*> (%)	R_{PU} (V)	R_{Re} (V)	$R = R_{Re}/R_{PU}$
90	120	101	110	1.09
85	120	96	106	1.1
80	120	90	100	1.11
75	120	84	94	1.12

6.9 DIRECTIONAL POWER PROTECTION

Figure 6.30 shows a Lucas-Nülle GmbH circuit diagram of directional power protection. The overhead transmission line receives a three-phase power supply and is loaded symmetrically at its end. A circuit breaker (power-switch module) is located before the transmission line for disconnecting the line from the power supply in the event of a fault. The directional time overcurrent relay receives the necessary measurement variables via a current transformer and a voltage transformer for each phase. Figure 6.31 shows a Lucas-Nülle GmbH layout plan test emulator at PVAMU.

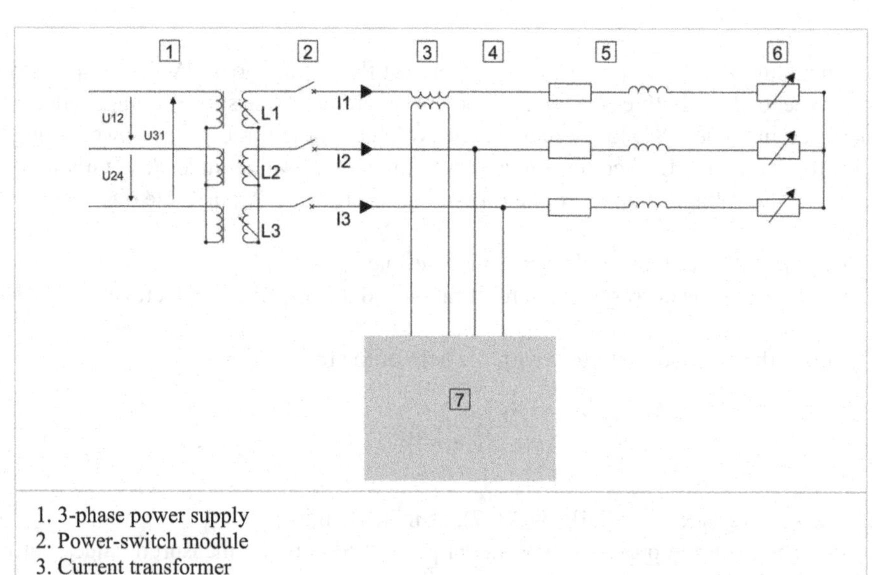

FIGURE 6.30 Lucas-Nülle GmbH circuit diagram of directional power protection.

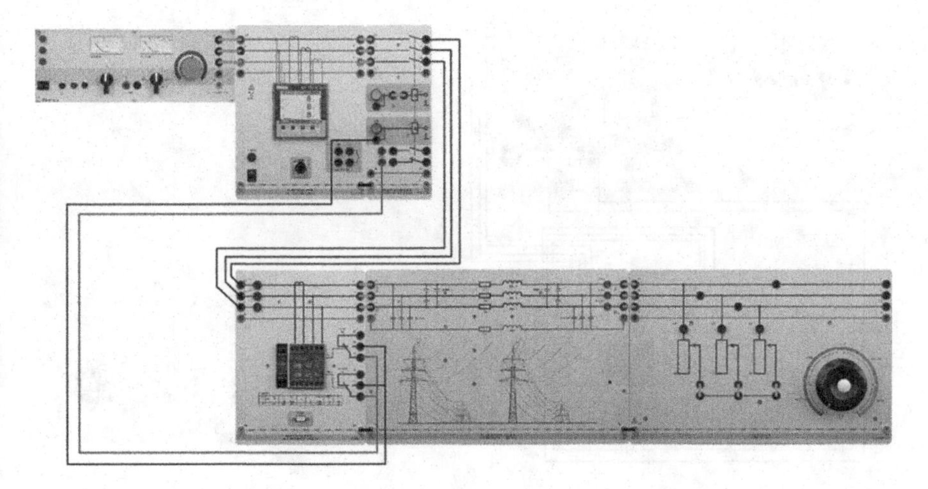

FIGURE 6.31 Lucas-Nülle GmbH layout plan test emulator at PVAMU.

6.10 TESTING FORWARD AND REVERSE POWER

Initial steps: Use the LN SCADA software for precise adjustment of values (refer to the chapter titled "Using LN SCADA Software"). This software indicates the power in watts (W), which is considered a percentage of the protective relay's reference power. Example: Un × In × power proportion (%) = 110 V × 1 A × 50% = 55 W; Un = nominal voltage of the XP2-R; In = nominal current of the XP2-R.

6.10.1 Test of Forward Power

For the connection in this section, the relay's DIP switches are set as shown in Table 6.11 (active setting = green background).

When the resistive load is at the maximum value, the voltage will be 110 V.

At current of 0.31, A, the apparent power will be 20 VA, and the percentage should be set for this power level on the potentiometer at 31%. The trip level will be 35 W at a delay time of 12.5 seconds.

TABLE 6.11
Show the Relay's DIP Switches Setting Using the Lucas-Nülle GmbH Test Emulator at PVAMU

Function	Un	Un	Un			*t* Return	t $RP_>$	*t* $P_>$
DIP Switch	1 (V)	2 (V)	3 (V)	4	5	6 (ms)	7	8
ON	110	230	400		Δ	40	×10	×10
OFF	100	100	100		Y	500	×1	×1

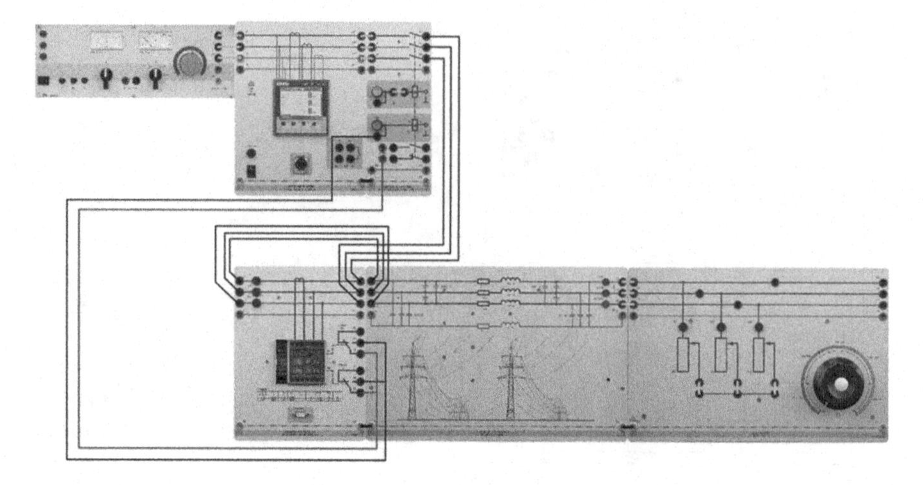

FIGURE 6.32 Lucas-Nülle GmbH layout for reverse power emulator at PVAMU.

6.10.2 Test of Reverse Power

When the resistive load is at maximum value, the voltage is 110 V.

Change the direction of the current flow by swapping the connections, as shown in Figure 6.32. This reverses the current and, therefore, the power direction for the protective relay.

The power of 10 W per phase corresponds to 6% of a gas turbine's.

Maximum permissible reverse power. The setting will be 16%. The power at which the relay is tripped is at a level of 17 W at a delay time of three seconds.

PROBLEMS

6.1. Determine the operation time of a 5 A, three seconds overcurrent relay with a plug setting of 125% and a time multiplier of 0.58. The supplying CT is rated 600:5 A, and the fault current is 3500 A.

6.2. The 40 MVA, 11/30 kV transformer shown in Figure 6.33 operates at 25% overload. The transformer's CB is equipped with a 1000/5 CT ratio, the feeder's CBs have a 500/5 CT ratio, the feeder relays are set at 125%, and

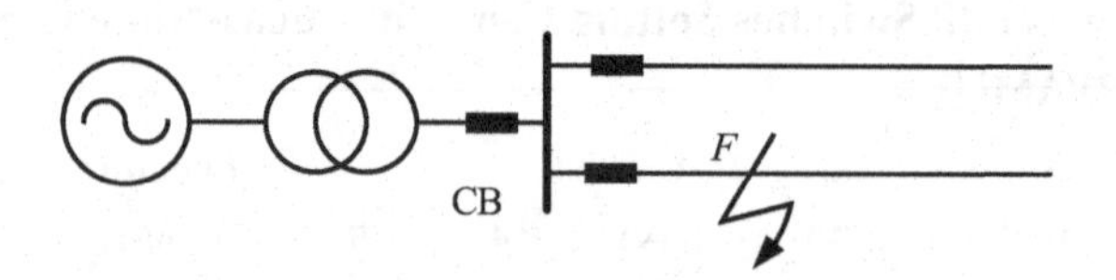

FIGURE 6.33 The power system configuration of Problem 6.2.

TMS = 0.35. Use the time–PSM given in Table 6.12 and a discrimination margin of 0.4 seconds for a three-phase fault current of 3300 A at point F to find (Table 6.13):

i. The operating time of the feeder relay.
ii. The minimum setting of the transformer relay.
iii. The TMS of the transformer.

TABLE 6.12
The Time–PSM Characteristics Table for TSM = 1.0

Time (seconds)	2.2	2.5	2.8	3.5	4.0	5.0	8.0	10.0	∞
PSM	20.0	15.0	10.0	6.4	5.0	4.0	3.2	2.0	1.0

TABLE 6.13
System Data

Generator	11 kV	40 MVA	$X = 0.2$ p.u.
Transformer	11/30 kV	60 MVA	$X = 0.12$ p.u.
Feeder (each)	30 kV	50 MVA	$R = 0.05$ p.u., $X = 0.25$ p.u.

6.3. The 25 MVA transformer shown in Figure 6.34 operates at 30% overload and feeds n 11 kV bus through a circuit breaker. The transformer's CB is equipped with a 1000/5 CT ratio, the feeder's CBs have a 400/5 CT ratio, the feeder relays are set at 125%, and TMS = 0.375. Use the time–PSM given in Table 6.12 and a discrimination margin of 0.4 seconds for a three-phase fault current of 5000 A at point F to find (Table 6.14):
i. The operating time of the feeder relay.
ii. The minimum setting of the transformer relay.
iii. The TMS of the transformer.

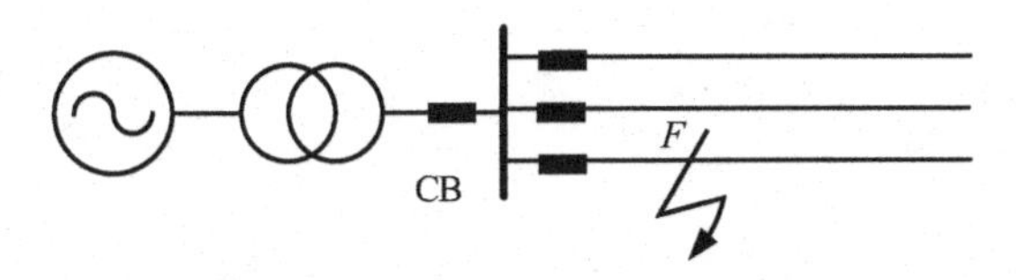

FIGURE 6.34 The power system configuration of Problem 6.3.

TABLE 6.14
System Data

Generator	11 kV	25 MVA	$X=0.1$ p.u.
Transformer	11/33 kV	60 MVA	$X=0.15$ p.u.
Feeder (each)	33 kV	50 MVA	$R=0.05$ p.u., $X=0.25$ p.u.

6.4. A 11 kV busbar has two incoming feeders, each fitted with 1000/5 A CT and an IDMTL (inverse definite minimum time-lag) relay having a plug and time setting of 150% and 0.4, respectively. The corresponding data for an outgoing feeder is 400/5 A, 175%, 0.3. For a three-phase 280 MVA fault along the outgoing feeder, calculate the difference between the operating times of the relay for:

   i. The two incoming feeders are in operation.
   ii. One incoming feeder is in operation (the other is switched out).
   iii. Show that the protection and a single incoming feeder will not operate to a load of 20 MVA.

6.5. It is required to provide time–current grading for the following system in Figure 6.35 (Table 6.15).

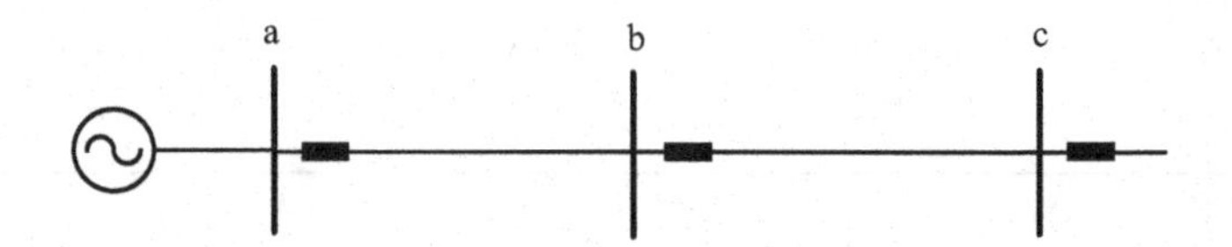

FIGURE 6.35 The power system configuration of Problem 6.5.

TABLE 6.15
System Data

Relay	Plug Setting (PS) (%)	CT Ratio	Fault Current (A)
a	150	400/5	5100
b	125	300/5	4200
c	100	200	3300

Use IDMTL characteristics for TMS = 1, shown in Figure 6.36.

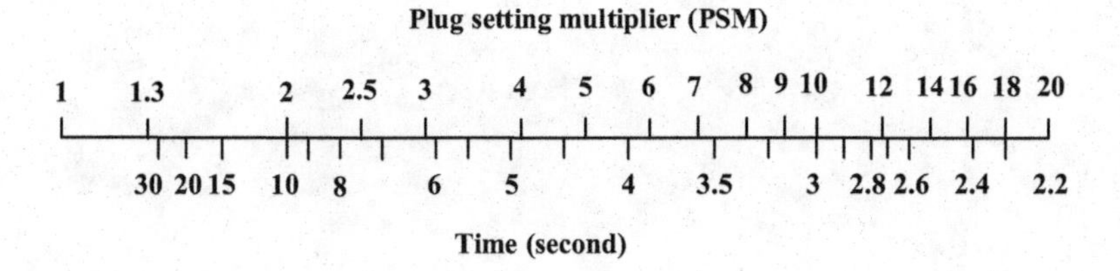

FIGURE 6.36 IDMTL characteristics.

6.6. For the relay R in the system shown, determine the current tap setting (CTS). If the maximum three-phase fault current is 2400 A and the TDS = 2.0 finds the operating time, the relay type is CO-8 (inverse type) (Figure 6.37).

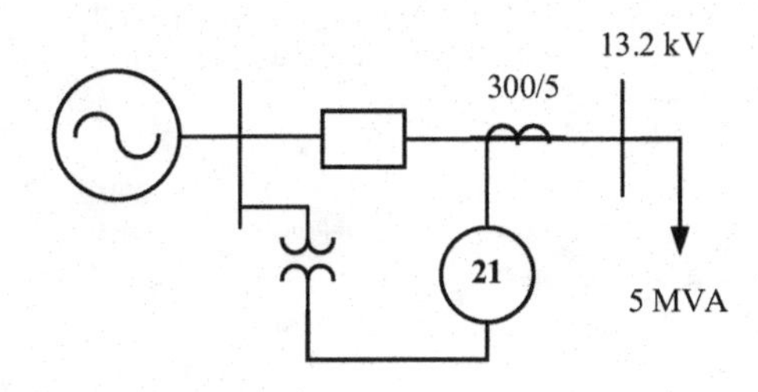

FIGURE 6.37 Connection of Problem 6.6.

7 Transmission Line Protection

7.1 INTRODUCTION

A transmission line is the most important and integral part of a power system. Due to more than 80% of disturbances or short-circuit faults in an overhead line, this section has become the most vulnerable part of the electrical system. Therefore, it is necessary to have designed a reliable protection system to protect against interference.

In order of ascending cost and complexity, the protective devices available for transmission line protection are:

i. fuses
ii. sectionalizes, reclosers
iii. instantaneous overcurrent
iv. inverse, time delay, overcurrent
v. directional overcurrent
vi. distance
vii. pilot.

The current chapter describes transmission lines protection, distance relay as impedance, reactance, and mho relay will be discussed in this chapter. The fundamentals of differential protection systems used to protect transmission lines will also be discussed. Protection of parallel lines (Parallel operation) and Parametrizing non-directional relays using the SCADA system are discussed. Directional time overcurrent relays and High-speed distance protection using SCADA technology are explained in this chapter.

7.2 DISTANCE RELAY

Distance relays represent the first option for replacing overcurrent relays when considered inadequate for a specific function. These relaying devices are not affected as much as overcurrent relays by relative source impedances and system conditions. Other advantages offered by distance relays are the integrated fault location function, the possibility of being applied as remote backup protection, and the wide variety of characteristics, which make the option for these devices the most adequate for certain applications

The ratio of voltage to current at the relay location governs the distance relay's performance. The operating time of the relay automatically increases with an increase in this ratio. The impedance or the circuit's reactance between the relay and the fault is proportional to the distance between them, provided the relay actuating quantities (VR and IR) are properly chosen. Figure 7.1 shows a distance relay configuration.

DOI: 10.1201/9781003394389-7

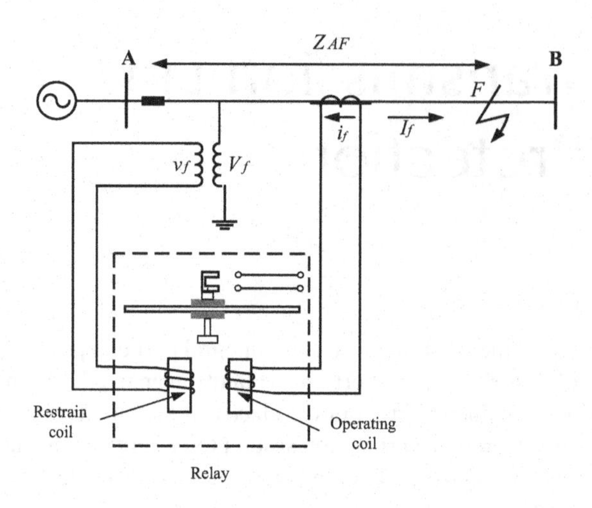

FIGURE 7.1 Distance relay configuration.

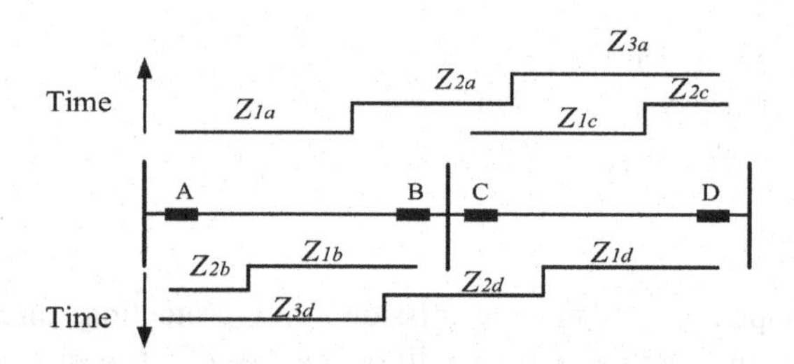

FIGURE 7.2 Two sections line system is protected by three zones protection scheme.

For the simple impedance relay shown, the relay operates for $Z = V/I$ below a particular value, whereas above this value of Z, the relay will restain. If possible, select a setting comparable along the time to be protected.

Strictly speaking, the impedance seen by the relay is not proportional to the distance between the relay and the fault in general because of the following reasons:

i. Presence of resistance at the fault location (arc resistance).
ii. Presence of loads or generating sources between the relay and the fault location.

Figure 7.2 shows a two sections line system protected by three zones protection scheme.

Z_{1a} corresponds to ~80% of the length of line AB and is a high-speed zone. It was extended only to 80% of the length to pass the impedance measurements' inaccuracies, especially when the current is offset.

The second Z_{2a} zone for relay A causes the remaining 100% of line AB plus 20% of the adjoining section CD.

The third Z_{3a} zone for relay A provides backup for fault in section CD.

Due to the complexity of the systems having some feeds from generating station, the need for faster clearing time as the fault level increases, and also of difficulty in grading time/overcurrent relay with an increasing number of switching stations, the use of high-speed Distance relay on the modern system has become imperative.

It is a non-unit form of protection, offering considerable economic and more advantages in medium and high voltage lines.

Being of the non-unit type distance scheme automatically provides backup to adjacent line sections. Selectivity is often achieved by a directional feature inherent to the distance relay or supplementary relay. Discrimination between equipment types with the same directional response on adjacent lines is by time grading but not cumulative towards the power sources.

7.3 SETTING OF DISTANCE RELAY

Distance relays are adjusted based on the positive sequence impedance between the relay location and fault position beyond which a given relay unit's operation should stop. The impedance or the corresponding distance is called the "reach of the relay." As the relays are energized through CTs and VTs, the primary value of impedance must be converted to secondary values

$$Z_R = \frac{V_R}{I_R} = \frac{V_s}{\text{VT ratio}} \times \frac{\text{CT ratio}}{I_s} = Z_{sys} \times \frac{\text{CT ratio}}{\text{VT ratio}} \tag{7.1}$$

$$\text{CT ratio} = \frac{\text{System phase current}}{\text{Relay phase current}} \tag{7.2}$$

$$\text{VT ratio} = \frac{\text{System line voltage}}{\text{Relay line voltage}} \tag{7.3}$$

Consider the simple radial line with a distance protection system installed at end A (the local end), while end B is called the remote end, as shown in Figure 7.3. These relays sense local voltage and current and calculate the effective impedance at that point. This means that the relay requires voltage and current information.

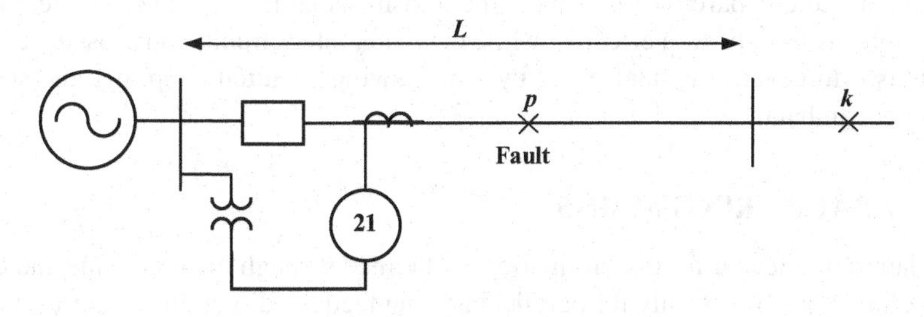

FIGURE 7.3 Radial power system provided by distance relay.

When the protected line becomes faulted, the effective impedance becomes the impedance from that point to the fault.

Assume a balanced three-phase fault at distance p:

For internal fault at point p:

$$Z_p = \frac{V_p}{I} < Z_L \quad \text{Relay will operate}$$

For external fault at point k:

$$Z_p = \frac{V_k}{I} < Z_L \quad \text{The relay will not operate}$$

Hence the distance relay action is to compare the local voltage with the local current, i.e., the secondary values of V and I in the voltage and current transformers, so as the quantity

7.4 DRAWBACK OF DISTANCE RELAY

The disadvantages of distance and directional overcurrent relay on the transmission line include;

i. The relays cannot disconnect instantly on both ends of the line if a fault occurs at the end of the line.
ii. Coordination is achieved by adjusting the time delay of the relay mounted on a channel next to the main and backup protection concept. As a result, termination disturbance will be slow in line with the relay's delay time that works on each protection zone.

To solve this, transmission lines can be protected by applying differential protection (for short transmission lines) and pilot relay protection (for long-distance transmission). The current differential protection is based on Kirchhoff's first law, whereas the impedance type is based on Kirchhoff's second law. When a fault occurs within the protected zone, the current flowing into the protected line is unequal to those flowing out from the protected line. Therefore, it requires a reliable communication channel to compare the currents at the transmission line terminals. Differential protection has been proven effective while evolving, inter-circuit, and crossing country faults. Moreover, it is unaffected by power swings, mutual coupling, and series impedance unbalances.

7.5 PARALLEL RING MAINS

The plain radial feeder for discrimination is obtained through the relay time and current setting adjustments only for parallel and ring feeders; directional features are to be incorporated to obtain proper discrimination.

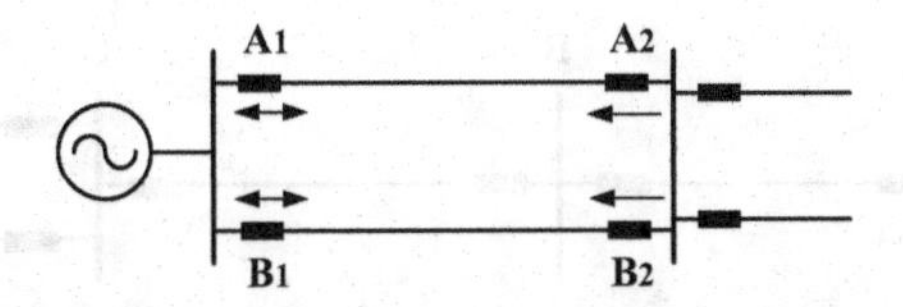

FIGURE 7.4 Parallel feeder system configuration.

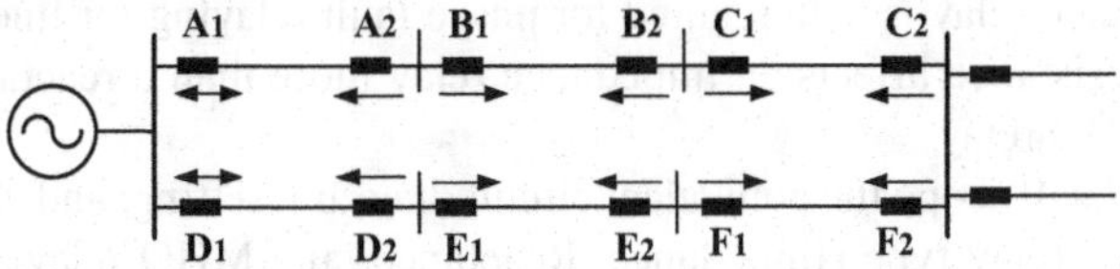

A1 A2 B1 B2 C1 C2
D1 D2 E1 E2 F1 F2

FIGURE 7.5 Ring system configuration.

For both the parallel feeder system shown in Figure 7.4, if the four relays are non-directional, then the supply will be disrupted completely for a fault anywhere in the system in both feeders. It is usual to have relays A_2 and B_2 as direction sensitive to ensure discrimination. They operate for faults occurring in the feeder in the direction indicated by the arrows (away from the bus). Furthermore, relays A_2 and B_2 should operate before non-directional relays A_1 and B_1.

Usually, a directional relay is required at one end of each feeder, with a non-directional relay at the other.

For the ring system shown in Figure 7.5, the supply side relay is non-directional. Relay settings are determined by considering the grading in one direction and then in the other, working backward toward the power source. The clockwise circuit, A_1 B_1 C_1 F_2 E_2 D_2 D_1. The counter-clockwise circuit D_1 E_1 F_1 C_2 B_2 A_2 A_1.

7.6 IMPEDANCE, REACTANCE, AND MHO RELAY

Distance relays are used for both phase fault and ground fault protection. Distance relays are high-speed and independent of changes in the magnitude of the short circuit current and are not so much affected by changes in the generation capacity. The reactance relays are used where arc resistance is likely comparable to the protective section's impedance. The arc resistance is given by

$$R_{arc} = \frac{8750l}{I^{1.4}} (\Omega)$$

Where l: length of the arc (ft). I: fault current. The reactance relay ignored the arc resistance and is dependable for a short line and protection against ground faults. The additional potential drop in an arcing fault prevents the impedance relay from operating on short lines.

The MHO type is best suited for phase relays for longer lines. MHO relays have the advantage over the two other relays of inherent directional characteristics.

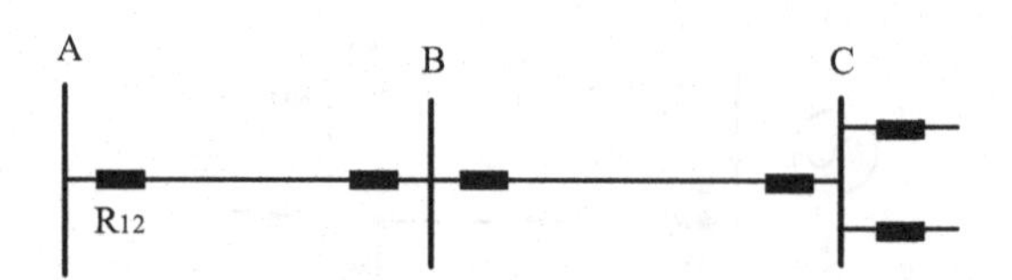

FIGURE 7.6 Power system supplied by distance relay (R_{12}).

The impedance relay is better suited for phase fault relaying for lines of medium length. ARC resistance affects an impedance relay more than a reactance relay but less than an MHO relay.

This section will scope the equivalent circuit diagram, setting, and characteristics of each distance relay type (Impedance, Reactance, and MHO relay). Suppose the power system shown in Figure 7.6 is protected by a distance relay.

7.6.1 Impedance Relay Protection Setting Diagram

Figure 7.7a shows an equivalent circuit diagram for the impedance relay; Figure 7.7b shows the three-zone setting for relay R_{12} on the secondary side as follows

$$Z_{\text{zone 1}} = 80\% Z_{AB} \times \frac{\text{CT ratio}}{\text{VT ratio}}$$

$$Z_{\text{zone 2}} = 120\% Z_{AB} \times \frac{\text{CT ratio}}{\text{VT ratio}}$$

$$Z_{\text{zone 3}} = \left[100\% \left(Z_{AB}\right) + 100\% \left(Z_{BC}\right)\right] \frac{\text{CT ratio}}{\text{VT ratio}}$$

To draw the impedance setting zone for two transmission line sections have the impedances of $Z_{AB} = 2 + j10\ \Omega$, and $Z_{BC} = 2.5 + j15\ \Omega$ respectively.

7.6.2 Reactance Relay Protection Setting Diagram

Figure 7.8a shows an equivalent circuit diagram for the reactance relay; Figure 7.8b shows the three-zone setting for relay R_{12} on the secondary side as follows

$$X_{\text{zone1}} = 80\%\ X_{AB} \times \frac{\text{CT ratio}}{\text{VT ratio}}$$

$$X_{\text{zone2}} = 120\% X_{AB} \times \frac{\text{CT ratio}}{\text{VT ratio}}$$

$$X_{\text{zone3}} = \left[100\% \left(X_{AB}\right) + 100\% \left(X_{BC}\right)\right] \frac{\text{CT ratio}}{\text{VT ratio}}$$

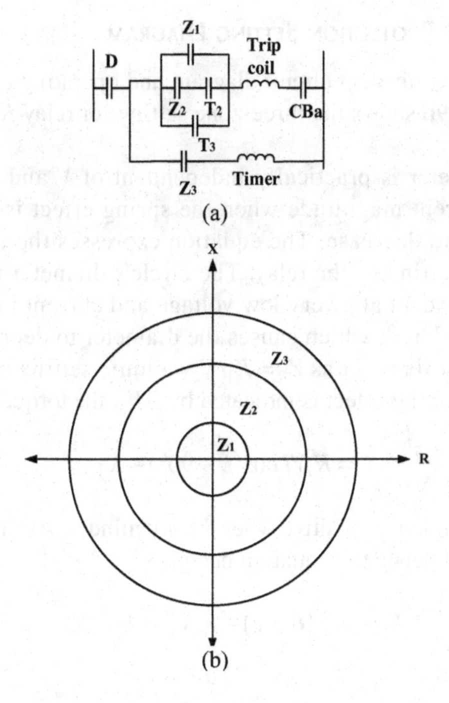

FIGURE 7.7 Impedance characteristics and contact circuit for three zones impedance relaying. (a) Contact circuit. (b) Impedance characteristics.

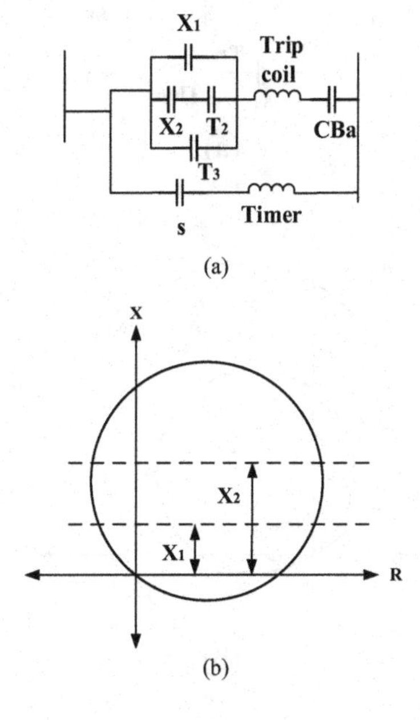

FIGURE 7.8 Reactance characteristics and contact circuit for three zones reactance relaying. (a) Contact circuit. (b) Impedance characteristics.

7.6.3 MHO Relay Protection Setting Diagram

Figure 7.9a shows an equivalent circuit diagram and operating characteristic for the MHO relay; Figure 7.9b shows the three-zone setting for relay R_{12} on the secondary side as follows.

The circle's diameter is practically independent of V and I, except at a very low voltage and current magnitude when the spring effect is considered, which causes the diameter to decrease. The equation expresses the circle's diameter as $Z_R = K_1/K_2$ = ohmic setting of the relay. The circle's diameter is practically independent of V and I, except at a very low voltage and current magnitude when the spring effect is considered, which causes the diameter to decrease. The equation expresses the circle's diameter as $Z_R = K_1/K_2$ = ohmic setting of the relay.

If the spring controlling effect is indicated by $-K_3$, the torque equation becomes,

$$T = K_1 VI \cos(\theta - 90°) - K_3$$

Where θ and τ are defined as positive when I lag behind V. At the balance point, the net torque is zero, and hence the equation becomes

$$K_1 VI \cos(\theta - \tau) - K_2 V^2 - K_3 = 0$$

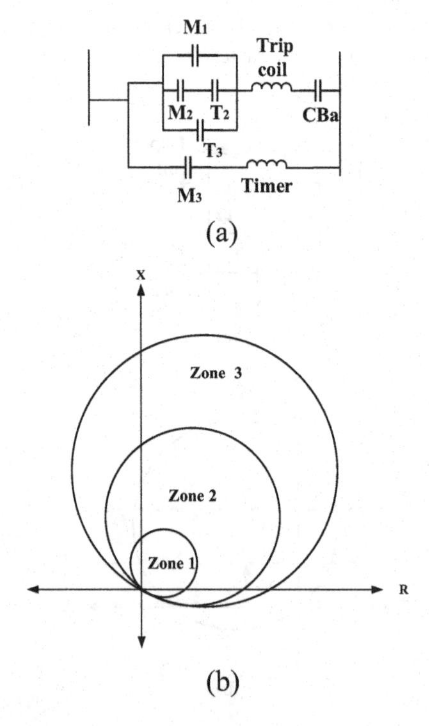

FIGURE 7.9 MHO characteristics and contact circuit for three zones MHO relaying. (a) Contact circuit. (b) Impedance characteristics.

$$\frac{K_1}{K_2}\cos(\theta-\tau)-\frac{K_3}{K_2\,VI}=\frac{V}{I}=Z$$

If the spring-controlled effect is neglected, mean $K_3=0$.

So

$$Z=\frac{K_1}{K_2}\cos(\theta-\tau)$$

The three-zone setting can be calculated from the following equation and taking the arc resistance to consider each zone setting and test whether the relay underreaches.

$$Z_{\text{zone1}}=80\%\ Z_{AB}\times\frac{\text{CT ratio}}{\text{VT ratio}}$$

$$Z_{\text{zone2}}=120\%\ Z_{AB}\times\frac{\text{CT ratio}}{\text{VT ratio}}$$

$$Z_{\text{zone3}}=\left[100\%\ (Z_{AB})+100\%(Z_{BC})\right]\frac{\text{CT ratio}}{\text{VT ratio}}$$

Example 7.1

For the 66 kV radial feeder shown in Figure 7.10, Calculate the zone 1 setting for the distance relay in primary ohms. CT ratio = 500/1, VT ratio = 130,000/120.

Solution

$$Z=(0.24+j0.8)\times 25=6+j\ 20\ \Omega\ \text{Primary}$$

$$\text{Zone1}=80\%(6+j20)\frac{\frac{500}{1}}{\frac{130,000}{120}}=2.215+j7.383\ \Omega\ \text{secondary}$$

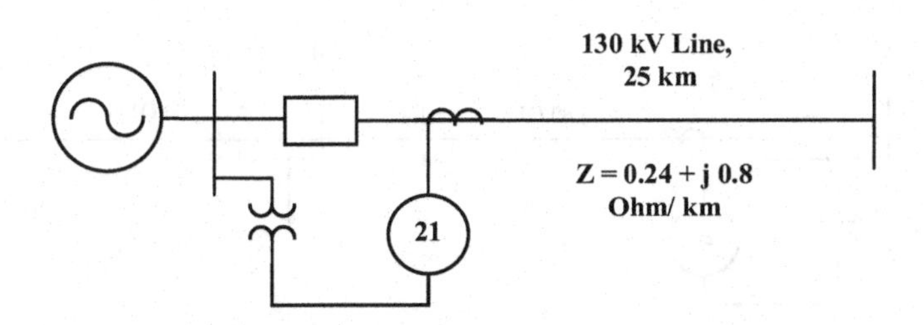

FIGURE 7.10 A 66 kV radial feeder system of Example 7.1.

Example 7.2

Figure 7.11 shows two simple sections of radial lines. We will consider the settings for line *AB* at bus *B*. The impedance angle for each line is 75°. The line length is 80 km, and the distance relay at bus *A* is fed by current transformers rated at 2000 A: 5 A and voltage transformers rated at 345/200 kV Y: 120/69 V Y. Find the settings of zone 1 and zone 2 of the relays.

Solution

SET ZONE 1 FOR 85%

$$\text{Zone 1 setting} = 0.85 \times 80 = 68\ \Omega,\ \text{primary}$$

$$\text{CT ratio} = \frac{2000}{5} = 400$$

$$\text{VT ratio} = 200{,}000/69 = 2900$$

$$\text{Relay setting for zone 1} = 68\frac{400}{2900}$$

$$= 9.38\ \text{relay ohms}$$

$$\text{Zone 2 setting: } 120\% - 150\% \ \text{Choose } 140\%$$

$$\text{Zone 2 setting} = 1.40 \times 80 = 112\ \text{ohm, primary}$$

$$\text{Relay setting for zone } 2 = 112\frac{400}{2900} = 15.44\ \text{relay ohms}$$

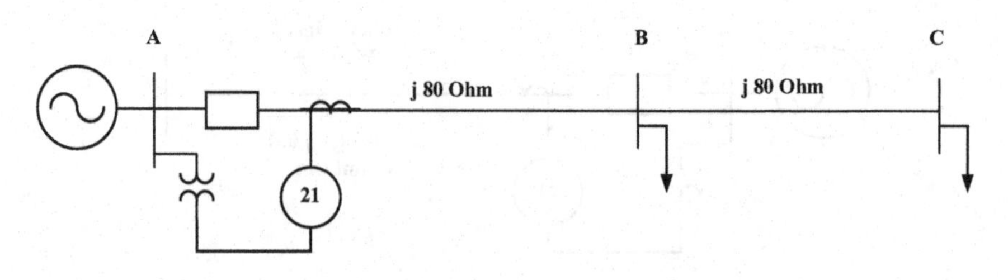

FIGURE 7.11 A radial lines system of Example 7.2.

Example 7.3

A section of 138 kV is shown in Figure 7.12, Line 1–2 is 64 km, and line 2–3 is 96 km long. The positive sequence impedance for both lines is 0.05 + *j*0.5 Ω/km. The maximum load current carried by line 1–2 under emergency conditions is 50 MVA. Design a three zones distance relying system specifying the relay setting for relay R_{12} (Figure 7.13).

Hint: Standard VT secondary voltage is 120 V line to line.

Solution

The positive sequence impedance of the two lines are:

$$Z_{12} = (0.05 + j0.5) \times 64 = 3.2 + j32\ \Omega$$

$$Z_{23} = (0.05 + j0.5) \times 96 = 4.8 + j48\ \Omega$$

$$\text{The maxmum load current} = \frac{50 \times 10^6}{\sqrt{3} \times 138 \times 10^3} = 209.2\ \text{A}$$

The suitable choice of CT ratio 200:5 A

The VT ratio is 138,000:120 V

The setting of zone no. 1

$$Z_{\text{zone 1}} = 80\%\ (3.2 + j32) \times \frac{\frac{200}{5}}{\frac{138{,}000}{120}}$$

$$= 3.2 + j32\ \Omega \text{ on the secondary side}$$

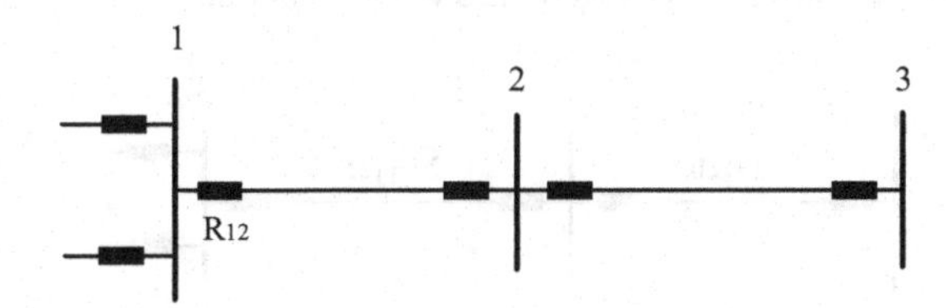

FIGURE 7.12 A radial lines system of Example 7.3.

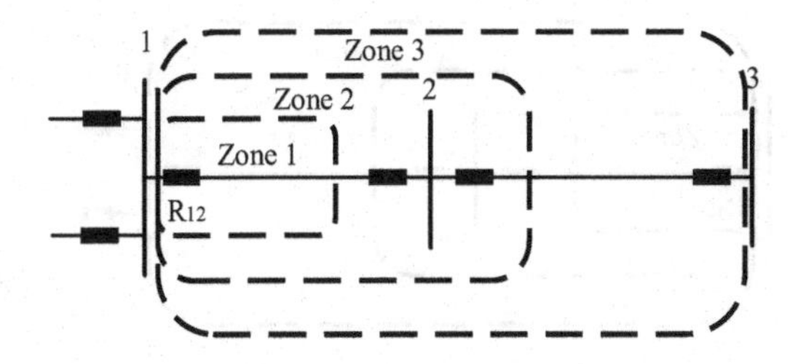

FIGURE 7.13 A radial lines system of Example 7.3 with zones.

The setting of zone no. 2

$$Z_{zone\,2} = 120\%\,(3.2 + j32) \times \frac{\frac{200}{5}}{\frac{138{,}000}{120}}$$

$$= 3.2 + j32\,\Omega \text{ on the secondary side}$$

The setting of zone no.3

$$Z_{zone\,3} = [100\%(3.2 + j32) + 100\%(4.8 + j48)] \times \frac{200/5}{138{,}000/120}$$

$$= 3.2 + j32\,\Omega \text{ on the secondary side}$$

Example 7.4

A section of 132 kV is shown in Figure 7.14, Line 1–2 is 64 km, and line 2–3 is 96 km long. The positive sequence impedance for both lines is $0.05+j0.5\ \Omega$/km. The maximum load current carried by line 1–2 under emergency conditions is 110 MVA (Figure 7.15).

Determine:

i. Maximum load current
ii. CT ratio.
iii. VT ratio.
iv. Impedance measure by the relay.
v. Load impedance based on secondary.
vi. Design a three zones distance relying system specifying the relay setting for relay R_{12}

Hint: Standard VT secondary voltage is 120 V line to neutral.

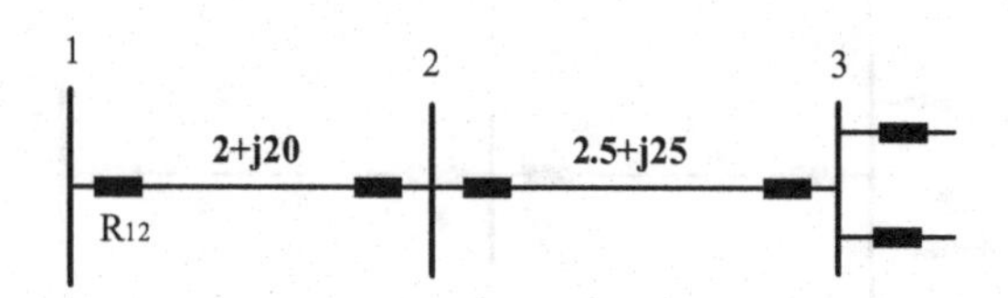

FIGURE 7.14 A radial lines system of Example 7.4.

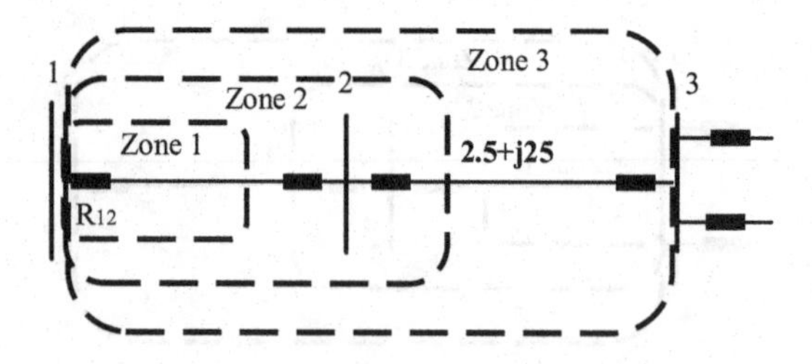

FIGURE 7.15 A radial lines system of Example 7.4 with zones.

Solution

The positive sequence impedance of the two lines are:

$$Z_{12} = 2 + j20\ \Omega$$

$$Z_{23} = 2.5 + j25\ \Omega$$

$$\text{The maxmum load current} = \frac{120 \times 10^6}{\sqrt{3} \times 132 \times 10^3} = 481.12\ \text{A}$$

The suitable choice of CT ratio is 500:5 A
The VT ratio is 132,000:120 V

$$Z_1 = 80\%\ (2 + j20) = 1.6 + j16\ \Omega$$

$$Z_2 = 120\%\ (2 + j20) = 2.4 + j24\ \Omega$$

$$Z_3 = \left[100\%\ (2 + j20) + 100\% (2.5 + j25)\right]$$
$$= 4.5 + j45\ \Omega \text{ on the secondary side}$$

$$Z_{load} = \left[\frac{\frac{132{,}000}{\sqrt{3}}}{481.12}\right] \times \frac{\frac{500}{5}}{\frac{132{,}000}{120}} = 14.4\ \Omega$$

The setting of zone no. 1

$$Z_{zone\ 1} = 80\%\ (2 + j20) \times \frac{\frac{500}{5}}{\frac{132{,}000}{120}}$$
$$= 0.145 + j1.45\ \Omega \text{ on the secondary side}$$

The setting of zone no.2

$$Z_{zone\ 2} = 120\%\ (2 + j20) \times \frac{\frac{500}{5}}{\frac{132{,}000}{120}}$$
$$= 0.218 + j2.18\ \Omega \text{ on the secondary side}$$

The setting of zone no.3

$$Z_{\text{zone 3}} = \left[100\% \, (2 + j20) + 100\% (2.5 + j25)\right] \times \frac{500/5}{132{,}000/120}$$

$$= 0.41 + j4.1 \, \Omega \text{ on the secondary side}$$

Example 7.5

A 380 kV transmission line shown in Figure 7.16, a directional impedance relay, is used for distance protection. With the data given in Table 7.1, for a three-phase fault

i. Determine the three zones setting for relay R_{12}.
ii. If the maximum current in line 1–2 during emergency loading is 1500 A at power factor 0.707 lagging, Show if the relay R_{12} trip such loading or not

CT ratio = 1200:5   VT ratio: 380,000/120

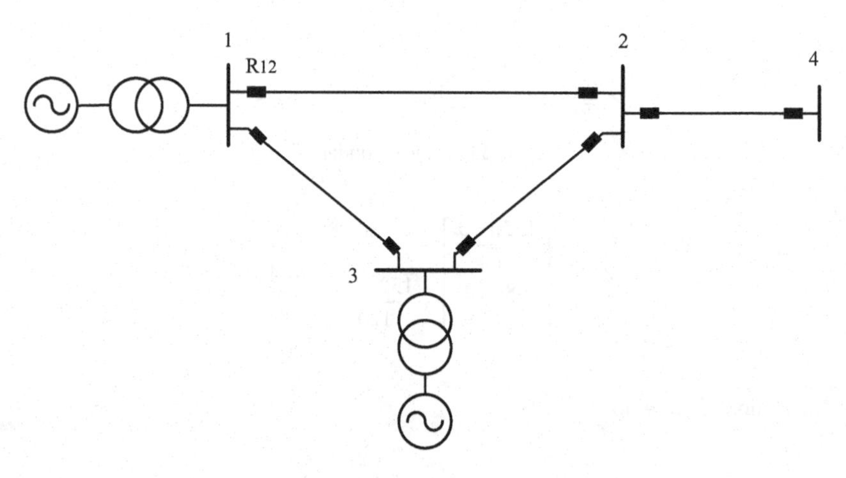

FIGURE 7.16 A power system of Example 7.5.

TABLE 7.1
Transmission Lines Impedance in Ohm

Line	Impedance (Ω)
1–2	$9 + j45$
2–3	$9 + j45$
2–4	$6 + j40$
1–3	$4 + j30$

Solution

The setting of zone no.1

$$Z_{\text{zone 1}} = 80\%(9 + j45) \times \frac{\frac{1200}{5}}{\frac{380,000}{120}}$$

$$= 0.545 + j2.728\ \Omega \text{ on the secondary side}$$

The setting of zone no.2

$$Z_{\text{zone 2}} = 120\%\ (9 + j45) \times \frac{\frac{1200}{5}}{\frac{380,000}{120}}$$

$$= 0.818 + j4.1\ \Omega \text{ on the secondary side}$$

The setting of zone no.3

$$Z_{\text{zone 3}} = \left[100\%\ (9 + j45) + 100\%(6 + j40)\right] \times \frac{1200/5}{380,000/120}$$

$$= 1.13 + j6.44\ \Omega \quad \text{on the secondary side}$$

$$Z_L = \frac{V}{I} = \frac{\frac{380,000}{\sqrt{3}}}{1500\angle -\cos^{-1} 0.707} = 146.26\ \angle 45° \Omega \text{ on primary side}$$

$$= 11\angle 45°\ \ \Omega \text{ on secondry side}$$

$\therefore Z_L > Z_{\text{zone 3}}$ So the load impedance location is out of three-zone characteristics. Therefore the relay will not detect the load.

Example 7.6

A 380 kV transmission line shown in Figure 7.17, a directional impedance relay, is used for distance protection, with the data given in Table 7.2, for a three-phase fault.

Determine the three zones setting for relay R_{12}.

If the maximum current in line 1–2 during emergency loading is 2000 A at power factor 0.88 lagging, Show if the relay R_{12} trip for such loading or not

CT ratio = 1000:5 VT ratio: 380,000/120

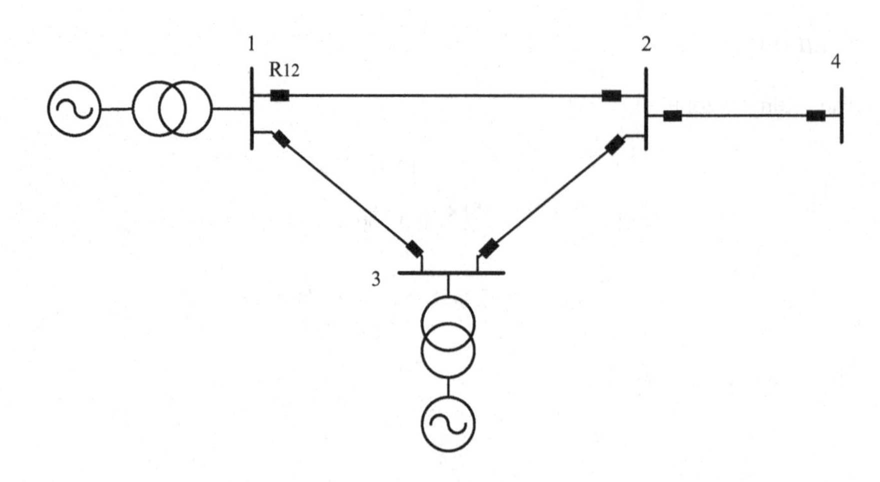

FIGURE 7.17 A power system of Example 7.6.

TABLE 7.2
Transmission Lines Impedance per km

Line	Impedance (Ω/km)	Long (km)
1–2	$0.01+j0.045$	100
2–3	$0.01+j0.045$	120
2–4	$0.012+j0.040$	120
1–3	$0.04+j0.030$	100

Solution

The impedances per phase are given in Table 7.3

The setting of zone no.1

$$Z_{\text{zone }1} = 80\%\left(1+j4.5\right)\times\frac{\frac{1000}{5}}{\frac{380{,}000}{120}}$$

$$= 0.05+j0.227\ \Omega \text{ on the secondary side}$$

The setting of zone no.2

$$Z_{\text{zone }2} = 120\%\left(1+j4.5\right)\times\frac{\frac{1000}{5}}{\frac{380{,}000}{120}}$$

$$= 0.075+j0.341\ \Omega \text{ on the secondary side}$$

TABLE 7.3
Transmission Lines Impedance

Line	Impedance (Ω/km)	Long (km)	Z (Ω)
1–2	$0.01+j0.045$	100	$1+j4.5$
2–3	$0.01+j0.045$	120	$1.2+j5.4$
2–4	$0.012+j0.040$	120	$1.44+j4.8$
1–3	$0.04+j0.030$	100	$4+j3$

The setting of zone no.3

$$Z_{\text{zone}3} = \left[100\%\,(1+j4.5)+100\%(1.44+j4.8)\right]\times \frac{\frac{1000}{5}}{\frac{380,000}{120}}$$

$$= 0.154 + j0.587\ \Omega \text{ on the secondary side}$$

$$= 0.606\angle 75.3^\circ\ \Omega \text{ on secondry side}$$

$$Z_L = \frac{V}{I} = \frac{\frac{380,000}{\sqrt{3}}}{2000\angle -\cos^{-1}0.88} = 109.69\angle 28.357^\circ\ \ \Omega \ \text{ on primary side}$$

$$= 6.928\ \angle 28.357\ \ \Omega \text{ on secondry side}$$

$\therefore Z_L > Z_{\text{zone}3}$ So the load impedance location is out of three-zone characteristics. Therefore the relay will not detect the load.

Example 7.7

Figure 7.18 shows a reactance relay situated at end A of 100 km. A single-circuit transmission line AB has a first zone setting of 80%. The positive phase sequence impedance of the line is $j0.5$ p.u. A single line to ground fault occurred at 70 km from end A. The current distribution is such that the actual fault arc current is $15-j15$ p.u., while the current flowing at the end A is $3-j10$ p.u. concerning the same reference. If the fault arc resistance is 0.07 p.u., will the relay see the fault in zone 1? If not, what is the % error in the relay reach? And where will it see the fault as % of the length?

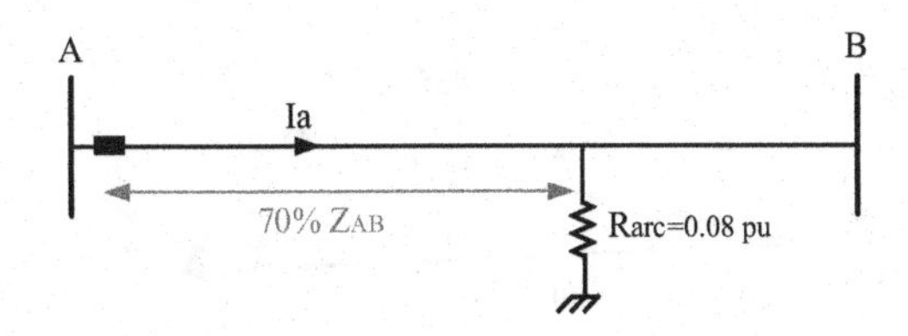

FIGURE 7.18 A single-circuit transmission line of Example 7.7.

Solution

$$Z_{AB} = j0.5 \text{ p.u.}$$

$$I_f = 15 - j15 \text{ p.u.}$$

$$I_a = 3 - j10 \text{ p.u.}$$

The voltage drop across AF

$$V_{AF} = I_a \cdot Z_{AF} = (3 - j10) \cdot (70\% \times j0.5) = 3.654\angle 16.69° \text{ p.u.}$$

The voltage drop through arc resistance

$$V_F = I_F \ R_{\text{arc}} = (15 - j15) \times 0.08 = 1.697\angle -45° \text{ p.u.}$$

The voltage from relay A to the fault location

$$V_A = V_{AF} + V_F = 3.654\angle 16.69° + 1.697\angle -45°$$

$$= 4.7\angle -1.84° \text{ p.u.}$$

To determine the error in reach from Figure 7.19.

$$\cos(16.69 + 45) = \frac{\text{Error}}{1.697}$$

$$\text{Error} = 0.804 \text{ p.u.}$$

$$\text{The percentage error in reach } = \frac{0.804}{3.654} \times 100 = 22\%$$

For 70% location of the fault from the relay located at *A* will be $70\% \times 22\% = 15.4\%$

The relay will see the fault at $70\% + 15.4\% = 85.4\%$

Therefore the relay will see the fault out of zone one (i.e., the relay under reach).

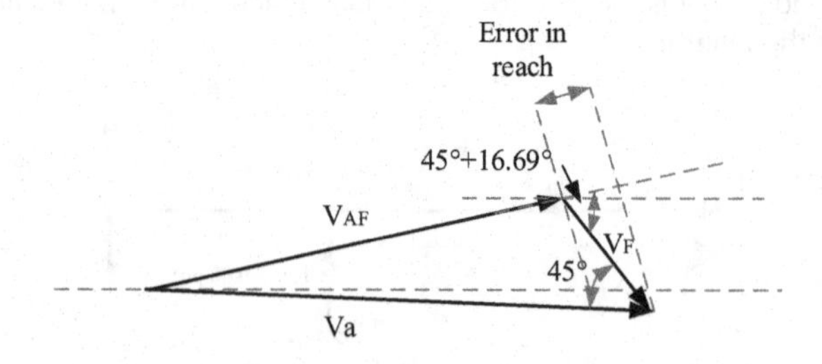

FIGURE 7.19 Phasor diagram of Example 7.7.

Example 7.8

A 138 kV sub-transmission system shown in Figure 7.20 considers the directional distance (MHO) relay located at *A*. Assume a single line to gram located at point F, at 70% distance from A. The magnitude of the fault current is 1.2 p.u. Assume that the fault arc length is 10.3 ft. The power, voltage, current, and impedance base values are 100 MVA, 138 kV, 418.4 A, and 190.4 Ω, respectively. Consider only two zones available and $Z_{AB} = 0.2 + j0.7$ p.u., determine (Figures 7.21 and 7.22)

i. The value of arc resistance at the fault location in Ohm and p.u.
ii. Value of line impedance, including the arc resistance.
iii. Line impedance angle without and with arc resistance.
iv. Graphically show whether or not the relay will clear the fault instantly.

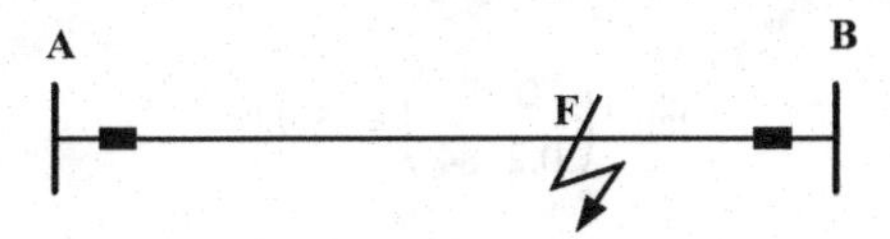

FIGURE 7.20 Sub-transmission system with the directional distance (MHO) relay.

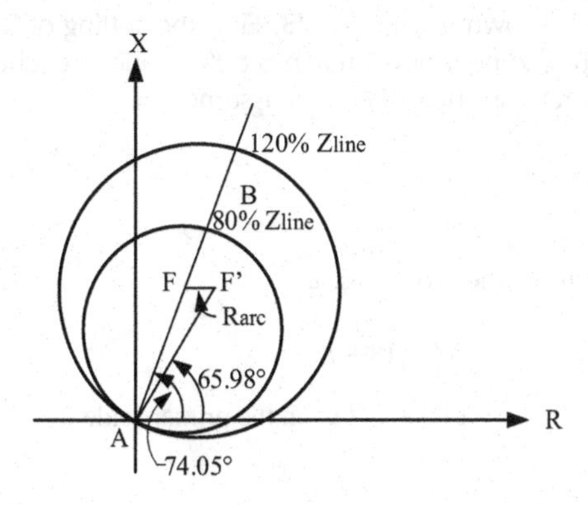

FIGURE 7.21 MHO characteristics of Example 7.8.

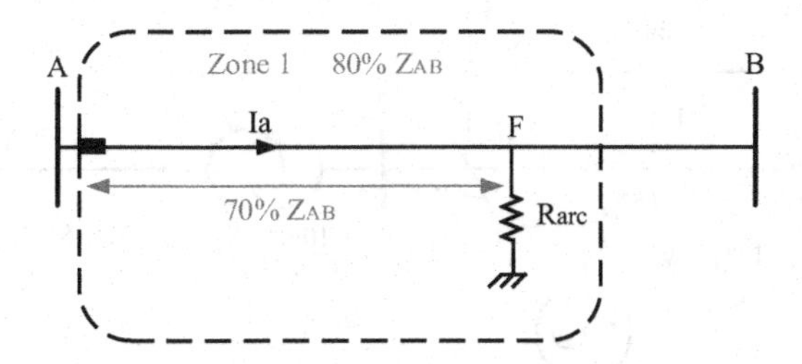

FIGURE 7.22 Sub-transmission system with 80% zone of the directional distance (MHO) relay.

Solution

i. The arc current is 1.2 p.u. or $I = 1.2 \times 418.4 = 502.08$ A

$$\text{The arc resistance } R_{arc} = \frac{8750\, l}{I^{1.4}} = \frac{8750 \times 10.3}{502.08^{1.4}} = 14.92\,\Omega \text{ or } 0.0784 \text{ p.u.}$$

ii. The impedance seen by the relay is

$$70\%\ Z_{line} + R_{arc} = 0.7(0.2 + j0.7) + 0.0784 = 0.2184 + j0.49 \text{ p.u.}$$

iii. The line impedance angle without arc resistance

$$\tan^{-1}\left(\frac{0.49}{0.14}\right) = 74.05°$$

The line impedance angle with arc resistance

$$\tan^{-1}\left(\frac{0.49}{0.2184}\right) = 65.98°$$

Example 7.9

The power system is shown in Figure 7.23. Find the setting of Breaker B_{12} for the first, second, and third zone. Zone 1 reaches 80%, zone 2 reaches 140%, CT ratio is 400/5 A, and VT ratio is 160,000/110 V; assume $I_1 = I_2$.

Solution

a. without injected generator at 65%

$$Z_1 = 80\%\ (18 + j55)$$

$$= 14.4 + j44\,\Omega \text{ on the primary side}$$

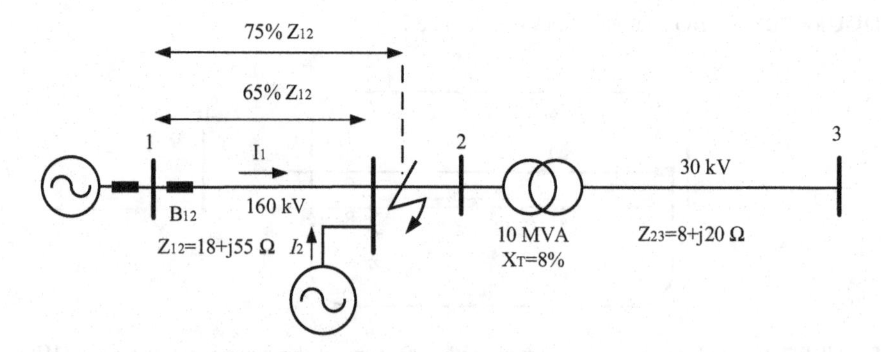

FIGURE 7.23 The power system of Example 7.9.

$$Z_{\text{transformer}} = 10\% \times \frac{160^2}{10} = 256\ \Omega$$

$$Z_2 = 140\%\ (18 + j55) + j256$$
$$= 25.2 + j333\ \Omega \text{ on the primary side}$$

$$Z'_{23} = \left(\frac{160}{30}\right)^2 (8 + j20) = 227.55 + j568.89\ \Omega$$

$$Z_3 = \left[100\%\ (18 + j55) + j256 + 100\%(227.55 + j568.89)\right]$$
$$= 245.55 + j879.89\ \Omega \text{ on the primary side}$$

$$Z_{\text{zone-1}} = (14.4 + j44) \times \frac{400/5}{160{,}000/110}$$
$$= 0.792 + j2.42\Omega \text{ on the secondary side}$$
$$= 2.746\angle 71.87°\Omega \text{ on the secondary side}$$

$$Z_{\text{zone-2}} = (25.2 + j333) \times \frac{400/5}{160{,}000/110}$$
$$= 1.386 + j18.315\ \Omega \text{ on the secondary side}$$
$$= 18.36\angle 85.67°\Omega \text{ on the secondary side}$$

$$Z_{\text{zone-3}} = (245.55 + j879.89) \times \frac{400/5}{160{,}000/110}$$
$$= 13.505 + j48.393\ \Omega \text{ on the secondary side}$$
$$= 50.24\angle 74.40°\Omega \text{ on the secondary side}$$

The fault at 75% of Z_{12}:

$$Z_{1f} = 75\%(18 + j55) \times \frac{400/5}{160{,}000/110}$$
$$= 0.7425 + j2.268\ \Omega \text{ on the secondary side}$$
$$= 2.386\angle 71.87°\ \Omega \text{ on the secondary side}$$

Therefore, the relay will detect the fault in zone 1 (Since $|Z_{1f}| < |Z_{\text{zone-1}}|$).

b. with the injected generator at 65% will be carrying I_1, and 15% will carrying $I_1 + I_2$

$$Z_1 = 65\%\ (18 + j55) + 15\% \left(\frac{I_1 + I_2}{I_1}\right)(18 + j55)$$

$$= 65\%\ (18 + j55) + 15\% \left(\frac{I_1 + I_2}{I_1}\right)(18 + j55)$$

$$= 17.1 + j52.25\ \Omega \text{ on the primary side}$$

$$Z_2 = 65\%\ (18 + j55) + 75\%\ (18 + j55)\left(\frac{I_1 + I_2}{I_1}\right) + j256\left(\frac{I_1 + I_2}{I_1}\right)$$

$$= 38.7 + j630.25\ \Omega \text{ on the primary side}$$

$$Z_3 = \left[65\%\ (18 + j55) + 35\%\ (18 + j55)\left(\frac{I_1 + I_2}{I_1}\right) + j256\left(\frac{I_1 + I_2}{I_1}\right) + 100\%(227.55 + j568.89)\left(\frac{I_1 + I_2}{I_1}\right)\right]$$

$$= 479.4 + j1724.03\ \Omega \text{ on the primary side}$$

$$Z_{\text{zone-1}} = (17.1 + j52.25) \times \frac{400/5}{160{,}000/110}$$

$$= 0.940 + j2.873\ \Omega \text{ on the secondary side}$$

$$= 3.022\angle 71.88°\ \Omega \text{ on the secondary side}$$

$$Z_{\text{zone-2}} = (38.7 + j630.25\ ) \times \frac{400/5}{160{,}000/110}$$

$$= 2.128 + j34.663\ \Omega \text{ on the secondary side}$$

$$= 34.72\angle 86.48°\ \Omega \text{ on the secondary side}$$

$$Z_{\text{zone-3}} = (479.4 + j1724.03) \times \frac{400/5}{160{,}000/110}$$

$$= 26.367 + j94.82\ \Omega \text{ on the secondary side}$$

$$= 98.41\angle 74.46°\ \Omega \text{ on the secondary side}$$

The fault at 75% of Z_{12}:

$$Z_{1f} = 65\%(18 + j55) \times \frac{400/5}{160{,}000/110} + 10\%\ (18 + j55)\left(\frac{I_1 + I_2}{I_1}\right) \times \frac{400/5}{160{,}000/110}$$

$$= 0.8415 + j2.571\ \Omega \text{ on the secondary side}$$

$$= 2.705\angle 71.87°\ \Omega \text{ on the secondary side}$$

Therefore the relay will detect the fault in zone 1 (since $|Z_{1f}| < |Z_{\text{zone-1}}|$).

7.7 FUNDAMENTALS OF DIFFERENTIAL PROTECTION SYSTEMS

It is difficult to apply overcurrent relays in some situations due to coordination problems and excessive fault clearance time problems. In these cases, differential protection is applied. The basic principle of differential protection is to measure the current at each end of the selected zone and operate where there is a difference. In Figure 7.24, the simple differential protection scheme is illustrated. The protected circuit can be equipment such as a transformer, generator, motor, and transmission line. Primary and secondary currents of the system are denoted by I_p and Are, respectively. In normal operating conditions, the secondary current flowing through CT_1 and CT_2 are the same. As a result, there is no current flowing through the differential relay.

Considering scenario one, there is a fault outside the zone, as shown in Figure 7.24. The fault current will pass through both CTs to the fault. Since the fault current passing through both CTs is the same, the secondary current of both will be the same. Therefore, the relay will not sense the fault. Even though the fault current is very high, based on the differential relay configuration, no current will flow through the relay path. However, the differential relay will operate during the internal fault.

Figure 7.25 shows the differential relay configuration with an internal fault. When there is an internal fault, the current will flow through both CTs but in the opposite direction. Hence there will be a difference in current, which will flow through the differential relay, and the relay will send a trip signal to the circuit breaker.

The problem with differential protection is the location of two current transformers. When the differential protection is applied to the equipment, such as the transformer, the CT location will be local. Still, in the transmission lines, the location will be remote, which might cause problems if an adequate pilot protection scheme is not applied.

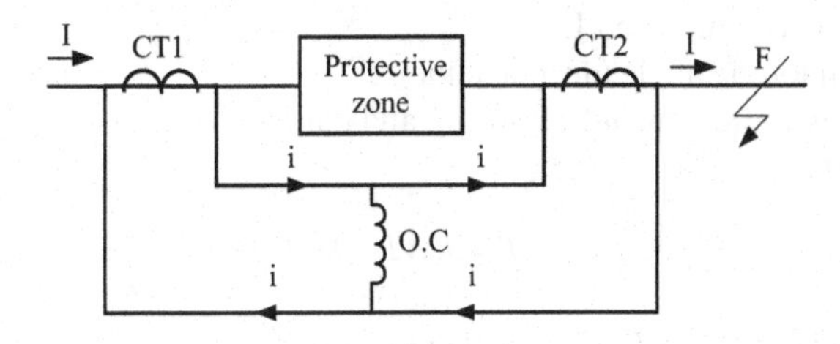

FIGURE 7.24 Simple differential protection scheme.

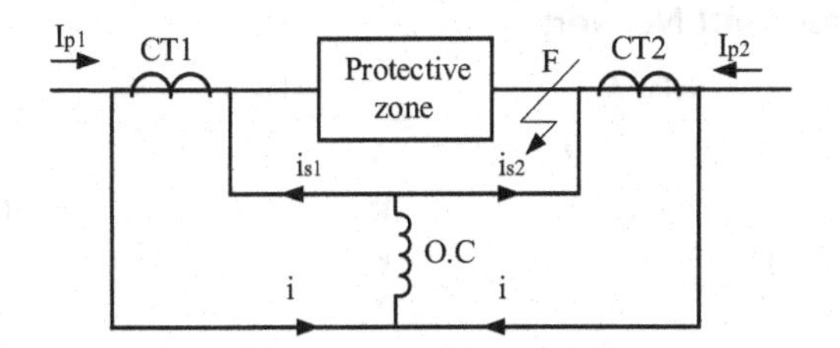

FIGURE 7.25 Differential relay configuration with an internal fault.

In summary, the circuit breaker will trip after getting a trip signal from the relay. The trip signal can be sent to either fault, which produces a high current in the system, or through manual switching contacts when the breaker needs to be out for maintenance.

7.8 DIRECTIONAL OVERCURRENT RELAY

Overcurrent relays are the simplest and cheapest form of transmission line protection. However, the correct coordination of these devices is the most difficult to achieve. These relays are also very susceptible to relative source impedances and system conditions.

When fault current can follow in both directions through the relay location, it is necessary to respond to the relay directional by adding directional control elements. These are power measuring devices in which the system voltage is used as a reference for establishing the relative

7.9 DIRECTION OR PHASE OF THE FAULT CURRENT

Directional relays are usually used in conjunction with other forms of relays, usually over the current type; when used as an overcurrent relay, the combination uses to select to respond to the fault current only in the protective zone's direction (Table 7.4).

The torque developed by the directional unit is

$$T = VI\cos(\theta - \tau) - K$$

Where V: RMS voltage fed to the voltage coil.

I: RMS current in the current coil.

θ: the angle between V and I.

τ: maximum torque angle (design value).

K: restraining torque, including spring and damper.

In particular $\theta - \tau = 0$

$$T = K_1 VI - K$$

Under threshold condition $T = 0,\ K_1 VI = K$

TABLE 7.4
Specify the Phase Shift Network

	Relay A		Relay B		Relay C	
Connections	V	I	V	I	V	I
90°	V_{bc}	I_a	V_{ca}	I_b	V_{ab}	I_c
60° no. 1	V_{ac}	$I_a - I_b$	V_{ba}	$I_b - I_c$	V_{cb}	$I_c - I_a$
60° no. 2	$-V_{cn}$	I_a	$-V_{an}$	I_b	$-V_{bn}$	I_c
30°	V_{ac}	I_a	V_{ba}	I_b	V_{cb}	I_c

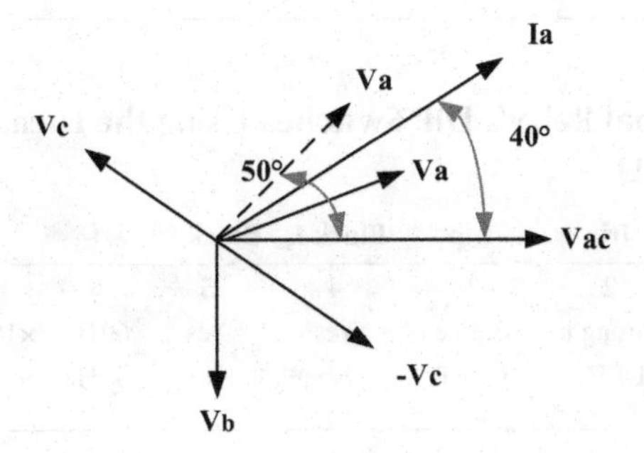

FIGURE 7.26 Directional relays 30° connection phasor diagram of Example 7.10.

Example 7.10

A directional overcurrent relay protects a line circuit with an impedance angle of 10°.

Use a 30° connection to specify the phase shift network used in the relay potential coil if phase a fault occurs. The relay potential coil has an impedance of $1000\angle 60°\ \Omega$ under the above fault condition V_a (Figure 7.26).

Shift in the leading direction by 20°, with the operating frequency of 50 Hz.

Solution

I_a leading V_{ac} by 40°

$$Z_{pc} = 1000\angle 60°\ \Omega = 500 + j860\ \Omega$$

For angle to be 50°

$$\tan 50° = \frac{X}{500}$$

$$X = 596\ \Omega$$

$$\text{Capacitive reactance require} = 860 - 596 = 264\ \ \Omega$$

$$\therefore \text{Phase shifting}\ C = \frac{1}{2\pi \times 50 \times 264} = 12.1\ \mu\text{F}$$

7.10 PROTECTION OF PARALLEL LINES (PARALLEL OPERATION)

The non-directional relay's DIP switches, as indicated in Table 7.5, use the Lucas-Nülle GmbH test emulator at PVAMU.

The directional relay's DIP switches as indicated in Table 7.6 (active setting = green background)

All components furnished are connected with a power plug to the supply voltage, and the resistive load is set to its maximum value. The protective relay settings will remain.

TABLE 7.5
The Non-Directional Relay's DIP Switches Using the Lucas-Nülle GmbH Test Emulator at PVAMU

Function	Trip	Trip	Trip	Block $I_>$	Block $I_{\gg}$	f	$tI_>$	$tI_>$
DIP Switch	1	2	3	4	5	6	7	8
ON	Inverse	Strong i.	Extreme i.	Yes	Yes	60 Hz	×10 seconds	×100 seconds
OFF	DEFT	DEFT	DEFT	No	No	50 Hz	× 1 seconds	×1 seconds

TABLE 7.6
The Directional Relay's DIP Switches Using the Lucas-Nülle GmbH Test Emulator at PVAMU

Function	Password		Reset	
DIP Switch	1	2	3	4
AN	Password		Manual	
OFF	Normal		Auto	

Any relay actions (attributable to excitation) should be ignored for the time being. Turn on the adjustable three-phase power supply and slowly raise the line-to-line voltage to 110 V. Decrease the resistive load until a total current of 0.75 A flows.

Using the multimeter, successively measure the current in phase 1 of lines I and II and the line-to-line voltage between phases L_2 and L_3 at the beginning and end of the lines. Enter the results in the table below.

Measurement results:

- Phase current I_I $(A_1) = 0.35$ A
- Phase current I_{II} $(A_2) = 0.37$ A
- Line-to-line voltage U_A $(V_1) = 110$ V
- Line-to-line voltage U_B $(V_2) = 84$ V

7.11 MINIMUM PICK-UP VALUE

Figure 7.27 shows Lucas-Nülle GmbH Layout plan for the Single-pole short circuit emulator at PVAMU. To determine the minimum pick-up value I_{PU}. The only line I will be operating for this purpose. The three-phase supply slowly increases the line-to-line voltage to 110 V.

The line I conducts the entire operating current of about 0.52 A.

When slowly decreasing the pick-up value $I_>$ of time overcurrent relay one until it responds (upper red LED comes on). When the fine adjustment potentiometer is to its

Phase current I_I (A1) = 0.35 A

Phase current I_{II} (A2) = 0.37 A

Line-to-line voltage U_A (V1) = 110 V

Line-to-line voltage U_B (V2) = 84 V

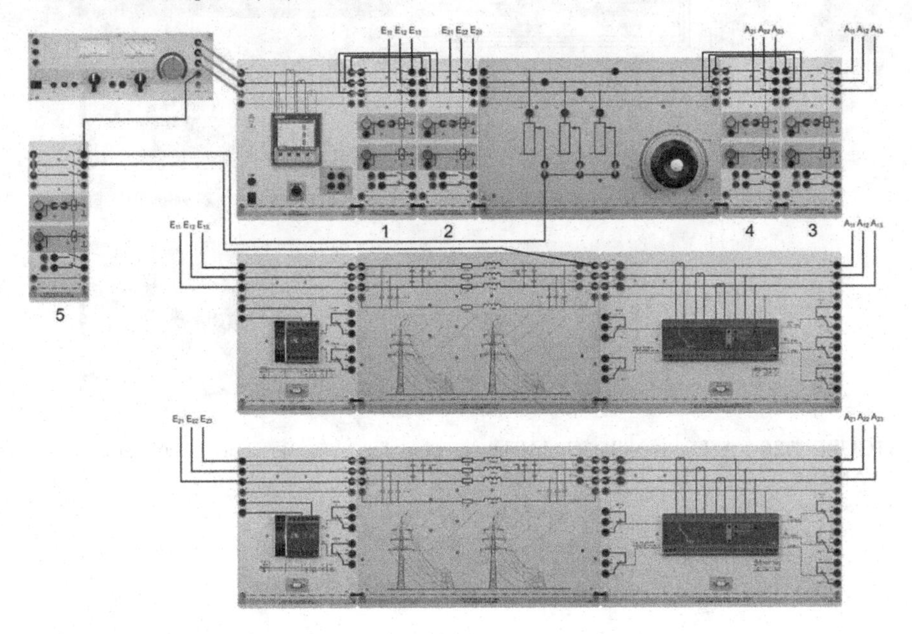

FIGURE 7.27 Lucas-Nülle GmbH layout plan for single-pole short circuit emulator at PVAMU.

right limit (upper red LED goes off again). The potentiometer back slowly until the relay is energized again,

- Both values $I_>$ and $I_>/I_n$
- Pick-up value $I_>=0.5$ A
- Pick-up value $I_>/I_n=0.05$ A

The total fault current resulting from the single-pole short circuit. Figure 7.28 shows a Lucas-Nülle GmbH Layout plan for Two-pole short circuit emulator at PVAMU.

The phase current measurements are:

- Phase $L_1=1.21$ A
- Phase $L_2=0.76$ A
- Phase $L_3=0.79$ A

Figure 7.29 shows a Lucas-Nülle GmbH Layout plan for three-pole short circuit emulator at PVAMU.

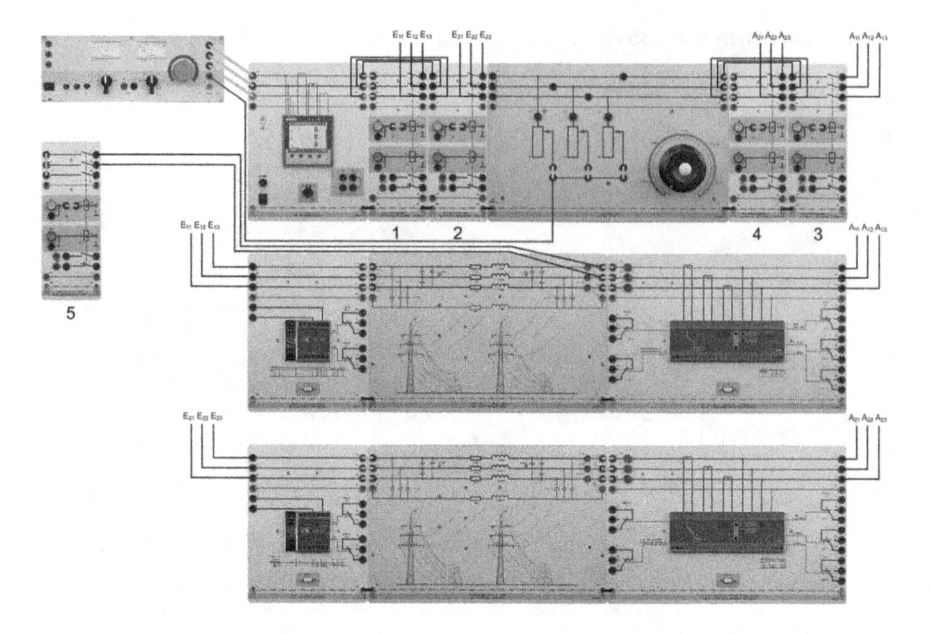

FIGURE 7.28 A Lucas-Nülle GmbH layout plan for two-pole short circuit emulator at PVAMU.

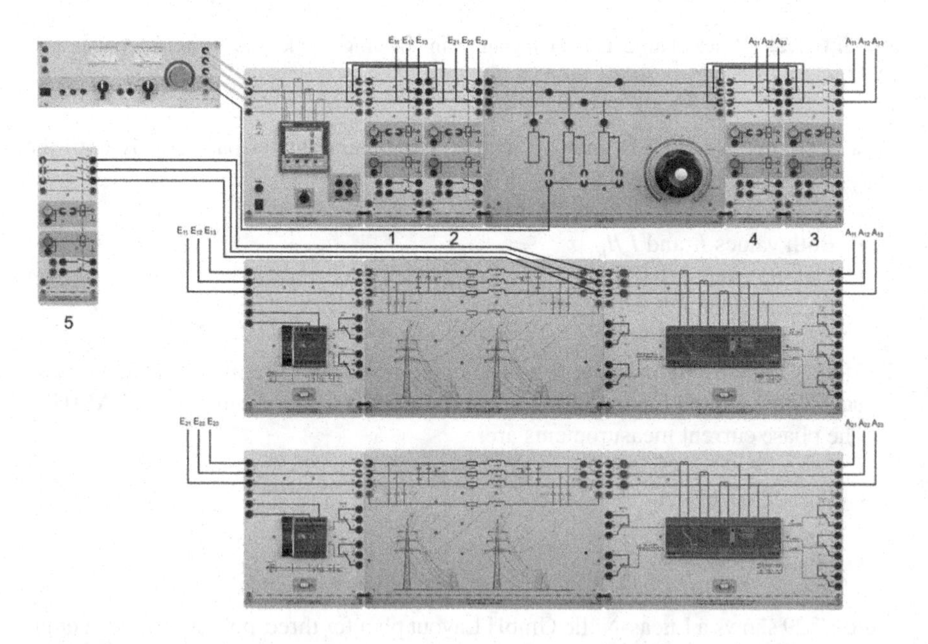

FIGURE 7.29 A Lucas-Nülle GmbH layout plan for three-pole short circuit emulator at PVAMU.

The phase current measurements are:

- Phase $L_1 = 1.23$ A
- Phase $L_2 = 1.26$ A
- Phase $L_3 = 1.25$ A

State the lowest fault current which can occur in a phase at $I_{K\,min} = 0.76$ A.

7.12 PARAMETRIZING NON-DIRECTIONAL RELAYS

The time overcurrent relays at the beginning of the line will be configured and monitored via the software. After that, the directional time over current relays will be set directly via the software.

7.13 TIME OVERCURRENT RELAYS

Before setting a protective relay, establish a connection between it and the computer.

- Open the accompanying SCADA software for better and more precise readings of the settings (**Using "LN SCADA Software"**).
- Precisely set the determined pick-up value $I_>$ via the potentiometers on both relays.
- Set the delay time $tI_>$ initially to 0 seconds (left limit).

7.14 DIRECTIONAL TIME OVERCURRENT RELAYS

The accompanying SCADA software directly sets the values (refer to the chapter titled **Using "LN SCADA Software"**).

The Setting values specified in Table 7.7 are required to determine the pick-up value $I_>$ for both directional relays. Figure 6.7 shows a Lucas-Nülle GmbH Layout plan for three-pole short circuit emulator at PVAMU.

TABLE 7.7
The Setting Values Required to Determine the Pick-Up Value Using the Lucas-Nülle GmbH Test Emulator at PVAMU

Parameter	Value
Pick-up value $I_>$	
Trip characteristic CHARI$_>$	DEFT
Trip delay $tI_>$ forward	0.03 seconds
Trip delay $tI_>$ reverse	0.03 seconds
Reset mode	0 seconds
Pick-up value $I_{\gg}$	EXIT
Trip delay $tI_{\gg}$ forward	EXIT
Trip delay $tI_{\gg}$ reverse	EXIT
Characteristic angle RCA	49
Switch failure protection	EXIT
Frequency	60 Hz
LEDs flash on energization	FLSH

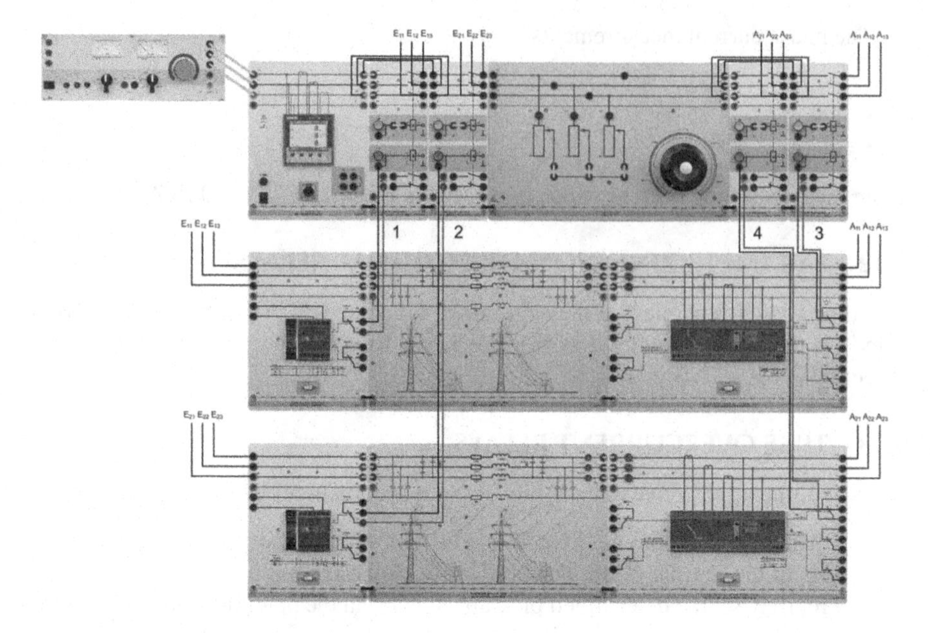

FIGURE 7.30 A Lucas-Nülle GmbH layout plan for connecting the power switches circuit emulator at PVAMU.

The relays have been configured as non-directional tripped in the same manner regardless of the direction of energy flow.

To modify the protective scheme setup, as shown in Figure 7.30, connect the relay outputs to the power switches.

7.15 HIGH-SPEED DISTANCE PROTECTION

With the aid of the line impedances, we will first identify the impedance zones. Let's look at Figure 7.31 for this reason. For instance, in this model, the stretch between stations A and B combines two lines, each 300 km (1864 miles) long. Consider an additional line segment (BC) that is 600 km (3728 miles) long to define the impedance zones protected by relay A. The reactance of the relevant line segment determines the range of each distance zone. The first zone with impedance Z_{AB} on line segment A-B is set to 85% of the distance with a delay time T_1 of 0.00 seconds, as is already known from the theoretical portion. Thus, the defect is limited by the intrinsic time of the component.

The other phases are chosen following the guidelines in Figure 7.31. Zones 1 and 2 should both respond by moving forward. Zone 3 needs to provide a non-directional backup defense.

With the help of the line impedances and Figure 7.31, zones 1–3 can be ascertained. Figure 7.32 shows the matching stepped characteristic. The indicated values correspond to the transformers' primary sides.

Sample calculation for zone 1: $\left(14.4\,\Omega + j\ 173.4\,\Omega\right) \times 0.85 = 12.2\,\Omega + j\ 147.4\,\Omega$

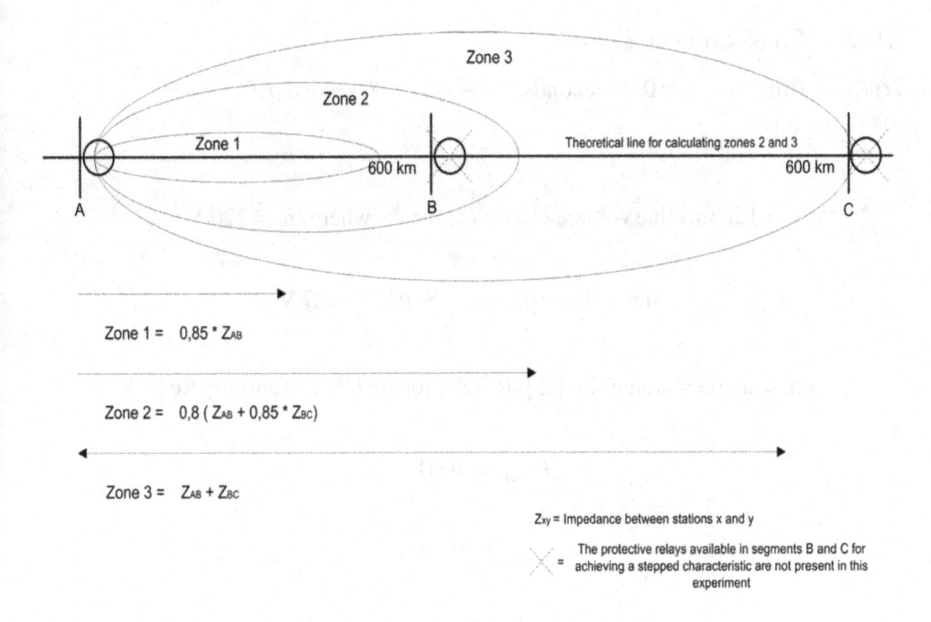

FIGURE 7.31 A Lucas-Nülle GmbH network plan.

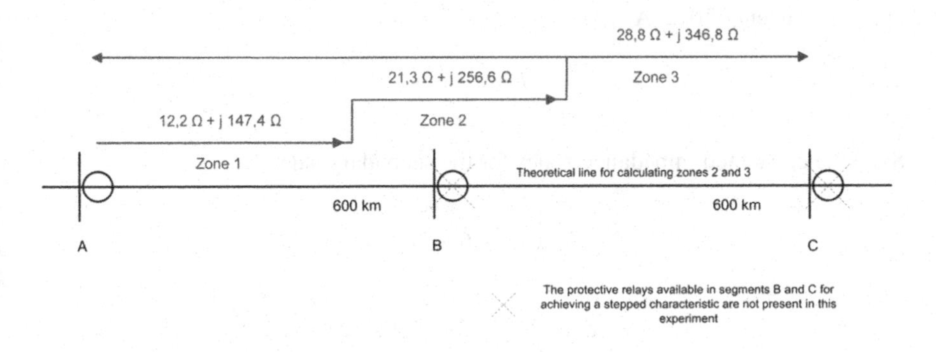

FIGURE 7.32 A Lucas-Nülle GmbH grading scheme.

7.16 FURTHER SETTINGS

During the parametrization of the protective device, some values need to be specified for the secondary side and must therefore be converted beforehand. When calculating voltages and amperages, this conversion is performed via the transformation ratios:

$$X_{\text{secondary}} = \frac{\text{Current transformation ratio}}{\text{Voltage transformation ratio}} \times X_{\text{primary}}$$

7.16.1 Characteristic Data

Tripping time $Z_1 - t_1 = 0.00$ seconds/$Z_2 - t_2 = 5.00$ seconds/$Z_3 - t_3 =$ 10.00 seconds

$$\text{Line-to-line voltage } U_{L12} = U_0 \times e^{j30^\circ} \text{ where } U_0 = 220 \text{ V}$$

$$\text{Star voltage } U_Y = U_0/\text{Sqrt}(3) = 127 \text{ V}$$

$$\text{Phase angle} = \arctan\left(\text{Im}\{Z\}/\text{Re}\{Z\}\right) \text{ for Im}\{Z\} > 0 \text{ and any Re}\{Z\}$$

$$U_{\text{prim}} = 400\,\text{kV}$$

$$I_{\text{prim}} = 2500\,\text{A}$$

$$\text{Angle} = 84^\circ$$

Using the chapter on fundamentals as help, ascertain the overcurrent excitation level at a load current of 0.52 A.

$$I = 0.25 \text{ A}$$

State the calculated impedance zones for the secondary side.

$$\text{Zone 1} = 7.6 + j92.1\,\Omega$$

$$\text{Zone 2} = 13.3 + j160.4\,\Omega$$

$$\text{Zone 3} = 16.8 + j216.8\,\Omega$$

Figure 7.33 shows an A Lucas-Nülle GmbH Layout for testing the trip characteristic one emulator at PVAMU.

For the next test, modify the setup as shown in Figure 7.34.

For the next test, modify the setup as shown in Figure 7.35.

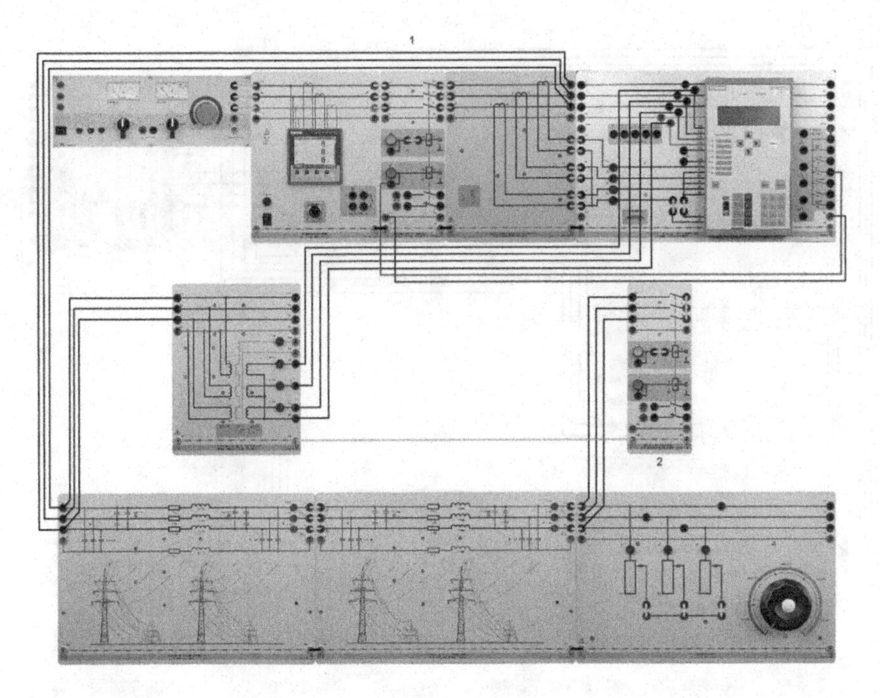

FIGURE 7.33 A Lucas-Nülle GmbH layout for the trip characteristic one emulator test at PVAMU.

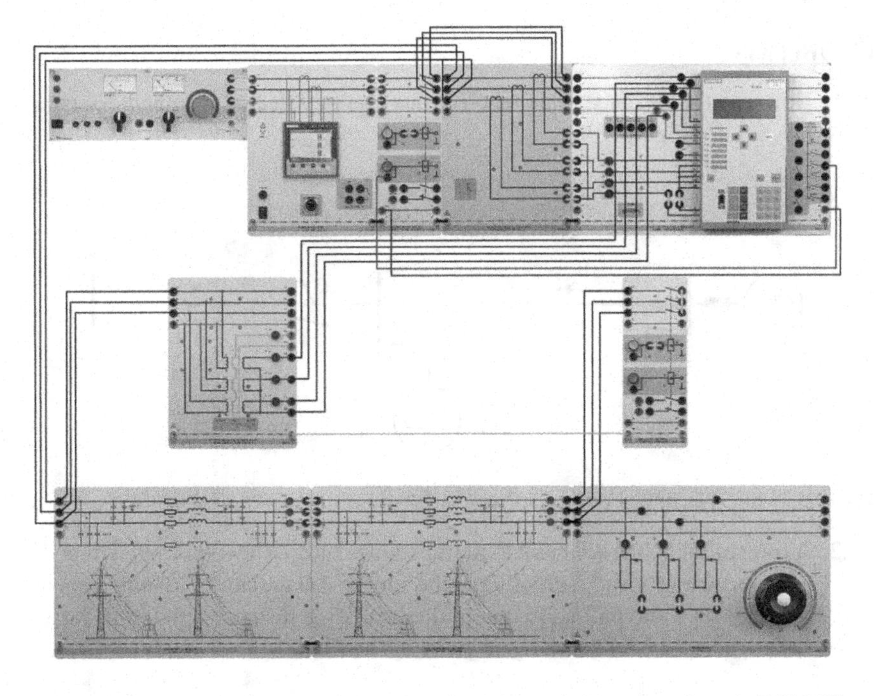

FIGURE 7.34 A Lucas-Nülle GmbH layout trip characteristic two emulator at PVAMU.

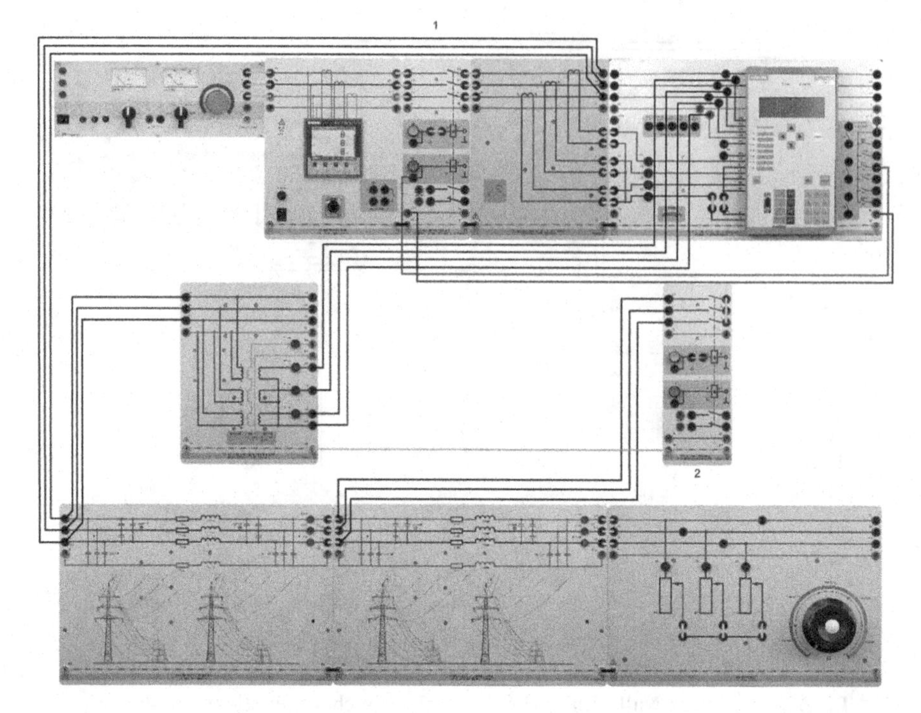

FIGURE 7.35 A Lucas-Nülle GmbH layout for the trip characteristic three emulator at PVAMU.

PROBLEMS

7.1. How can the distance relay reach at bus one be affected if infeed exists in bus two for Figure 7.36?

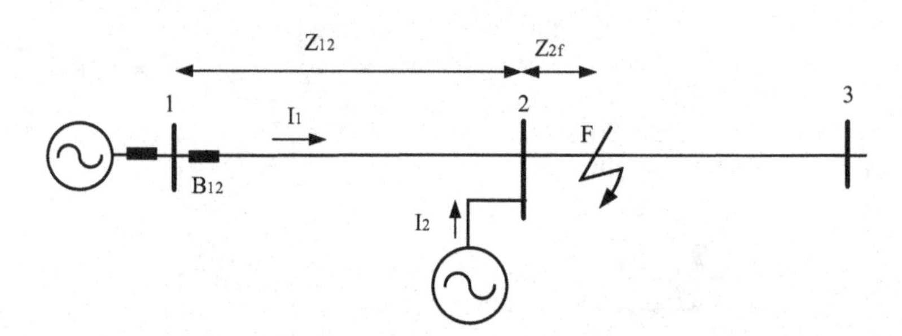

FIGURE 7.36 The power system of Problem 7.1.

7.2. A transmission line AB has a total impedance of Z_L. The total source impedance feeding at A and B are Z_a, and the source voltage can be assumed equal in magnitude and phase. To obtain the line AB, a distance relay impedance measured by the relay (V_{an}/I_a) is given by

$$Z_L + R\left(2 + \frac{Z_L}{Z_a}\right)$$

Assume that both sources are star connection solidly to the ground and all plant has the same impedance to all three sequence currents.

If the relay is a reactance relay set to 80% of the total line impedance, and $Z_L = 16\angle 50°\ \Omega$, Show if the relay will trip for $R > 3$ or not.

7.3. What are the protection system components, and what are the basic concepts of protective relay application?

Q) Show, with the aid of diagrams, a sample method of protecting generators against inter-turn faults.

7.4. The line impedance for the power system shown in Figure 7.37 is $Z_{12} = Z_{23} = 32 + j35\ \Omega$ reach for the zone 3 B_{12} impedance relay is set for 100% of line 1–2 plus 120% of line 2–4

i. A bolted three-phase at bus 4 shows that the apparent primary impedance seen by the B_{12} relay is

$$Z_{Apparent} = Z_{12} + Z_{24}\left(\frac{I_{32}}{I_{12}} + 1\right)$$

Where $\frac{I_{32}}{I_{12}}$ is line 2–3 to line 1–2 fault current ratio.

ii. If $\frac{I_{32}}{I_{12}} > 0.2$ does the B_{12} relay see the fault on bus 4?

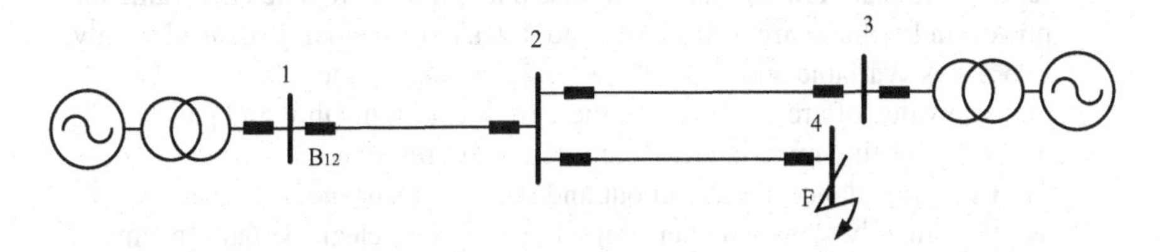

FIGURE 7.37 The power system of Problem 7.4.

7.5. A section of 110kV is shown in Figure 7.38, line 1–2 is 100km, and line 2–3 is 80km long. The positive sequence impedance for both lines is $0.02 + j0.2\ \Omega$/km. The maximum load current carried by line 1–2 under emergency conditions is 30 MVA. Design a three zones distance relying system specifying the relay setting for relay R_{12}.

Hint: Standard VT secondary voltage is 120V line to line.

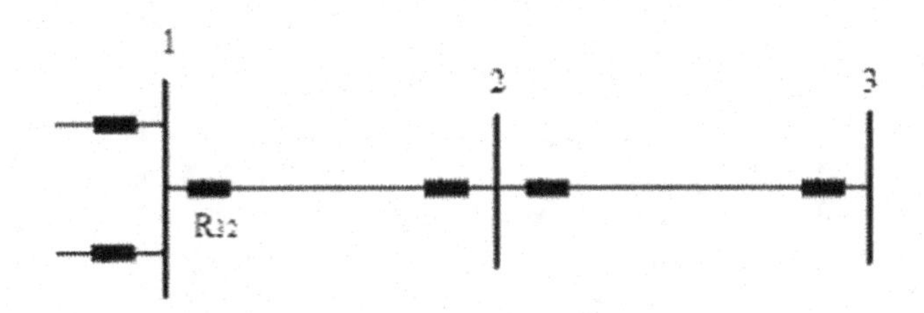

FIGURE 7.38 The power system of Problem 7.5.

7.6. Figure 7.39 shows a reactance relay situated at end A of 120 km, A single-circuit transmission line AB has a first zone setting of 80%. The positive phase sequence impedance of the line is $j0.4$ p.u. A single line to ground fault occurred at 70% from end A. The current distribution is such that the actual fault arc current is $10 - j10$ p.u., while the current flowing at the end A is $1 - j5$ p.u. concerning the same reference. If the fault arc resistance is 0.05 p.u., will the relay see the fault in zone 1? If not, what is the % error in the relay reach? And where will it see the fault as % of the length?

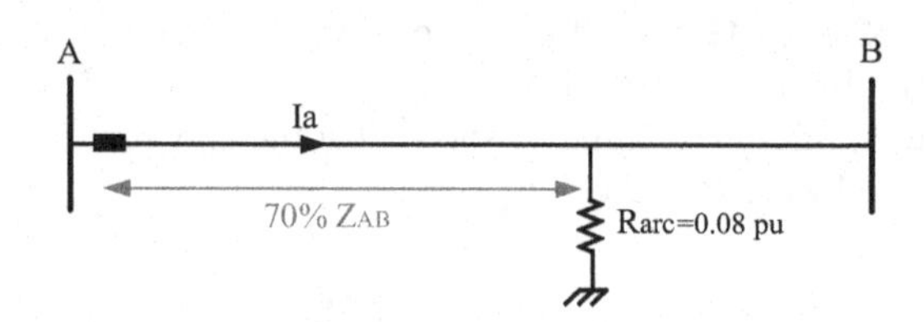

FIGURE 7.39 The power system of Problem 7.6.

7.7. For a 130 kV sub-transmission system shown in Figure 7.40, consider the directional distance (MHO) relay located at A. Assume a single line to gram located at point F, at 70% distance from A. The magnitude of the fault current is 1.25 p.u. Assume that the fault arc length is 10 ft. The base value for power and voltage are 100 MVA, and 130 kV, respectively. Consider only two zones available and $Z_{AB} = 0.2 + j0.7$ p.u., determine

i. The value of arc resistance at the fault location in Ohm and p.u.
ii. Value of line impedance, including the arc resistance.
iii. Line impedance angle without and with arc resistance.
iv. Graphically show whether or not the relay will clear the fault instantly.

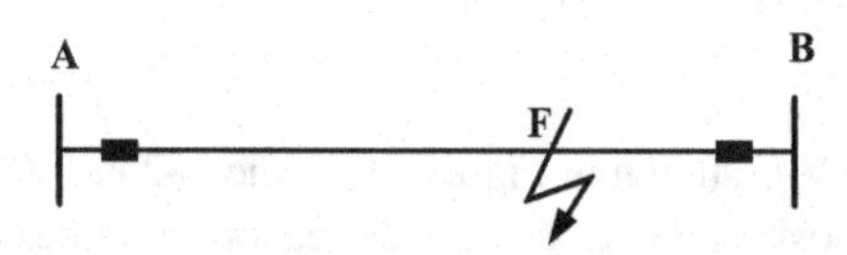

FIGURE 7.40 The power system of Problem 7.7.

8 Transformer Protection

8.1 INTRODUCTION

The type of protection used for power transformers depends upon their size, voltage, rating, and nature of their application. For small-size (<2 MVA) transformers, protection with fuses may be adequate, whereas, for a transformer of >10 MVA, a differential relaying with harmonic restraint may be used.

There are some peculiar issues with the inherent characteristics of power transformers, which are not present in the safety of transmission lines, turbines, motors, or other power system devices. The effect of internal electrical faults is transformer faults, i.e., short circuits, the most common one being the phase-to-ground fault. Turn-to-turn faults are much less common. The physical extent of a transformer is limited to within a substation, unlike a transmission line. Thus, differential relaying, the most suitable type of protection available, can be used to protect transformers.

Fuses, overcurrent relays, differential relays, and pressure relays may secure the transformer and be controlled with winding temperature measurements and chemical analysis of the gas above the insulating oil for incipient trouble.

This chapter presents transformer protection, types, connection, and mathematical models for each type of device.

8.2 TRANSFORMER FUNCTIONS

8.2.1 Transformer Size

Usually, transformers with a capacity below 2500 kVA are shielded by fuses. The transformer can be secured with fuses with ratings between 2500 and 5000 kVA. From the point of view of sensitivity and synchronization with protective relays on the transformer's high and low sides, instantaneous and time-delay overcurrent relays may be more suitable. An induction disk overcurrent relay connected in a differential configuration is commonly used between 5000 and 10,000 kVA. A harmonic restraint and percentage differential relay are recommended above 10 MVA pressure, and temperature relays with this transformer size are commonly applied.

The protection method used for power transformers depends on the transformer ratings. Their ratings show transformers are usually categorized as follows: The protection method used for power transformers depends on the transformer ratings. Transformers are usually categorized according to their ratings, as in Table 8.1.

Transformers below 5000 KVA (Category I & II) are protected using fuses. Fuses and MV circuit breakers often protect up to 1000 kVA (distribution transformers for 11 and 33 kV).

DOI: 10.1201/9781003394389-8

TABLE 8.1
Transformers Rating Categories

	Transformer Rating (kVA)	
Category	**Single Phase**	**Three Phase**
I	5–500	15–500
II	501–1667	501–5000
III	1668–10,000	5001–30,000
IV	>10,000	>30,000

For transformers 10 MVA and above (Category III & IV), differential relays are commonly used to protect them. Current differential relays are applied for transformers as main protection. Regarding backup protection, distance protection, overcurrent (phase current, zero sequence current) protection, or both are mainly applied.

8.2.2 Location and Function

In addition to the transformer's size, the particular safety application's decision is significantly affected given the transformer's role within the power network. If the transformer is an integral part of the bulk power system, it would need more sophisticated relays in design and redundancy.

A single differential relay and overcurrent backup will normally suffice if it is a distribution station step-down transformer. The high *X/R* ratio of the fault path will require harmonic restraint relays if the transformer is close to a generation source to accommodate the higher magnetic in-rush currents.

8.2.3 Voltage

In general, higher voltages need more sophisticated and expensive protective equipment due to the harmful impact on system efficiency of delayed fault clearing and the high cost of transformer repair.

8.2.4 Connection and Design

The safety systems can differ considerably between two- or three-winding transformers and autotransformers. A three-phase transformer's winding connection will make a difference in the selected safety scheme, whether delta or wye. The existence of tertiary windings, grounding used, tap changers, or phase-shifting windings is also significant.

We will comment on the effect of these and other factors on the relaying systems of choice as we develop safety ideas for transformers.

8.3 FAULTS ON POWER TRANSFORMER

Faults on power transformers can be classified into three categories:

i. Faults in the auxiliary equipment, which is part of the transformer:
   Detecting a fault in auxiliary equipment is necessary to prevent the main transformer windings' ultimate failure. The following can be considered an auxiliary equipment
   a. Transformer oil.
   b. Gasoline.
   c. Oil pumps and forced air fans.
   d. Core and windings insulation.
ii. Faults in the transformer windings and connections:
   Electrical faults that cause immediate serious damage are detected by unbalance current or voltage may be divided into the following:
   a. Faults between the adjacent turn of coils, such as phase-to-phase faults on the HV and LV external terminals, the windings themselves, or short circuits between turns of HV and LV windings.
   b. Faults to the ground or across complete windings, such as phase-to-earth faults on the HV and LV external terminals or the windings.
iii. Overload and external short circuit:
   Overloads can be sustained long, limited only by the permitted temperature rise in the windings and cooling medium. Excessive overloading results in the deterioration of insulation and subsequent failure. It is usual to monitor the winding and oil temperature conditions, and an alarm is initiated when the permitted temperature limits are exceeded.

8.4 MAIN TYPES OF TRANSFORMER PROTECTION

The main types of transformer protection are

A. Electrical Protection
   1. Differential protection (87T).
   2. Over current protection (51).
   3. Earth fault and restricted earth fault protection.
B. Nonelectrical Protection
   1. Buchholz relay.
   2. Oil pressure relief device.
   3. Oil temperature (F49).
   4. Winding temperature (F49).

8.4.1 PERCENTAGE DIFFERENTIAL PROTECTION

Figure 8.1 shows the single-phase, two-winding power transformer. The algebraic sum of the ampere-turns of the primary and secondary windings must equal the magnetomotive force (MMF) needed to set up the transformer center's working flux

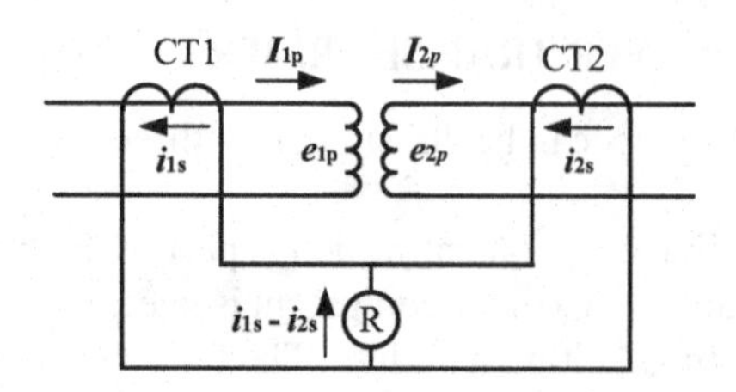

FIGURE 8.1 Differential relay connections.

during normal operation of transformer. The MMF is negligible (often <0.5% of the MMF generated by the load current) because of the very small air gap in the transformer core, and hence for a normal power transformer

$$N_1 \ i_{1p} = N_2 \ i_{2p} \tag{8.1}$$

Where $N_1, \ N_2$ number of power transformer primary and secondary winding turns and i_{1p}, i_{2p} primary and secondary power transformer current.

If we use current transformers with turn ratios of 1: n_1 and 1: n_2 on the primary and secondary sides, respectively, the currents in the secondary windings of the current transformers are connected under normal conditions by the currents

$$N_1 \ n_1 \ i_{1s} = N_2 \ n_2 \ i_{2s} \tag{8.2}$$

where i_{1s} and i_{2s} secondary current for current transformers on the primary and secondary of the power transformer.

If we select the CTs appropriately, we may make $N_1 \ n_1 = N_2 \ n_2$ and then, for a normal transformer, $i_{1s} = i_{2s}$. However, if an internal fault develops, this condition is no longer satisfied, and the difference between i_{1s} and i_{2s} becomes much larger, proportional to the fault current. The differential current I_d.

$$I_d = i_{1s} - i_{2s} \tag{8.3}$$

Provides the fault current with a highly sensitive calculation. If, as shown in Figure 8.1, an overcurrent relay is attached, it will provide the power transformer with excellent protection.

Until a workable differential relay can be introduced, some practical problems must be considered. First, it may not be possible to obtain the CT ratios that will satisfy the condition ($N_1 \ n_1 = N_2 \ n_2$) on the primary and secondary sides on the primary and secondary sides, as we must choose standard ratio CTs. The problem is somewhat alleviated because most relays have different tap locations for each CT input to the relay, supplying auxiliary CTs that can correct any deviation from the desired ratios. Using auxiliary CTs to accomplish the same purpose is a much less suitable technique.

In either case, some residual ratio mismatch exists, even with these changes, which under normal conditions leads to a small differential current I_d. Second, the two CTs' transformation errors may vary from each other, resulting in substantial differential current or external failure when there is normal load flow. Lastly, if the power transformer is fitted with a tap changer, it will introduce a main transformer ratio adjustment when the taps are changed.

In the overcurrent relay, these three impacts allow a differential current to flow, and without creating a ride, the relay design must handle these differential currents. Because each of these triggers leads to a differential current equal to the real flowing current in the transformer's primary and secondary windings, a percentage differential relay is an excellent solution to this problem.

Including the fault current to the dc component, acting magnetizing characteristics of the CTs (different saturation) may cause CTs to transform the primary current differently even under fault conditions. This is more pronounced in the power transformer. To overcome these difficulties, Merz Price protection is modified by biasing the relay. This is commonly known as biased differential/protection or percentage differential/protection.

The relay consists of an operating coil (OC) and a restraining coil (RC). The OC is connected to the midpoint of RC. The operating current is a variable quantity because of the RC. If it is possible to make the value $|i_1 - i_2|$ dependent upon the average i_1 and i_2. A relay can be designed such that the operation principle for it

$$|i_1 - i_2| > \frac{K|i_1 + i_2|}{2} \rightarrow \text{Trip} \tag{8.4}$$

$$|i_1 - i_2| < \frac{K|i_1 + i_2|}{2} \rightarrow \text{Block} \tag{8.5}$$

where $|i_1 - i_2|$ the OC current and $\frac{|i_1 + i_2|}{2}$ the RC current.

In a percentage differential relay, the differential current must exceed a fixed percentage of the "through" current in the transformer. The through current is defined as the average of the primary and the secondary currents:

$$I_r = \frac{i_{1s} + i_{2s}}{2} \tag{8.6}$$

where I_r is a current derived from the electromechanical relay's nature of creating a restraint current torque on the moving disk, whereas the differential current generates the operating torque. The relay works when

$$I_d \geq k I_r \tag{8.7}$$

The slope of the percentage differential characteristic is where k. K is normally expressed as a percent value: 10%, 20%, and 40% generally. A relay with a 10% slope is much more responsive than a relay with a 40% slope. While lowercase symbols were used for the currents, indicating the instantaneous values, it should be clear that the corresponding relationships still exist between the currents' root mean square (RMS) values. Thus, for the RMS currents, Equations 8.1–8.5 also hold.

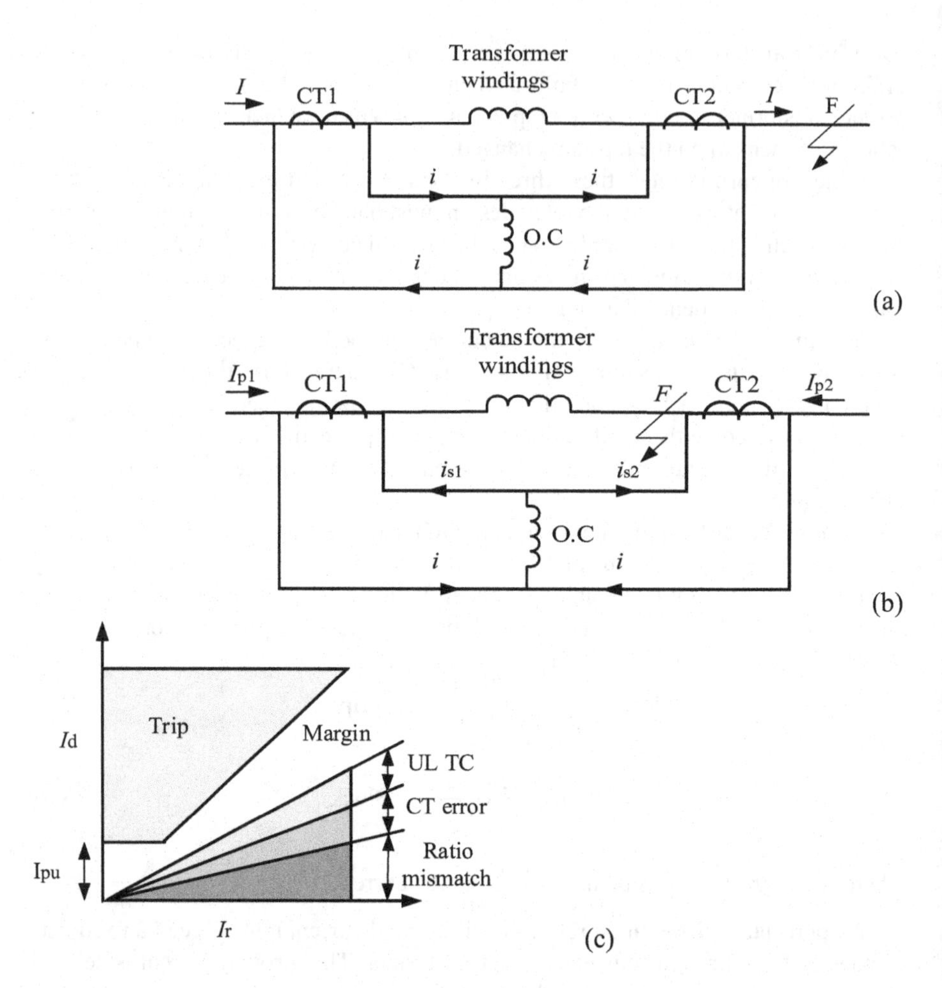

FIGURE 8.2 Differential relay protective scheme with (a) external fault, (b) internal fault, and (c) percentage differential relay characteristic.

Figure 8.2 shows a functional percentage differential protective scheme and characteristics. The slope of the relay defines the trip region. The three differential current sources during normal transformer operation are shown as the safety margin used when arriving at the slope. The relay has a limited pickup current setting, i.e., unless the differential current is above this pickup value, the relay does not work. Usually, the pickup setting is very low: standard values are 0.25 A secondary. This accounts for any residual CT errors of transformer load current at low values.

To obtain differential protection of the three-phase transformer, the three-phase transformers' protection requires that the three phases' primary and secondary currents be compared individually. The key difference between three-phase transformer safety and three single-phase transformers is the need to deal with a wye–delta transformation. The main and secondary winding currents are in phase under standard load conditions, but the line currents on the three-phase transformer's wye and delta sides are 30% out of phase.

Since CTs are typically linked on the line and not on the delta side in the winding, this phase shift creates a standing differential current, even though the main transformer's turns ratio is correctly regarded. The difficulty is resolved by connecting the CTs to undo the effect of the wye–delta phase shift created by the main transformer.

The CTs are linked in the delta on the power transformer's wye side, and the CTs are linked in the wye on the delta side of the power transformer. Recall that there are two forms of a delta relation that can be made.

The delta currents lag 30° behind the primary currents, and the other in which delta currents lead 30° behind the primary currents. The link that compensates for the phase shift generated by the power transformer must, of course, be used. Carefully testing the current flow in the differential circuits when the power transformer is usually loaded is the best way to achieve the right connections. When the CT connections are right, there will be no (or little) current in the differential circuit under these conditions.

Changing the turns ratios of the CTs and the phasing consideration discussed earlier is also important. The delta link on the power transformer's wye side generates relay currents that are numerically balanced with the relay currents produced by the wye-connected CTs. Therefore, the delta CT winding currents must be $(1/\sqrt{3})$ times that of the wye CT currents.

The differential protection for the power transformer of the biased type is used to override the dissemination caused by

i. Mismatch of the CTs on both sides of the power transformer due to different currents and voltage.
ii. Use the load tap-changing gear on large transformers, which under normal operation affects the transformer ratio.
iii. Transformer magnetizing in-rush current, such current may attain peak values corresponding several times the transformer full-load current aid decay relatively slowly.

8.4.2 Overcurrent Protection

The external faults or steady-state load currents must be differentiated from the currents generated by internal faults, as in all safety applications with overcurrent relays. External faults that are not immediately cleared or steady-state heavy loads affect overheating of the transformer windings and weaken the insulation. This will render internal flashovers vulnerable to the transformer. Time delay over overcurrent relays can be provided to protect transformers against internal faults. Arcing, potential fire, and magnetic and mechanical forces result from a sustained internal fault, resulting in structural damage to windings, tanks, or bushings with a corresponding danger to staff or surrounding equipment. High-side fuses, instantaneous and time-delay overcurrent relays, or differential relays may protect transformers.

8.4.2.1 Protection with Fuses

To cover transformers with ratings above 2.5 MVA, fuses are not used. The fundamental principle is close to that used in other fuse applications when choosing fuses for the high-voltage side of a power transformer. The ability to interrupt the fuse must surpass the maximum short-circuit current that the fuse would be asked to interrupt. The fuse's continuous rating must exceed the maximum load of the transformer.

The fuse level should typically exceed 150% of the full load. The minimum melt characteristic of a fuse means that if conditions to the right of (greater than) the function are obtained, the fuse will be impaired.

On the low side of the power transformer, the fuse's minimum melt characteristic must coordinate with (i.e., should be well separated from) the protective devices. The ambient temperature, previous loading, and reclosing adjustment variables should be considered when considering the coordination. These factors impact the fuse's previous heating and melt at various times than suggested by the criteria for a "cold" fuse.

The magnetizing current of the transformer does not cause the fuse to be fried. This calls for a current of longer length that must be lower than the characteristic of minimum melt. The fuse "speed ratio" is the ratio between the minimum melt current values at two widely spaced times: 0.1 and 100.

A lower velocity ratio would mean a more sloping characteristic, and such a fuse would have to be set with a smaller sensitivity for the proper degree of coordination. A fuse with as high a speed ratio as possible is ideal. Finally, to prevent the arrester from flickering overdue to the fuse operation, if a current-limiting fuse is used, the lightning arresters on the line side of the fuse should have a rating equal to or higher than the overvoltages that the fuse can produce.

8.4.2.2 Time-Delay Overcurrent Relays

Time-delay overcurrent relays provide security against unreasonable overload or persisting external fault. Usually, the pickup setting is 115% of the maximum overload acceptable. This margin covers the existing transformers (CTs), relays, and calibration ambiguity. The overcurrent relays with time delay must communicate with the low side. Devices of security. This can include low-voltage bus relays for phase-to-phase failures, phase-directional relays for parallel transformers, and timers for low-voltage breakers for breaker failure relays.

8.4.2.3 Instantaneous Relays

The use of instantaneous relays is subject to many constraints; some depend on the relay's nature. Of course, the relay must not be worked on in-rush or for low-side faults in both situations. In a transformer, the peak magnetizing current can be as high as eight to ten times the peak full-load current. Since the relay can see low-side faults when they are completely offset, one must consider these faults.

Some relay designs refer to the current instantaneous value, including the DC offset and electromechanical plunger-type relays. With maximum DC offset, such a relay must be set above the low-side fault currents. Disk-type relays respond only to the current wave's AC portion. Depending on their nature, solid-state or computer-based relays may or may not respond to DC offset.

8.4.3 Earth Fault and Restricted Earth Fault Protection

This type of protection is specific to transformers with at least one directly earthed or resistance-earthed winding. The protection is specialized to protect against winding faults to earth. The overcurrent units' connections can be only in the neutral, residual phase, or a differential connection, including all phases and the earth. These overcurrent units can be set much lower than the phase overcurrent units because of the load current's cancellation. Harmonic restraint may be required if non-differential connections are used. Generally, they will give a lower setting but only protect the earthed winding.

Conventional earth fault protection using overcurrent elements fails to adequately protect transformer windings.

This is especially the case with an impedance-earthed neutral for a star-connected winding. The degree of protection is greatly improved by applying minimal earth fault protection (or REF protection).

This is a unit safety function for one transformer winding. As shown in Figure 8.3, it can be of the high-impedance form or the biased low-impedance type. The residual current of three-line current transformers is balanced against a current transformer's output in the neutral conductor for the high-impedance type. The three-phase and neutral currents become the bias inputs to a differential element in the biased low-impedance version.

In the region between current transformers, the system is operational for faults and faults on the star winding. For all faults outside this region, the system will remain stable.

Restricted earth fault defense is also applied even when the neutral is firmly earthed. Since fault current remains at a high value until the last turn of the winding, the virtually complete cover is obtained for earth faults. Compared with the efficiency of devices that do not calculate the neutral conductor present, this is an enhancement. The safety of the earth's fault applied to a delta-connected or unearthed star winding

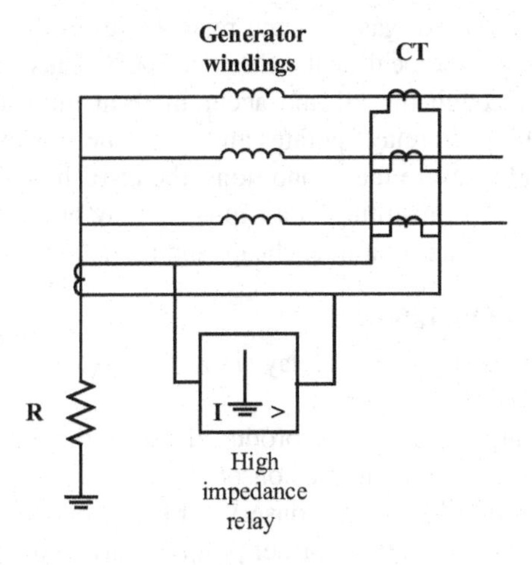

FIGURE 8.3 Restricted earth fault protection for a star winding.

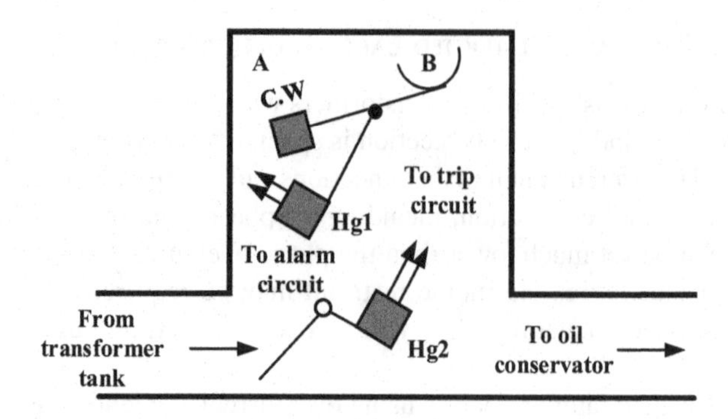

FIGURE 8.4 Buchholz relay mounting arrangement.

is fundamentally limited as no zero-sequence components can be transferred to the other windings via the transformer.

With restricted earth fault protection, both transformer windings can be secured separately, providing high-speed protection against earth faults for the entire transformer with relatively simple equipment. A high-impedance relay provides stability for rapid operation and phase fault.

8.4.4 Buchholz Relay

Buchholz protection usually is provided on all transformers. The Buchholz relay is contained in a cast housing connected in a pipe to the conservatory, as in Figure 8.4. A typical Buchholz relay will have two sets of contacts.

One is arranged to operate for slow gas accumulations, the other for a bulk displacement of oil in the event of a heavy internal fault. An alarm is generated for the former, but the latter is usually direct-wired to the CB trip relay.

Buchholz relay is a type of gas-actuated relay. Core insulation failures and poor electrical connection create local heat, which at 350°C causes the oil decomposes into gases that rise through the oil and accumulate at the transformer's tap. The Buchholz relay gas-actuated relay operates an alarm when a specified amount of gas has accumulated. Under severe fault conditions, the in-rush of gases and oil through the pipe moves the lever *L*, resulting in a trip. The relay provides a good method of detecting incipient faults and is simple, cheap, and robust.

8.4.4.1 Principle of Operation

The steps of operation of a Buchholz relay are as follows:

i. A Buchholz relay will detect gas produced within the transformer.
ii. An oil surge from the tank to the conservator.
iii. A complete loss of oil from the conservator (very low oil level).
iv. Fault conditions within a transformer produce carbon monoxide, hydrogen, and a range of hydrocarbons.

v. A small fault produces a small volume of gas deliberately trapped in the gas collection chamber (A) built into the relay. The oil level will be lowered, and the oil in the bucket (B) will tilt the counterweight (CW). Thus, switch Hg1 operates an alarm circuit to send an alarm.
vi. A large fault produces a large gas volume, which drives a surge of oil toward the conservatory. This surge moves a flap (P) in the relay to operate switch Hg2 and send a trip signal to open the main circuit breaker.
vii. The device will also respond to a severe reduction in the oil level due to the tank's oil leakage.

8.4.5 Oil Pressure Relief Devices

The commonly used "breakable disk," typically located at the end of the oil vent pipe protruding from the top of the transformer tank, is the simplest type of pressure relief system. The oil rush causes the disk to burst because of a severe failure, allowing it to drain rapidly.

Reducing and decreasing the pressure increase prevents the tank's explosive breach and fire risk. Outdoor transformers submerged in oil are usually installed in a storage pit to catch the spilled oil, reducing the risk of contamination.

The drawback of the shredded disk is that the remaining oil in the tank is exposed to the atmosphere after ripping. This is avoided by the snap pressure relief valve, a more powerful mechanism, which opens to allow the oil to drain if the pressure reaches the set level, but automatically shuts off when the internal pressure drops below that level. The valve can work within a few milliseconds if the pressure is abnormally high and provides a fast release when sufficient contacts are mounted.

This system is usually used in power transformers rated at 2 MVA or higher and can be used for distribution transformers rated as low as 200 kVA.

8.4.6 Oil Temperature (F49)

An oil level monitor for transformers containing the oil conservator(s) (expansion tank) is monitored. Usually, on both sides of the alarm, it has a screen. One contact for a warning of the maximum oil level and one contact for a minimum oil level warning.

The upper oil temperature gauge transformer contains an upper oil thermometer with a liquid thermometer bulb in the top pocket adapter. One to four contacts on the upper oil thermometer are sequentially closed at a higher temperature.

Figure 8.5 shows the construction of an upper oil thermometer of the capillary type, with the lamp located on top of the switch in a "pocket" surrounded by oil. The lamp is attached via a capillary tube to the measurement bottom inside the main unit. Via mechanical connections, bellows shift the indicator, activating the contacts at specified temperatures. Suppose the overhead oil temperature is much lower than the rolling temperature, especially shortly after a sudden overload. It means that the upper oil thermometer does not have effective overheating protection.

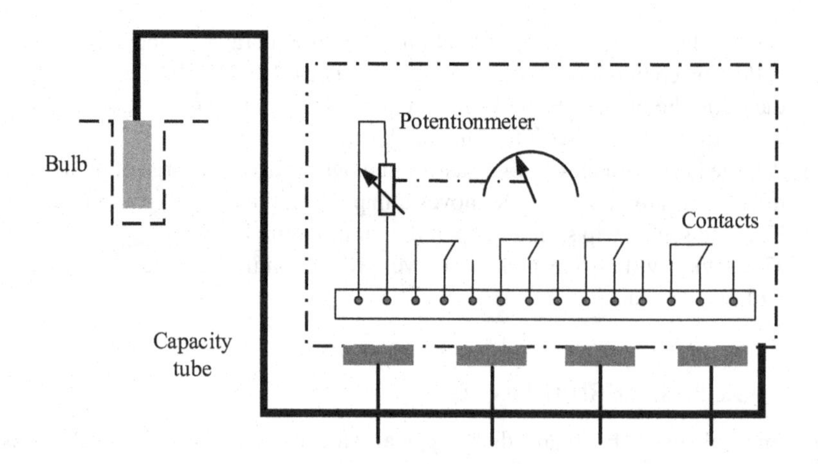

FIGURE 8.5 Capillary type of top-oil temperature measurement device.

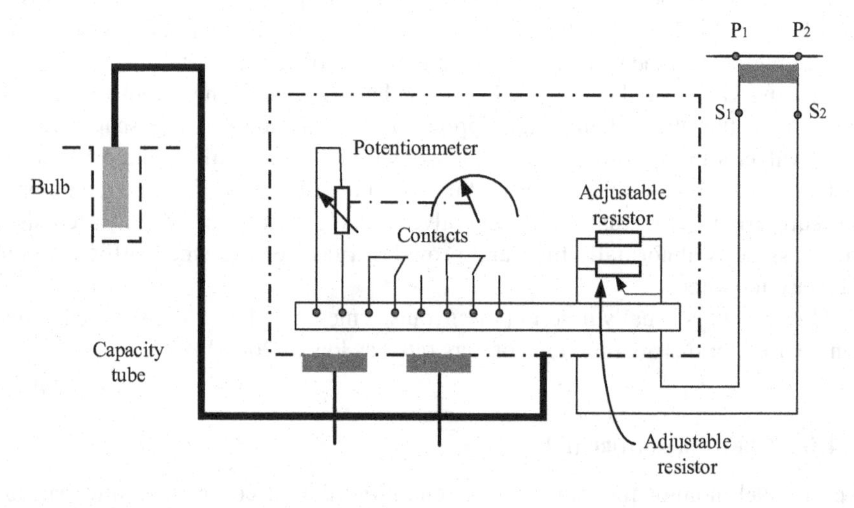

FIGURE 8.6 Capillary type of winding thermometer.

8.4.7 Winding Temperature (F49)

The winding thermometer shown in Figure 8.6 will respond to the overhead oil temperature and the load current's heating effect. And the thermometer winding will create an image of the toughest part of the coil. The overhead oil and its mantle temperature are measured by expanding the measurement further with a current signal proportional to the coil's load current.

This signal is taken from the current transformer inside the sleeve of this coil. This current will lead to the resistor component in the main unit. This resistor gets hot, and as a result of the current flowing through it, it heats the measuring bellows, which increases the pointer's movement. The temperature bias is proportional to the resistance of the electrical (resistor) heating element.

The result of the thermal operation provides data for adjusting resistance and hence temperature bias. The bias must correspond to the difference between the hot spot and overhead oil temperatures. The time constant for heating the pocket must match the time constant for heating the coil.

The temperature sensor then measures the temperature, equal to the coil's temperature, if the bias equals the temperature difference and the time constants are equal. With four contacts installed, the lower two levels are commonly used to start the fans or pumps for forced cooling, the third level to start the alarm, and the fourth step to trip the circuit breakers of the load or disconnect the power transformer, or both together. The latter usually controls forced cooling if the transducer is equipped with an upper oil thermometer and winding thermometer.

8.5 VOLTAGE BALANCE RELAY

In this type, the CTs at the two ends are connected in opposition. The relays are connected in series with the pilot wires. These types of connections are suitable for short-feeder protection. Figure 8.7 shows a schematic for a voltage balance relay.

8.6 TRANSFORMER MAGNETIZING IN-RUSH

The magnetizing flux currents in the power transformers are consequential in a sudden change in the magnetizing voltage. In addition to being usually considered the result of activating the transformer, the magnet impulse may also be due to the following:

1. As a result of an external error.
2. Changes like an external error.
3. Voltage recovery after removing an external malfunction.
4. Out-of-phase synchronization for a nearby generator.

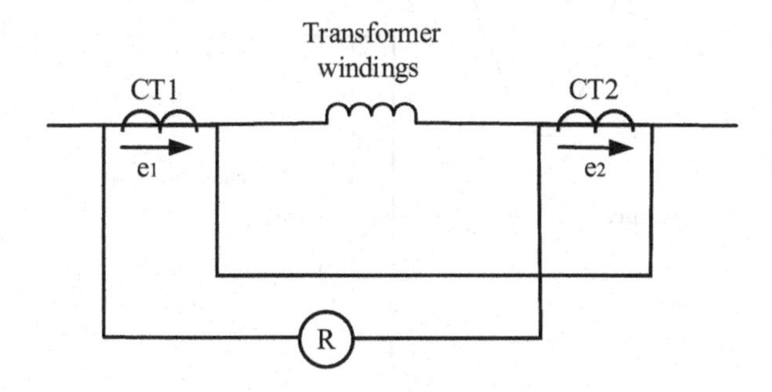

FIGURE 8.7 Voltage balance relay.

The magnetizing current disturbs the balance between the currents at the transformer terminals and is determined by the differential relay as a "wrong" differential current. However, the relay must remain constant during impulse conditions.

Given the transformer's life, tripping during the flow state is a very undesirable condition (breaking pure inductive load current generates high overvoltage, which may jeopardize transformer insulation and be the indirect cause of internal fault).

The magnetizing in-rush current cannot be easily detected; thus, this current characteristic is complex and difficult to analyze. It depends on many factors, such as connecting the supply to the power transformer, residual flux, the type of winding connection, core material, and the relation between current and voltage.

8.6.1 The Magnitude of Magnetizing In-Rush Current

Figure 8.8 shows a derivation of the in-rush current waveform from the saturation curve; when the transformer is initially energized, a transient in-rush current creates the transformer's magnetic field. Given that a transformer is energized from sinusoidal voltage, steady-state flux is an integral of the voltage shown as follows

$$\varphi = \frac{1}{N}\int \sin(\omega t)dt = -\frac{1}{\omega N}\cos(\omega t) \tag{8.8}$$

where N is the number of turns of the coil and is the angular frequency of the voltage signal, the instantaneous magnitude of the core flux at the moment of activation is defined as the residual flux $\varnothing_r$, and the amount of voltage-induced sinusoidal flux compensation depends on the voltage wave point at which the transformer is energized.

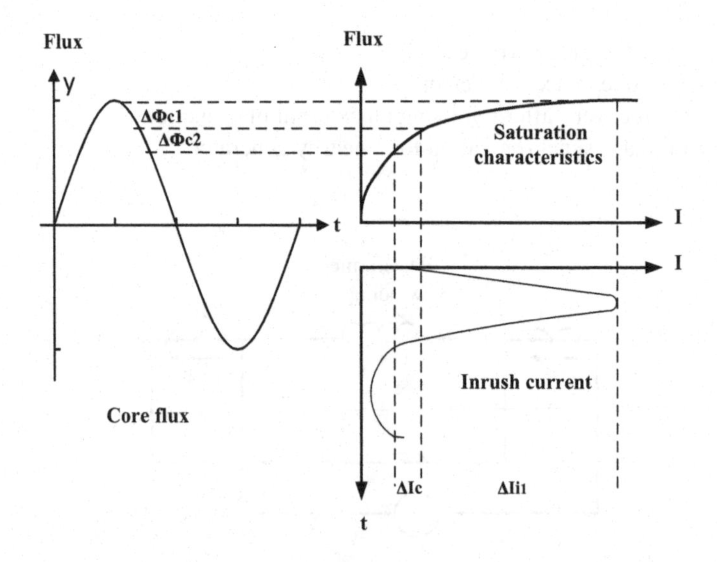

FIGURE 8.8 Derivation of the in-rush current waveform from the saturation curve.

Assuming that the peak normal core flux is called $\varnothing_m$, the peak core flux reaches two $\varnothing_m + \varnothing_r$, causing the core to saturate. The worst case is that the transformer is energized at the voltage wave's zero-crossing point with residual flux $\varnothing_m$. In this case, the saturation is greater, and a large in-rush current is induced.

Figure 8.8 shows the impulse formation at each point along with the fundamental flow wave (above the left diagram), it has a corresponding point on the saturation curve (upper right diagram, Figure 8.8), which is plotted against the flowing current (lower diagram, Figure 8.8 in the evolution of time).

The flux of the magnetic core is mapped to the flux current across the saturation curve. The impulse current is not a sine waveform because the saturation curve is not linear. The first half of the in-rush current cycle has a peak at the maximum flow, asymmetric with the second half of the cycle.

The flowing current is generated in the first half of the cycle, mostly by the current flow path above the saturation curve's knee point. In contrast, most of the second half of the current cycle is determined from the path below the knee point; the flux current is deformed due to magnetic saturation. Moreover, as shown in Figure 8.8, two equal increases of core flux in a sine wave, $\Delta\Phi_{c1}$, and $\Delta\Phi_{c2}$ generate significantly different height increases in the flux current, such as ΔI_{i1} and ΔI_{i2}, respectively.

At fault currents, the increase has the same height due to the linear designation. This is the main reason for the difference between impulse and fault currents in the characteristics of their changes in slope and symmetry. However, these features can be extracted by the MM-based charts.

This is done by giving a typical flow current waveform in Figure 8.9. A large, long-acting DC component can be observed, which has large peak values at the beginning and decays significantly after a few tenths of a second but only completely decays after several seconds. The shape, size, and duration of the flow rate depend on several factors, including:

i. The size of the transformer.
ii. Magnetic properties and resonance of the core.
iii. The impedance of the energizing system.
iv. The Phase angle when the transformer is *switched on*.

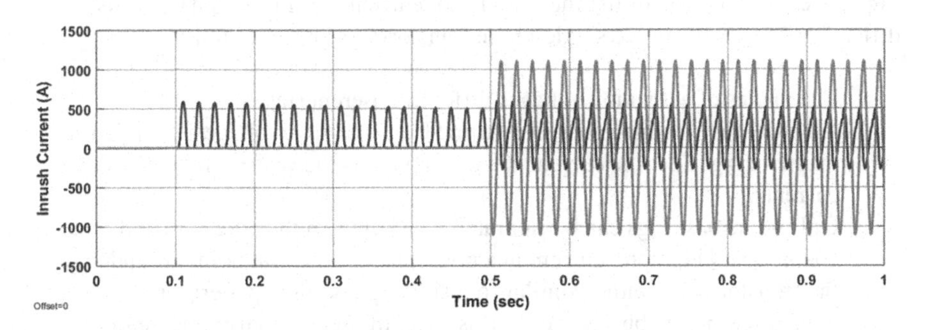

FIGURE 8.9 A typical in-rush current waveform.

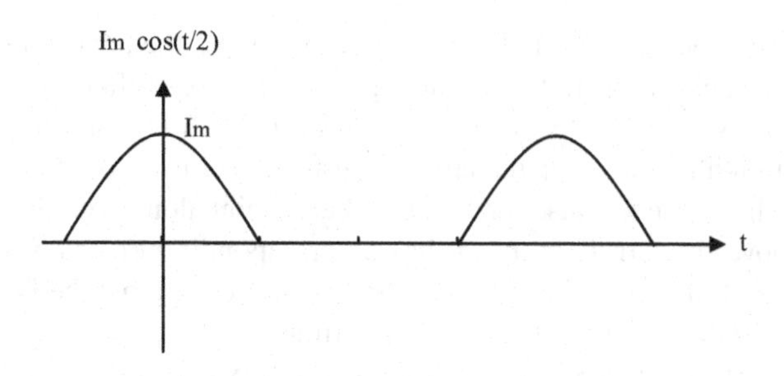

FIGURE 8.10 An idealized in-rush current waveform for spectral analysis.

8.6.2 Harmonics of Magnetizing In-Rush Current

To obtain the frequency spectrum of the impulse waveform in the single-phase transformer, an analytical approximation of the impulse waveform is given in Figure 8.10; the waveform between α and -α is the impulse current due to the saturation of the transformer air core between α and $2\pi - \alpha$ the discontinuous angle of the waveform, known as the gap. The angle α is used to facilitate modeling the actual flux current. The harmonic amplitude of n is calculated as follows

$$A_n = \frac{I_m}{\pi}\left[\frac{1}{n+1}\sin\{(n+1)\alpha\} + \frac{1}{n-1}\sin\{(n-1)\alpha\} - 2\cos\left(\frac{\alpha}{2}\right)\sin(n\alpha)\right] \quad (8.9)$$

where I_m is the peak value of the burst current. Figure 8.10 shows an idealized in-rush current waveform for spectral analysis. The second harmonic always predominates due to the large DC offset. However, the amount of the second harmonic may drop below 20%.

The second harmonic's minimum content mainly depends on the core's magnetic property's knee point. The lower the saturation flux density, the greater the amount of the second harmonic. Modern transformers built with enhanced magnetic materials have high knee points; therefore, their in-rush currents display a relatively low amount of the second harmonic.

Some difficulties arise when protecting transformers due to the second harmonic, which is the primary limiting criterion for installing differential relays during impulse conditions where the measured flux currents of a three-phase transformer differ significantly from the single-phase transformers above, as follows:

1. The angles of activation voltages differ in different stages.
2. When a delta-connected coil is turned on, the line voltage is applied as magnetizing. The line current at a given point is the vector sum of the other currents.
3. Only some base legs can be saturated depending on the base type and other conditions. Therefore, current in a given phase or neutral point grounded to the transformer is either similar to a single-phase flow pattern or becomes a distorted but wobbly wave. In this case, the second harmonic amount is greatly reduced, which causes problems with a differential relay.

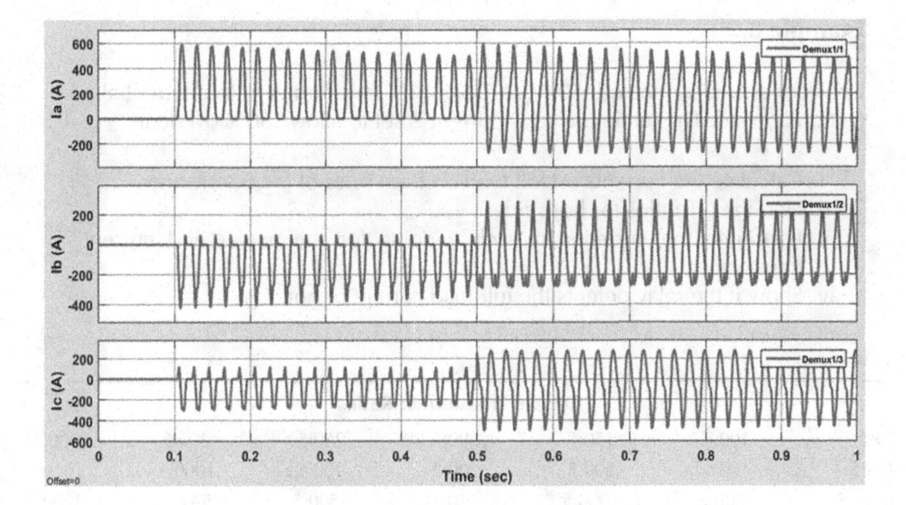

FIGURE 8.11 The in-rush currents in a three-phase transformer.

Figure 8.11 shows the waveforms for activating a three-phase transformer, the currents in phases A and B showing a typical corrected impulse pattern. In contrast, phase C is an oscillating waveform due to the displacement of the greatly diminishing direct current in the large volume, the flowing current.

The waveform is likely to be distorted due to the saturation of the CT.

In this case, the CT secondary current shows a certain distortion level, with a significantly reduced amount of the second harmonic, and in the most adverse conditions, the gap disappears.

Example 8.1

A 10 MVA, 132/6.6 kV, delta/star power transformer, what should be the CT'S ratio on both sides, and how should they be connected for a differential protection scheme to circulate a current of 5 A in the pilot wire?

Solution

The CTs on the HT Δ winding should be connected in Y, and this LTY winding should be connected in Δ

$$\text{The line current on H.T side} = \frac{10\times10^6}{\sqrt{3}\times132\times10^3} = 43.73\,\text{A}$$

CTs on the HT side ratio is 43.73 : 5 A

$$\text{The line current on L.T side} = \frac{10\times10^6}{\sqrt{3}\times6.6\times10^3} = 874.75\,\text{A}$$

CTs on the LT side ratio is $874.75 : \frac{5}{\sqrt{3}}$ A

Example 8.2

Consider a Δ/Y connected, 50 MVA, 33/132 kV transformer with differential protection applied. The minimum relay current is set to allow 150% overload.

i. Calculate the currents on full load on both sides of the transformer.
ii. Find the CT ratios for both sides.
iii. Distribute the currents in each winding for the power and current transformer.
iv. Show if the relay detects the full-load current or not.

Current Transformer Ratios

50:5	100:5	150:5	200:5	250:5	300:5	400:5
450:5	500:5	600:5	800:5	900:5	1000:5	1200:5
1500:5	1600:5	2000:5	240:5	2500:5	3000:5	3200:5
4000:5	5000:5	6000:5				

Solution

i. Calculate the relay currents on full load

$$I = \frac{S}{\sqrt{3}\,\mathrm{V}}$$

Δ-Side:

$$I_\Delta = \frac{50\times 10^6}{\sqrt{3}\times 33\times 10^3} = 874.77\ \mathrm{A}$$

Y-Side:

$$I_Y = \frac{50\times 10^6}{\sqrt{3}\times 132\times 10^3} = 218.6\ \mathrm{A}$$

ii. The nearest CT ratio on the *Y*-side at a load current of

$$I_\Delta = 874.77\ \mathrm{A} \quad \mathrm{is}\, 900/5\,\mathrm{A}.$$

iii. The nearest CT ratio in the Δ side at a load current of

$$I_Y = 218.6\,\mathrm{A} \quad \mathrm{is} \quad 250/5\,\mathrm{A}.$$

iv. The output current of CTs at full load:

$$\text{At } \Delta\text{-side of the power transformer} = 874.77 \times \frac{5}{900} = 4.86\,\text{A}$$

$$\text{At } Y\text{-Side of the power transformer} = 218.6 \times \frac{5}{250} \times \sqrt{3} = 7.573\,\text{A}$$

$$\text{Relay current at full load} = |7.573 - 4.86| = 2.71\,\text{A}$$

v. Minimum relay current setting to permit 50% overload

$$= 150\% \times 2.71 = 4.065\,\text{A}$$

vi. With the secondary current of 5 A and 150% CTs, the overload current will be 7.5 A, and since the overload current of 4.065 is <7.5 A, the relay will not operate.

Figure 8.12 shows the wiring diagram configuration.

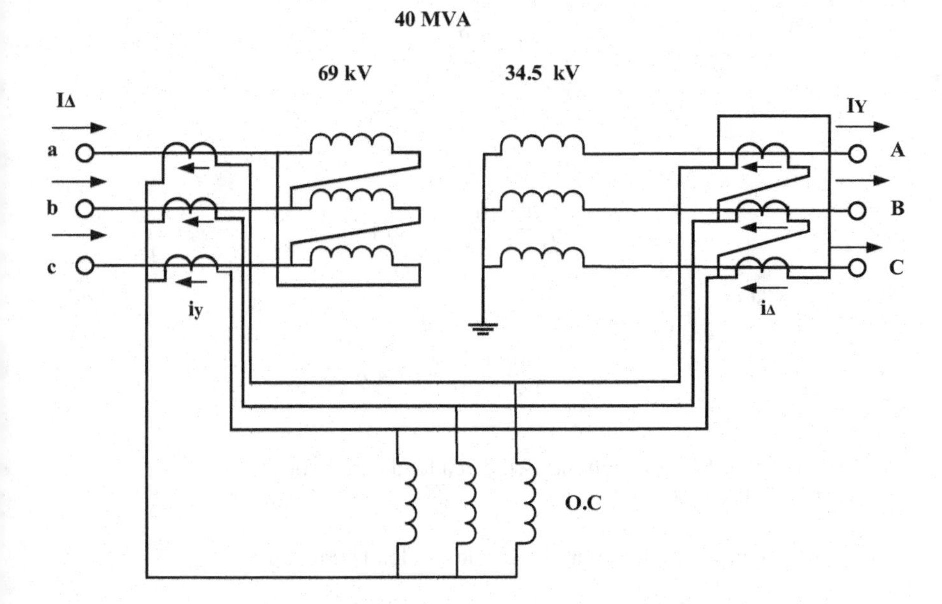

FIGURE 8.12 Wiring diagram of Example 8.2.

Example 8.3

Consider a Y/Δ connected, 30 MVA, 11/33 kV transformer with differential protection applied. The minimum relay current is set to allow 125% overload.

i. Calculate the relay currents on full load.
ii. Find the CT ratios for both sides.
iii. Find the pilot wires' current under load conditions and operating coil current.
iv. If the *line-to-line fault* of 2000 A occurs on the 33 sides between lines B and C out of the protective zone, distribute the currents in each winding for power and current transformer.
v. Show if the relay detects the fault.

Current Transformer Ratios

50:5	100:5	150:5	200:5	250:5	300:5	400:5
450:5	500:5	600:5	800:5	900:5	1000:5	1200:5
1500:5	1600:5	2000:5	2400:5	2500:5	3000:5	3200:5
4000:5	5000:5	6000:5				

Solution

Solution

i.

$$I = \frac{S}{\sqrt{3}\ V}$$

Y-Side:

$$I_Y = \frac{30\times10^6}{\sqrt{3}\times33\times10^3} = 1574\ \text{A}$$

Δ-Side:

$$I_\Delta = \frac{30\times10^6}{\sqrt{3}\times33\times10^3} = 524\ \text{A}$$

ii. The nearest CT ratio on the Y-side at a load current of
iii. $I_\Delta = 524$ A is 600/5 A.

The nearest CT ratio in the Δ side at a load current of

$$I_Y = 1574\,\text{A is}\,1600/5\,\text{A}.$$

iv. The pilot wires current (relay side)

$$i_\Delta = 1574 \times \frac{5}{1600} = 4.92\,\text{A}$$

$$i_1 = \sqrt{3} \times i_\Delta = \sqrt{3} \times 4.92 = 8.52\,\text{A}$$

$$i_y = 524 \times \frac{5}{600} = 4.366\,\text{A} = i_2$$

Operating current

$$i_{\text{o.c}} = |i_1 - i_2| = |8.52 - 4.366| = 4.15\,\text{A}$$

Figure 8.13 shows the distribution load current.

v. During the out of the protective zone, the distribution currents are given in Figure 8.14

$$i_\Delta = 6000 \times \frac{5}{1600} = 18.75\,\text{A}$$

$$i_y = 2000 \times \frac{5}{600} = 16.66\,\text{A}$$

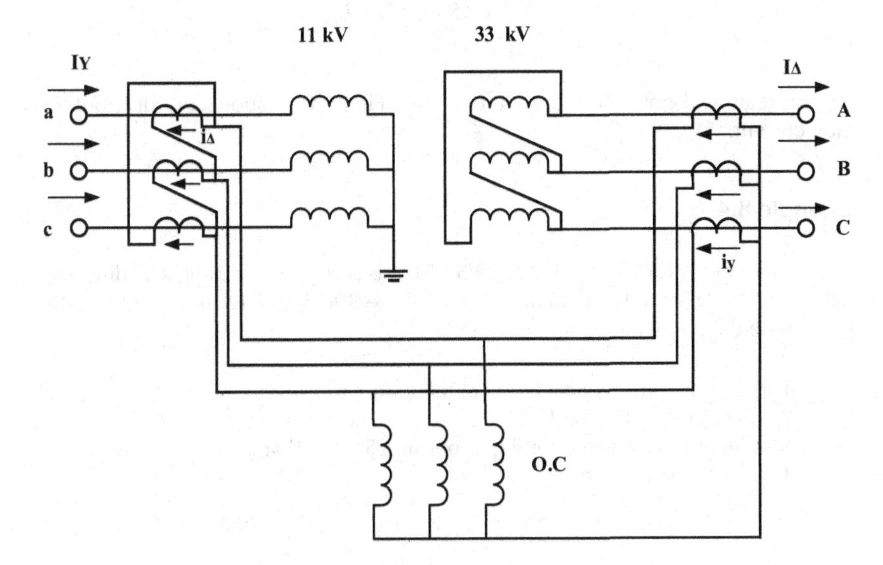

FIGURE 8.13 The distribution load current of Example 8.3.

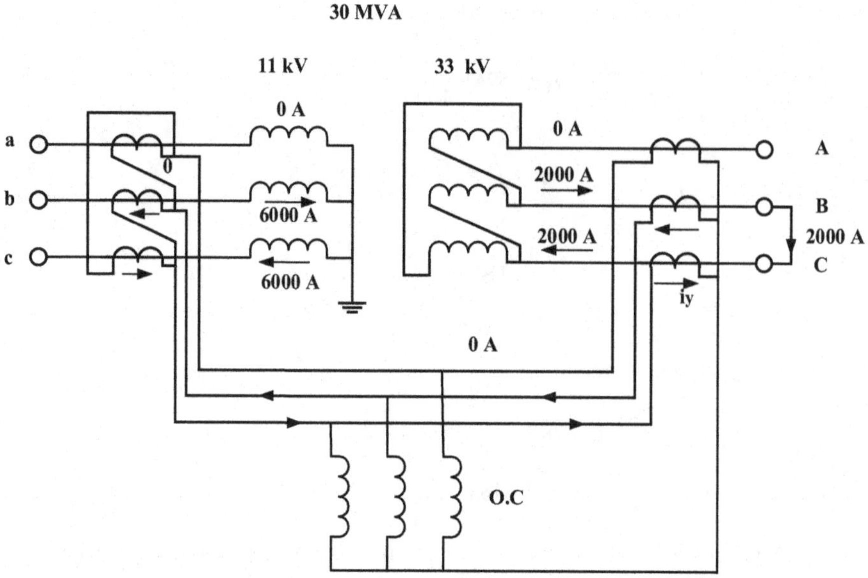

FIGURE 8.14 The distribution fault current of Example 8.3.

vi. Operating current

$$i_{o.c} = |i_1 - i_2| = |18.75 - 16.66| = 2.08\,\text{A}$$

$$i_s = 125\% \times 5 = 7.5\,\text{A} > i_{o.c}$$

So, the relay will not detect the fault current. Figure 8.14 shows the distribution fault current.

Example 8.4

A Δ/Y power transformer, 40 MVA, 69/34.5 kV, is protected by using a % differential relay. The CTs on the delta and star sides are 300/5 and 1200/5, respectively. Determine the following

i. The output current of both CTs at full load.
ii. Relay current at full load.
iii. Minimum relay current setting to permit 25% overload.
iv. The auxiliary turn ratio.

Solution

i. The output current of both CTs at full load.

$$I = \frac{S}{\sqrt{3}\ V}$$

Δ-Side:

$$I_\Delta = \frac{40 \times 10^6}{\sqrt{3} \times 69 \times 10^3} = 334.7\ \text{A}$$

Y-Side:

$$I_Y = \frac{40 \times 10^6}{\sqrt{3} \times 34.5 \times 10^3} = 669.4\ \text{A}$$

So, the output current of CTs at full load:

$$\text{At } \Delta\text{-Side of the power transformer} = 334.7 \times \frac{5}{300} = 5.578\ \text{A}$$

$$\text{At } Y\text{-side of the power transformer} = 669.4 \times \frac{5}{1200} \times \sqrt{3} = 4.83\ \text{A}$$

ii.

$$\text{Relay current at full load} = |5.578 - 4.83| = 0.748\ \text{A}.$$

iii.

$$\text{Minmum relay current setting to permit 25\% overload}$$

$$= 125\% \times 0.748 = 0.935\ \text{A}$$

iv.

$$\text{The auxiliary CT turn ratio} = \frac{5.578}{4.83} = 1.154.$$

Figure 8.15 shows a wiring diagram configuration.

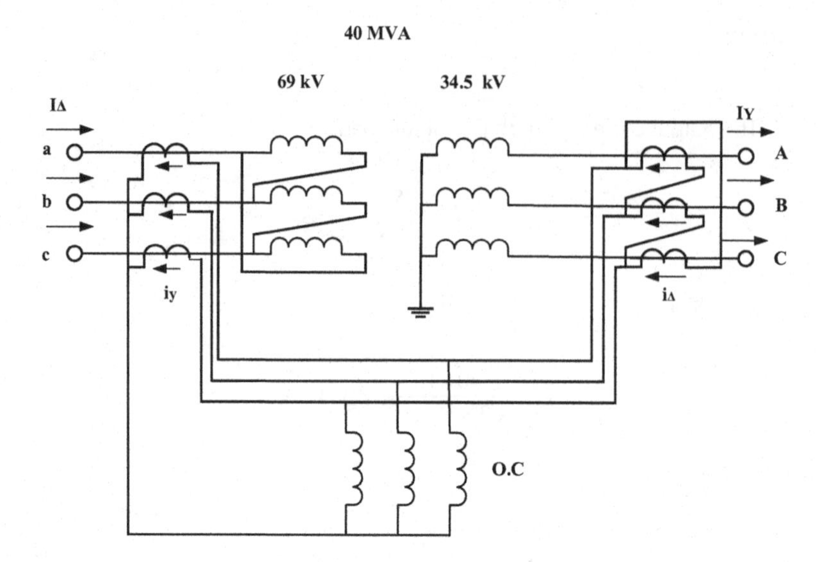

FIGURE 8.15 The distribution fault current of Example 8.4.

8.7 MODELING OF POWER TRANSFORMER DIFFERENTIAL PROTECTION

This section presents the Simulink models that have been developed for the simulation of the power system. The power transformer's differential relay simulates both internal and external faults, as well as the in-rush current that occurs when the transformer is energized.

8.7.1 Differential Protection Difficulties

Three major difficulties usually handicap the traditional differential defense. They allow a false trip signal to be emitted by the differential relay without any fault being present. To make the differential relay run properly, these problems must be solved.

During initial energization,

i. In-rush present,
ii. Mismatch and saturation of CTs,
iii. Due to the tap changer, the transformation ratio varies.

8.7.1.1 In-rush Current During Initial Energization

On the transformer's primary side, the transient magnetizing in-rush or the thrilling current occurs if the transformer is turned on (energized) and the instantaneous voltage value is not 900. The first peak of the flux wave is higher than the peak of the steady-state flux. This current occurs as an internal fault, and the differential relay detects it as a differential current.

The magnetizing current's first peak value can be as high as many times that of the maximum load current's peak. Many factors affect the magnitude and length of the magnetizing in-rush current, some of which are: The instantaneous value of the voltage waveform at the moment of closing CB.

i. The value of the residual (remnant) magnetizing flux.
ii. The sign of the residual magnetizing flux.
iii. The type of iron laminations used in the transformer core.
iv. The saturation flux density of the transformer core.
v. The total impedance of the supply circuit.
vi. The physical size of the transformer.
vii. The maximum flux-carrying capability of the iron core laminations.
viii. The input supply voltage level.

The consequence of the in-rush current on the differential relay is that the transformer is triggered incorrectly without any established type of fault. According to the differential relay principle, the relay contrasts the currents from both sides of the power transformer.

The in-rush current is, however, still flowing on the primary side of the power transformer. So, the differential current would have a large value because of the current on only one hand. Therefore, the relay must be built to understand that this current is a natural phenomenon and not to move because of this current.

8.7.1.2 False Trip Due to CT Characteristics

In reproducing their primary currents on their secondary side, the differential relays' efficiency depends on the CTs' accuracy. The CTs' primary ratings, located on the power transformer's high- and low-voltage sides, do not exactly fit the rated currents of the power transformer in many situations. Because of this disparity, a mismatch of CTs occurs, which generates a slight false differential present depending on the amount of this mismatch.

This quantity of the differential current is often sufficient to operate the differential relay. Therefore, correction of the CTs' ratio must be performed to solve CTs' mismatch using multi-tap interposing CTs.

The saturation issue is another problem that the perfect operation of CTs could face. False differential current occurs in the differential relay when saturation happens to one or more CTs at different stages. This differential current may cause the differential relay to mal-operate. The worst case of CT saturation could generate the DC portion of the primary side current, in which DC offset and extra harmonics are present in the secondary current.

8.7.1.3 False Trip Due to Tap Changer

The on-load tap-changer (OLTC) is mounted on the power transformer to automatically control the transformer's output voltage. Wherever there are strong fluctuations in the voltage of the power system, this device is necessary. Just one point of the tap-changing spectrum can be contrasted with the transformation ratio of the CTs. Therefore, if the OLTC is altered, the differential relay operating coil disbalances current flows. This interference activates CT discrepancies. This current will be considered a fault current that enables a trip signal emitted by the relay.

8.8 PERCENTAGE DIFFERENTIAL RELAY MODELING

The model of percentage differential relay, as shown in Figure 8.16, is designed to protect of power transformer of 132/33 kV, 63MVA, for different faults in inter-zone or outside of zone according to the following steps:

Step 1: The relay feed from the current measurement from both sides of the power transformer instead of the current transformer because there isn't a current model.

The behavior of current measurement is similar to the current transformer after adding two functions in our model, the first function multiplies the output current of current measurement by the gain (*k* factor, which represents CTs ratio for each side) to transform the primary current to secondary current, and the second function using delay function to compensate phase shift between the primary and secondary of the power transformer instead of using proper CT connections in case of the power transformer vector group (Y-D).

The traditional way to provide compensation is to connect a wye CT for a delta winding and a delta CT for a wye winding. This connection compensates for the delta–wye phase shift of the power transformer vector group (Y-D1) or (Y-D11).

Step 2: Figure 8.17 shows the differential relay (discrete model). The relay work according to equation ($I_d \geq K\ I_r$), where I_d represents the differential current (measured current) equal to subtract current on both sides of the transformer ($I_d = I_1 - I_2$), which compares with a multiply ($K\ I_r$), where *k* represents the minimum slope of the singular curve of a differential relay, which was chosen dependent on the different factors related to mismatching current transformers ratio on both sides of the power transformer, second

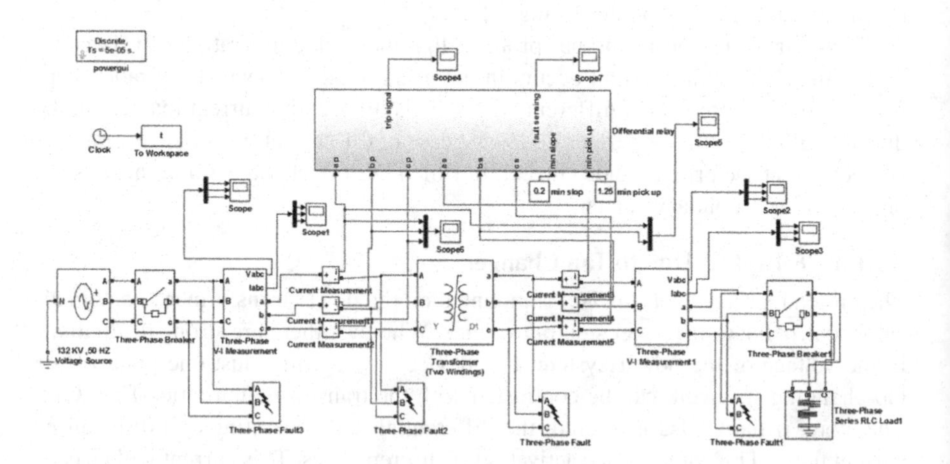

FIGURE 8.16 Differential relay protection of transformer using MATLAB/Simulink.

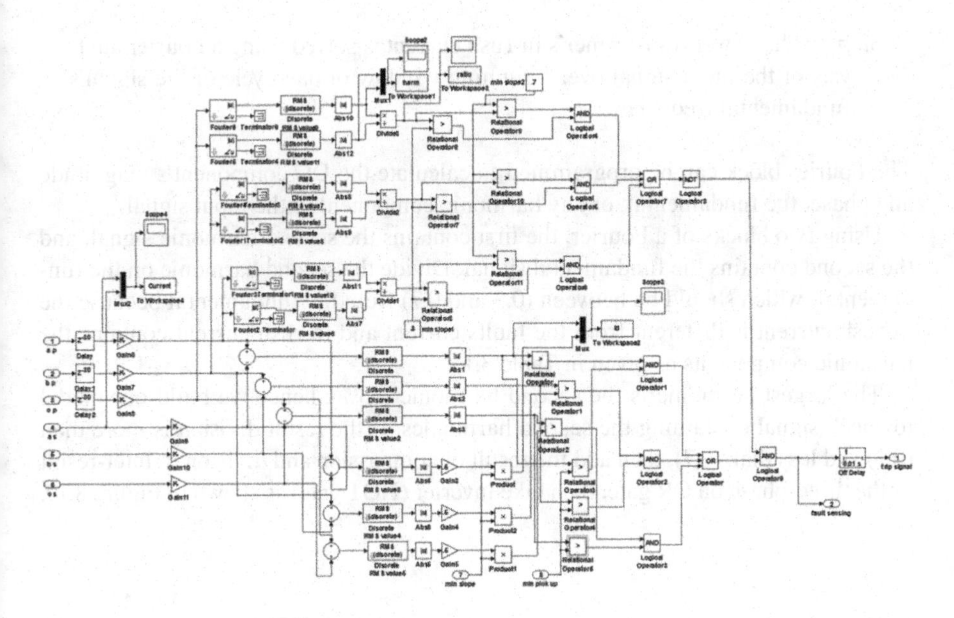

FIGURE 8.17 Differential relay (discrete model).

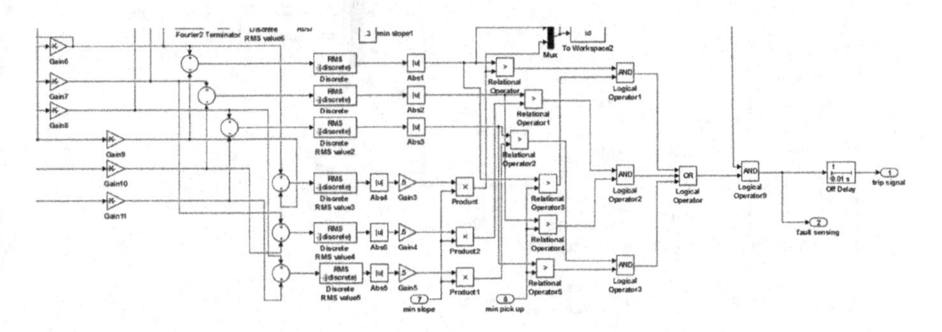

FIGURE 8.18 Amplitude part of a differential relay (discrete model).

using the on-load tap changer in the power transformer, and third unequal error in current transformers.

The restraint current ($I_r = (I_1 + I_2)/2$) represents the average current of both sides of the power transformer. According to the differential relay curve, the following conditions should be satisfied to obtain a trip.

Condition 1: The differential current Id (measured current) is greater than the restraint current ($I_d > K\,I_r$).

Condition 2: The differential current *Id* (measured current) is greater than the minimum pick up ($I_d > I_P$) where the minimum pick up value was chosen from values between (0.2 *and* $0.4 \times I_N$).

The circuit represents the amplitude part shown in Figure 8.18.

Step 3: The power transformer's in-rush current is solved using a Fourier analysis of the input signal over a running window of one cycle of the signal's fundamental frequency.

The Fourier block can be programmed to calculate the DC component's magnitude and phase, the fundamental, or any harmonic component of the input signal.

Using two blocks of a Fourier, the first contains the second harmonic signal, and the second contains the fundamental signal. Divide the second harmonic on the fundamental, which should be between (0.3 and 0.7) from the fundamental because the in-rush current is different from the faults current and in-rush current contains the harmonic components as given in Table 3.1.

The largest harmonic is the second harmonic (63%); hence we build our model to check signal containing the second harmonics. If the result division is more than (0.3) and less than (0.7), then add the result of compassion and division to inter-result of the three-phase on OR gate, then take inverter (NOT gate) as shown in Figure 8.19.

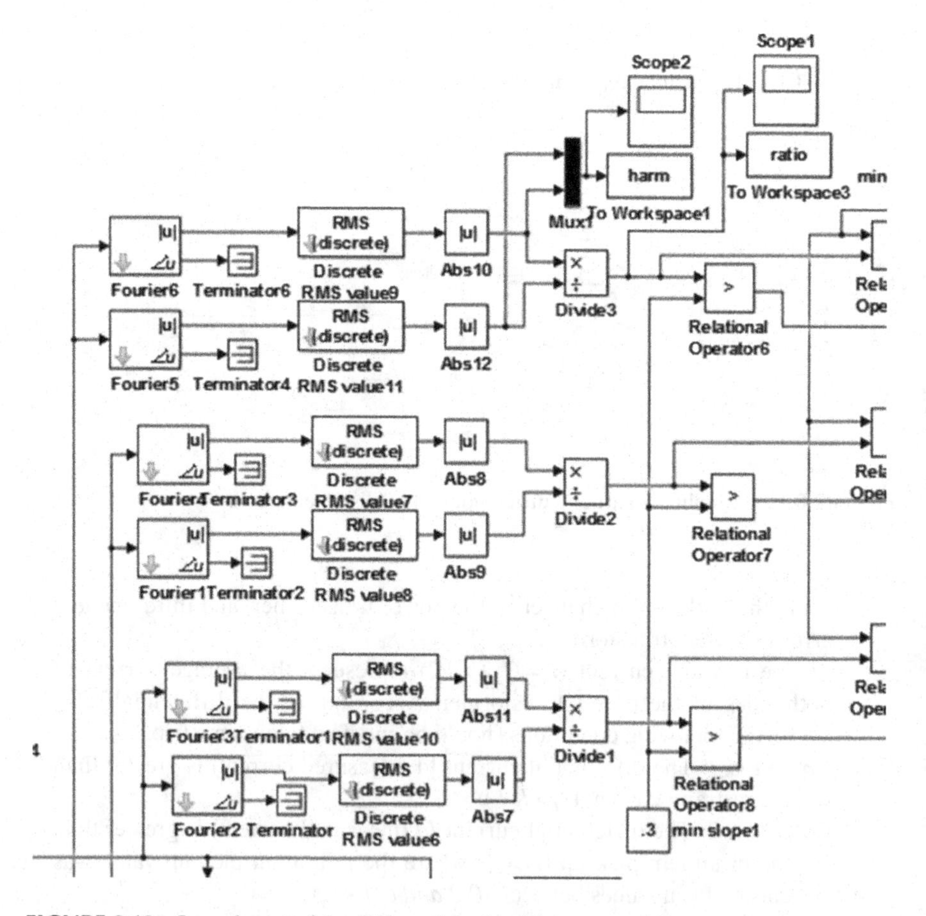

FIGURE 8.19 In-rush part of the differential relay (discrete model).

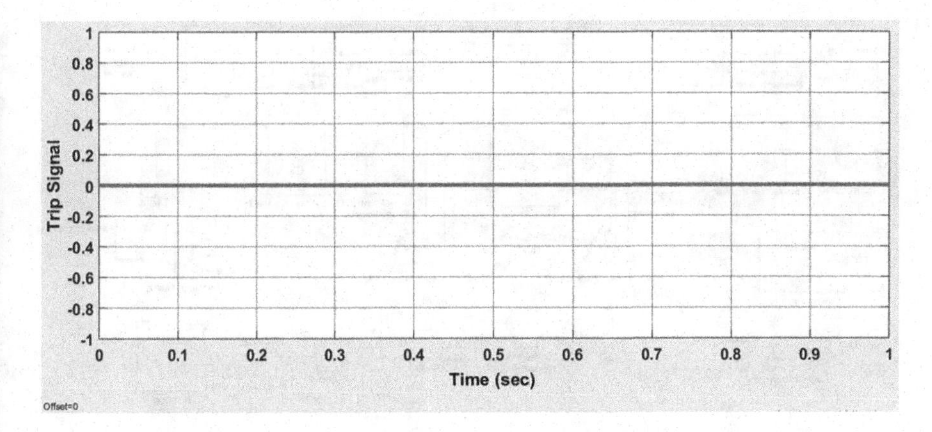

FIGURE 8.20 The trip signal in case differential current in-rush current.

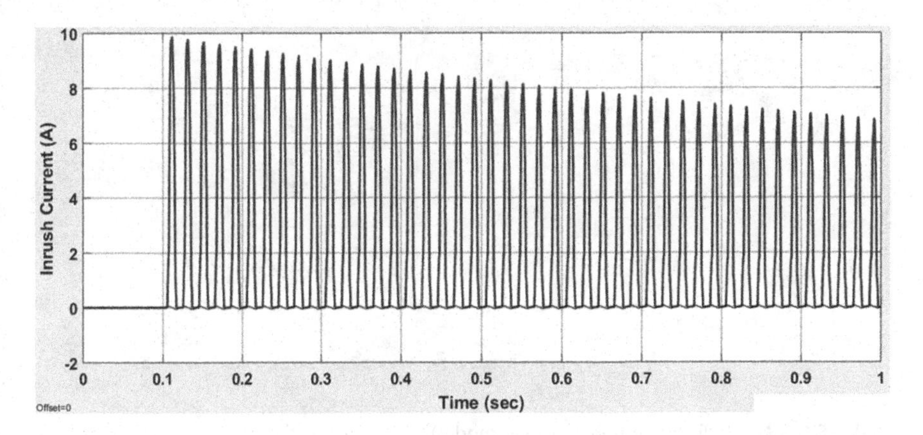

FIGURE 8.21 The differential current in-rush current.

Figure 8.19 shows the in-rush part of the differential relay (discrete model). Figure 8.20 shows a trip signal in case of differential current in-rush current, while Figure 8.21 shows the differential current in-rush current.

8.9 PHASOR MODEL

Applying the phasor simulation method to a simple linear circuit is interested only in the changes in magnitude and phase of all voltages and currents when switches are closed or opened, as shown in Figure 8.22. Figure 8.23 shows a differential relay (phasor model).

Step 1: The relay feed from *V–I* measurement from both sides of the power transformer instead of the current transformer because there isn't a current model.

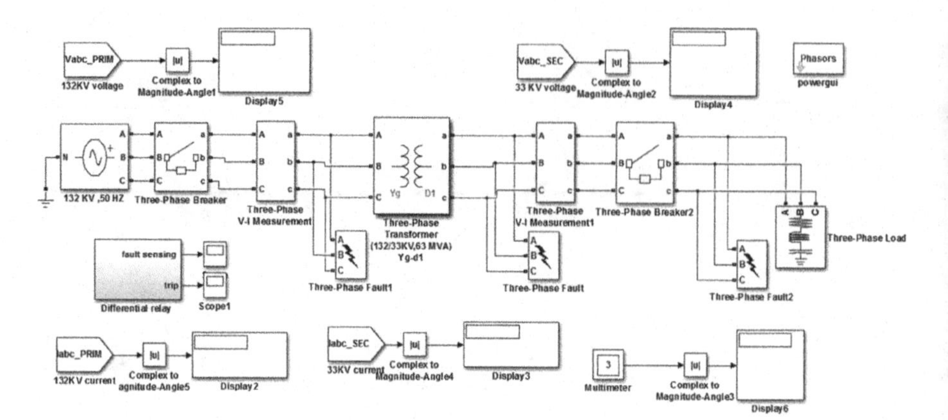

FIGURE 8.22 Phasor model of the differential relay.

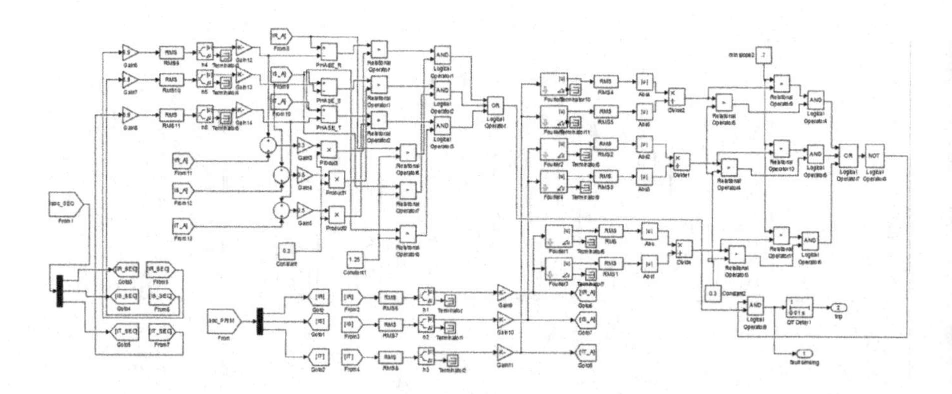

FIGURE 8.23 Differential relay (phasor model).

The behavior of current measurement is similar to the current transformer after multiplying the output current of *V–I* measurement by the gain (*k* factor, which represents CTs ratio for each side) to transform the primary current to the secondary current. *From block,* receive data (current) by tagging the required current side and send without wiring to *Goto block* by tagging the required side, then take RMS and the absolute value of each side.

Step 2: The relay work according to equation ($I_d \geq K\ I_r$), where I_d represents the differential current (measured current) equal to subtract current in both sides of the transformer ($I_d = I_1 - I_2$), which compares with a multiplied ($K\ I_r$), where *k* represents the minimum slope of the differential relay's singular curve, which was chosen dependent on the different factors related to the first mismatching current transformers ratio on both sides of the power transformer, second using on-load tap changer in power transformer, and third unequal error in current transformers.

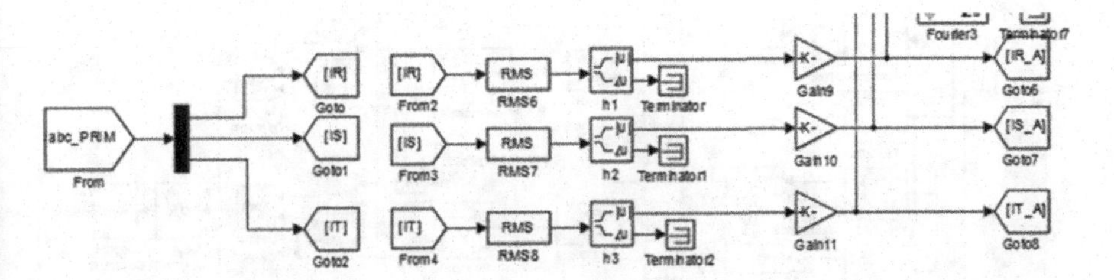

FIGURE 8.24 Amplitude part of the differential relay (phasor model).

The restraint current ($I_r = (I_1 + I_2)/2$) represents the average current of both sides of the power transformer. According to the differential relay curve, the following conditions should be satisfied to obtain a trip.

Condition 1: The differential current I_d (measured current) is greater than the restraint current ($I_d > K\, I_r$).

Condition 2: The differential current I_d (measured current) is greater than the minimum pick up ($I_d > I_P$), where the minimum pick up value was chosen from values between (0.2 *and* $0.4 \times I_N$).

Step 3: The power transformer's in-rush current is solved using a Fourier analysis of the input signal over a running window of one cycle of the signal's fundamental frequency.

The Fourier block can be programmed to calculate the DC component's magnitude and phase, the fundamental, or any harmonic component of the input signal.

Use two blocks of a Fourier, and the first contains the second harmonic signal. The second contains the fundamental signal by dividing the second harmonic on the fundamental, which should be between (0.3 and 0.7) from the fundamental because the in-rush current is different from the faults current that in-rush current contains the harmonic components.

The largest harmonic is the second harmonic (63%); hence we build our model to check the signal if it contains the second harmonics. If the result division is more than (0.3) and less than (0.7), then add the result of compassion and division to inter-result of the three-phase on OR gate, then take inverter (NOT gate) as shown in Figure 8.24.

The harmonic comparator shows in Figure 8.25 that the value of the second harmonic is higher than 0.3 of the fundamental component.

8.10 THREE-PHASE TO GROUND FAULT AT THE LOADED TRANSFORMER

In this case, the algorithm's security is checked by generating a three-phase ground fault. After switching CB1 to 0.1 seconds, by connecting the three phases, A, B, and C of the power transformer's secondary side to the field, an internal fault is generated

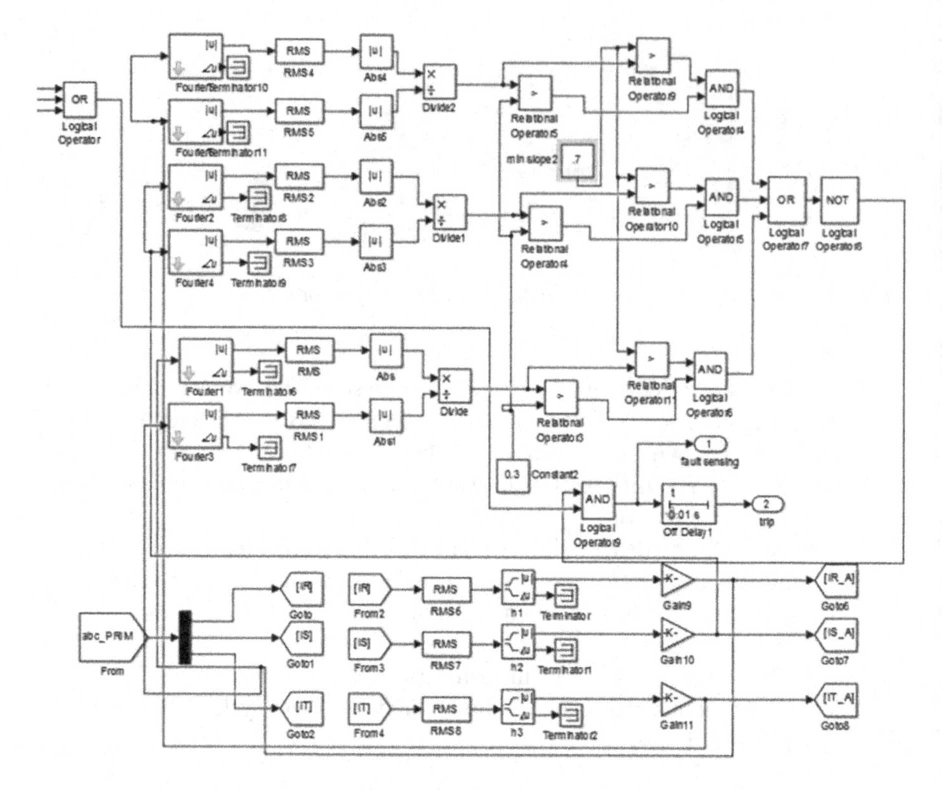

FIGURE 8.25 Inrush part of the differential relay (phasor model).

at 0.5 seconds on the power transformer's secondary side. In this scenario, a substantial increase in the primary current occurs due to faults within the safe zone at 0.5 seconds.

Using the harmonic and amplitude comparators, the relay observed this rise and identified it as an internal fault. The transformer is, thus, disconnected from the grid. It is also evident from Figure 8.26 that a trip signal after 0.57 ms was emitted by the relay. After the fault, which can be regarded as a very good transformer isolation speed, the relay released a trip signal after 0.57 ms. After the fault, which can be considered a very good speed to isolate the transformer. Figure 8.27 shows a three-phase current of the out-zone fault at 0.5 seconds for a loaded transformer, and Figure 8.28 shows a three-phase current of the protective zone for a loaded transformer. Figure 8.29 shows a three-phase fault in the protective zone for the loaded transformer. A three-phase fault in the protective zone for the loaded transformer is given in Figure 8.30. The single phase to the protective zone's ground fault for a loaded transformer is given in Figure 8.31. Finally, Figure 8.32 shows a line-to-line fault in the protective zone for the loaded transformer.

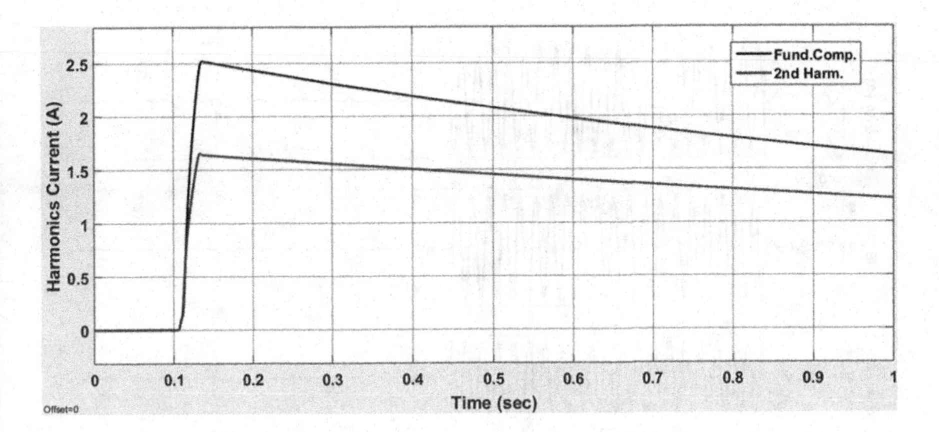

FIGURE 8.26 Relay trip signal.

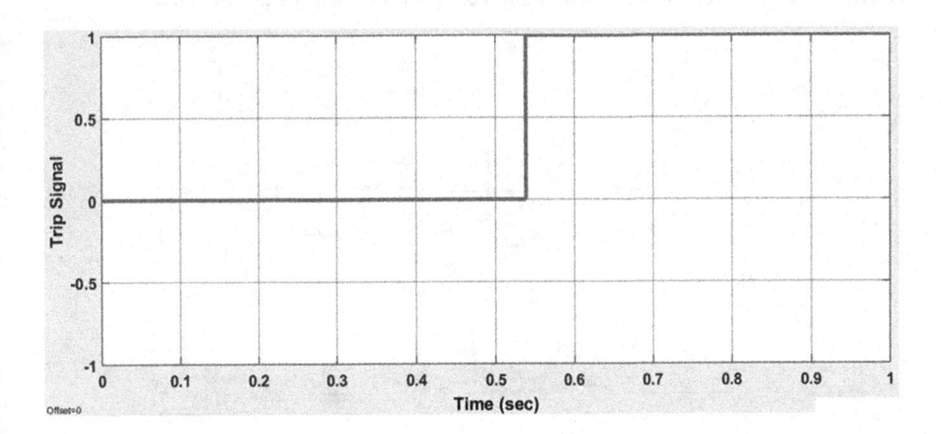

FIGURE 8.27 Three-phase current of out-zone fault at 0.5 seconds for a loaded transformer.

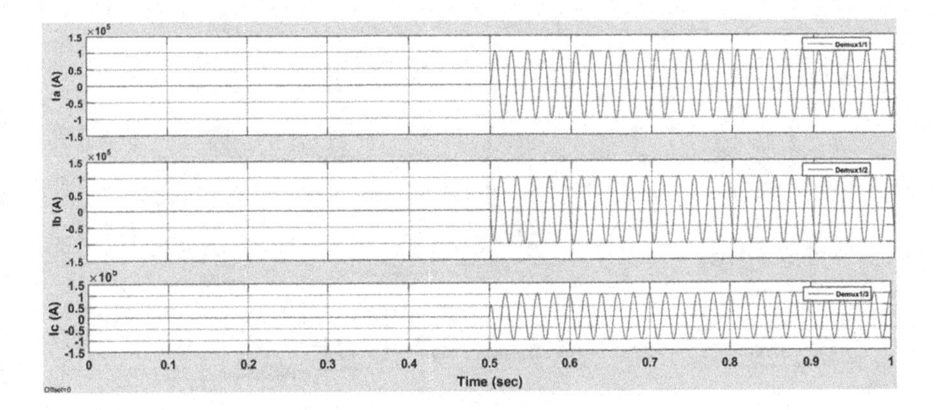

FIGURE 8.28 Three-phase current of protective zone for a loaded transformer.

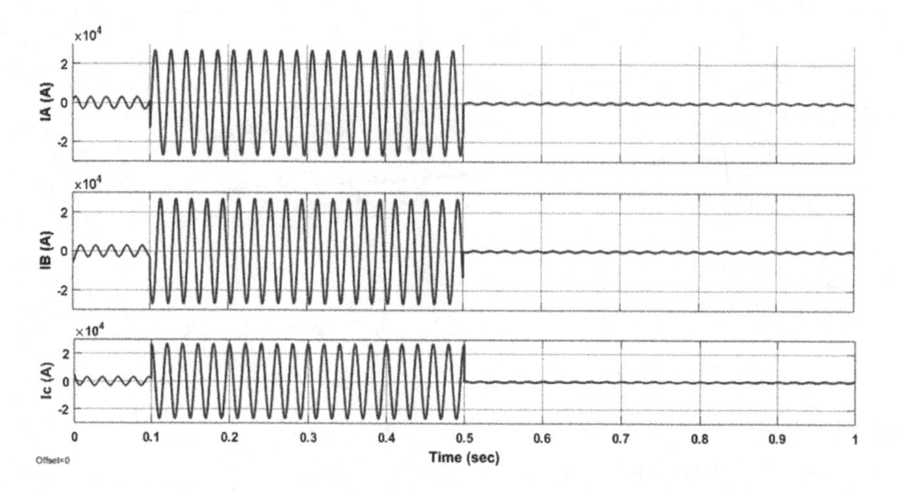

FIGURE 8.29 Three-phase current of protective zone for a loaded transformer.

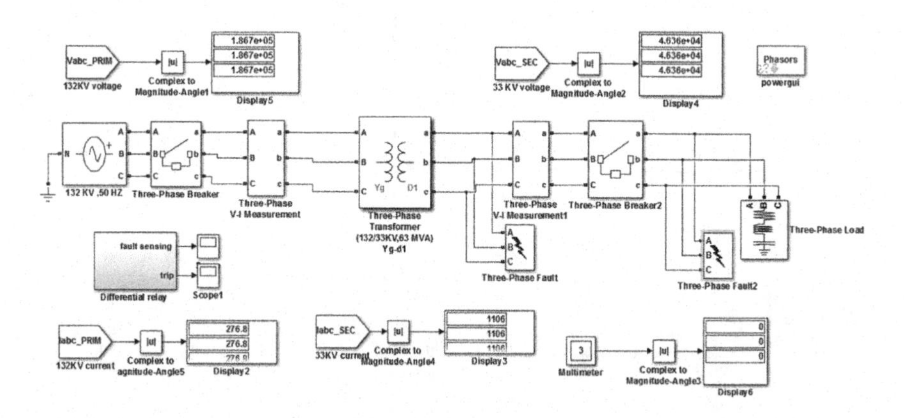

FIGURE 8.30 Three-phase fault of protective zone for a loaded transformer.

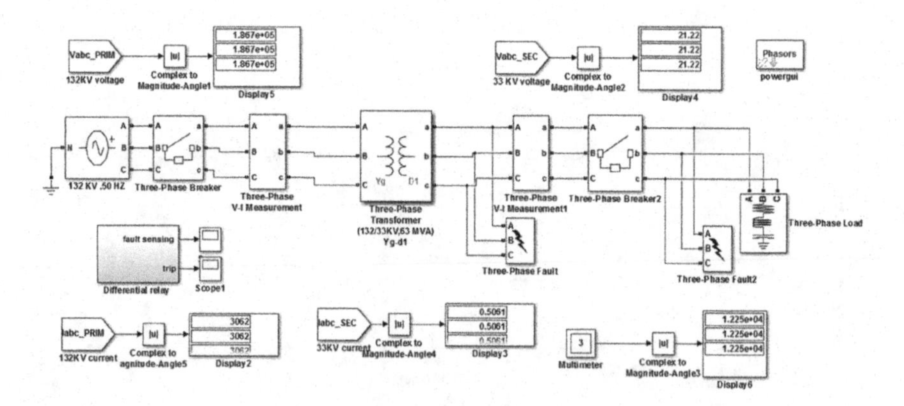

FIGURE 8.31 Three-phase fault of protective zone for a loaded transformer.

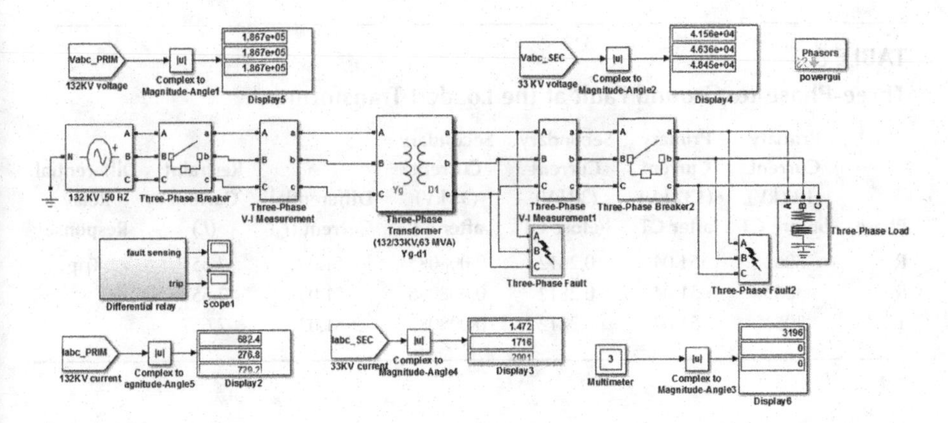

FIGURE 8.32 Single-phase to the ground fault of protective zone for a loaded transformer.

8.11 MAGNETIZING IN-RUSH CURRENT

Only the in-rush current flows through the power transformer's primary circuit in this simulation scenario. No current passes through the power transformer to the secondary side when the primary side CB1 is closed at 0.1 seconds.

In this case, logic (0) was released by the harmonic calculation section. The amplitude comparator that the differential current equals the in-rush current is released by the amplitude comparator logic (1). No trip signal is released for this logic coordination (1, 0). The result is given in Table 8.2.

8.12 THREE PHASES TO GROUND FAULT AT THE LOADED TRANSFORMER

In this case, the algorithm's security is checked by generating a three-phase ground fault. After switching CB1 to 0.1 seconds, by connecting the three phases, A, B, and C of the power transformer's secondary side to the field, an internal fault is generated at 0.5 seconds on the power transformer's secondary side. In this situation, there is

TABLE 8.2
Magnetizing In-rush Current

Phase	Primary Current (132 kV) before CT	Primary Current (132 kV) after CT	Secondary Current (33 kV) before CT	Secondary Current (33 kV) after CT	Differential Current (I_d)	Restraint Current (I_r)	Differential Relay Response
R	0.4153	0.006921	0.02694	0.0007746	0.006147	0.00115	No trip
S	0.4153	0.006921	0.02694	0.0007746	0.006147	0.00115	
T	0.4153	0.006921	0.02694	0.0007746	0.006147	0.00115	

TABLE 8.3
Three-Phase-to-Ground Fault at the Loaded Transformer

Phase	Primary Current (132 kV) before CT	Primary Current (132 kV) after CT	Secondary Current (33 kV) before CT	Secondary Current (33 kV) after CT	Differential Current (I_d)	Restraint Current (I_r)	Differential Relay Response
R	3062	51.04	0.2812	0.008085	51.03	7.657	Trip
S	3062	51.04	0.2812	0.008085	51.03	7.657	
T	3062	51.04	0.2812	0.008085	51.03	7.657	

a substantial increase in the primary current due to faults within the safe zone at 0.5 seconds.

Using the harmonic and amplitude comparators, the relay observed this rise and identified it as an internal fault. The transformer is, thus, disconnected from the grid. A trip signal after the relay emitted 0.57 ms. As given in Table 8.3, the result is after the fault, which can be considered a very good velocity to isolate the transformer.

8.13 PHASE-TO-GROUND EXTERNAL FAULT AT THE LOADED TRANSFORMER

This situation is similar to case 2, where the occurrence outside the safe zone (at the secondary side) of the fault current led to a rise in the fault currents on both sides of the power transformer. The relay, therefore, deemed this condition to be a significant increase in load currents.

The second harmonic value decreased to <0.3 of the fundamental portion after the external fault at 2 seconds. Consequently, logic (1) was released by the harmonic calculation component, but logic (0) was released by the amplitude comparator because the differential current was almost zero. Consequently, no trip signal is released for this logic coordination (0, 1). The result is given in Table 8.4.

TABLE 8.4
Phase-to-Ground External Fault at the Loaded Transformer

Phase	Primary Current (132 kV) before CT	Primary Current (132 kV) after CT	Secondary Current (33 kV) before CT	Secondary Current (33 kV) after CT	Differential Current (I_d)	Restraint Current (I_r)	Differential Relay Response
R	395.7	6.595	1826	52.51	−45.91	8.865	No trip
S	154.6	2.576	1011	29.07	−26.5	4.747	
T	410.6	6.843	1102.7	31.7	−24.86	5.782	

TABLE 8.5
Two-Phase-to-Ground Fault at a Loaded Transformer

Phase	Primary Current (132 kV) before CT	Primary Current (132 kV) after CT	Secondary Current (33 kV) before CT	Secondary Current (33 kV) after CT	Differential Current (I_d)	Restraint Current (I_r)	Differential Relay Response
R	3062	51.04	0.4753	0.01367	51.03	7.658	Trip
S	1332	22.21	0.0269	0.000774	22.2	3.331	
T	1730	28.84	923	26.54	2.303	8.307	

8.14 TWO-PHASE-TO-GROUND FAULT AT THE LOADED TRANSFORMER

In this case, phases A and B to ground fault are developed to test the algorithm's security. After flipping CB1 to 0.1 seconds, by connecting phases A and B to the ground of the power transformer's secondary side, an internal fault is formed at 0.5 seconds on the power transformer's secondary side. In this situation, there is a substantial increase in the primary current due to faults within the safe zone at 0.5 seconds.

Using the harmonic and amplitude comparators, the relay observed this rise and identified it as an internal fault. The transformer is, thus, disconnected from the grid. It is also clear that the relay emitted a trip signal after 0.53 ms. As given in Table 8.5, the result is after the fault, which can be considered a very good place to isolate the transformer.

PROBLEMS

8.1. Using a 20 MVA, 66/6.6 kV, delta/star power transformer, determine the CT's ratio on both sides and how they should be connected for a differential protection scheme circulate a current of 5 A in the pilot wire.

8.2. Using a 30 MVA, 132/6.6 kV, star/delta power transformer, determine the CT'S ratio on both sides and how they should be connected for a differential protection scheme circulate a current of 5 A in the pilot wire.

8.3. Consider a Δ/Y connected, 40 MVA, 33/66 kV transformer with differential protection applied. The minimum relay current is set to allow 125% overload.

i. Calculate the currents on full load on both sides of the transformer.
ii. Find the CT ratios for both sides.
iii. Distribute the currents in each winding for the power and current transformer.
iv. Show if the relay detects the full-load current or not.

8.4. Consider a Y/Δ connected, 40 MVA, 11/66 kV transformer with differential protection applied. The minimum relay current is set to allow 125% overload.

i. Calculate the relay currents on full load.
ii. Find the CT ratios for both sides.
iii. Find the pilot wires' current under load conditions and operating coil current.
iv. If the *line-to-line fault* of 3000 A occurs on 66 sides between lines B and C out of the protective zone, distribute the currents in each winding for power and current transformer.
v. Show if the relay detects the fault.

8.5. Consider a Y/Δ connected, 30 MVA, 11/66 kV transformer with differential protection applied. The minimum relay current is set to allow 150% overload.

i. Calculate the relay currents on full load.
ii. Find the CT ratios for both sides.
iii. Find the pilot wires' current under load conditions and operating coil current.
iv. If the *single line-to-ground fault* of 3000 A occurs on 66 sides between lines B and C out of the protective zone, distribute the currents in each winding for power and current transformer.
v. Show if the relay detects the fault.

8.6. A Δ/Y power transformer, 45 MVA, 69/132 kV, is protected by the use % differential relay. The CTs on the delta and star sides are 300/5 and 400/5, respectively. Determine the following

i. The output current of both CTs at full load.
ii. Relay current at full load.
iii. Minimum relay current setting to permit 25% overload.
iv. The auxiliary turn ratio.

9 Generator, Motor, and Busbar Protection

9.1 INTRODUCTION

Synchronous generators used in industrial and commercial applications are typical of the non-unit type (directly connected to the bus vice through a step-up transformer), with ratings varying from 48 to 13.8 kV and 5–30 MVA.

In medium and large power stations, large generating units and large power stations have to be connected to large grids for secure and safe operation and full exploitation of their benefits. The generators are operated in unit connection exclusively. In the connection, the generator is linked to the busbar of the higher voltage level via a step-up transformer or several parallel transformer units; the transformers electrically isolate the generators. A circuit breaker can be connected between the generator and the transformer.

Although generators are subject to numerous types of hazards, this chapter will briefly discuss the types of internal faults and various abnormal operating and system conditions. Additional protective schemes, such as overvoltage, out-of-step, synchronization, and the like, should also be considered depending on the generator's cost and relative importance.

The current chapter deals with generator protection and generator fault types. It also presents motor and busbar protection.

9.2 GENERATOR FAULT TYPES

The generator is protected against two types of faults:

i. Stator fault.
ii. Rotor fault.

The stator faults are

i. Phase to phase fault.
ii. Phase to ground fault.
iii. Inter-turn fault.

The danger of these faults is that they may damage the limitations due to heat generated at the point of fault.

The rotor faults are

i. Ground fault.
ii. Inter-turn fault.

DOI: 10.1201/9781003394389-9

TABLE 9.1
Generator *k*-Values

Type of Machine	Permissible I^2t
Salient pole generator	40
Synchronous condenser	30
Cylindrical rotor generator (indirectly cooled-air)	30
Cylindrical rotor generator (directly cooled H_2)	10

The field system is not grounded normally; therefore, a single ground fault does not give any fault current. A second earth fault will short circuit part of the field windings and gives rise to unbalanced forces on the rotor, which results in excess pressure on bearing and shaft distortion

If the fault is not removed quickly, it is necessary to know the exitance of the first fault occurrence.

The abnormal running conditions involve

i. Overloading.
ii. Overspeed.
iii. Unbalanced loading.
iv. Overvoltage.
v. Failure of prime-mover.
vi. Loss of excitation.

Unbalanced loading results in negative sequence current circulation in the stator windings, giving rise to a rotating magnetic field at double synchronous speed regarding the rotor. This will induce a voltage of double frequency in the rotor, and depending on the degree of unbalance, this can result in the rotor overheating.

An unbalanced negative sequence current overcurrent relay (ANSI Device No. 46) is recommended for this unbalanced. Table 9.1 gives a list of typical *k*-values for different types of generators.

Overloading the stator will overheat the stator windings, which may damage the insulation.

In hydro units, a sudden loss of load results in overspeeding.

The generator causes overvoltages over speed or faulty voltage regulator operation.

Failure of the prime-mover results in motoring; thus, the generator draws power from the system. This may lead to the rotor's acceleration and cause a dangerous mechanical condition if allowed to persist.

The loss of excitation may result in a loss of synchronism and slightly increased generator speed as the machine's power input remains unchanged.

Therefore, the machine behaves as an induction generator and draws its excitation current from the system, equal to the fall load-rated current. This leads to stator windings overheating and the rotor-to-body overheating because of current-induced rotor bodies due to slip speed.

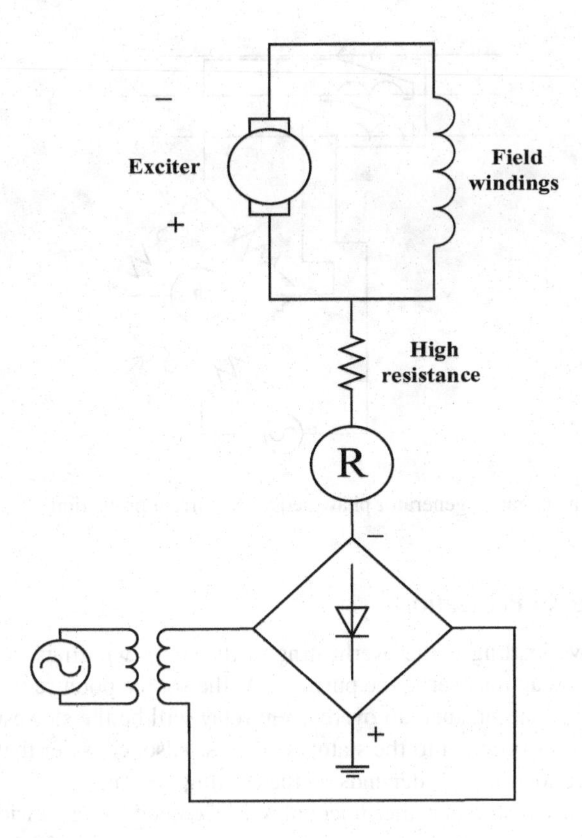

FIGURE 9.1 Protecting the rotor against the earth's fault.

This situation should not continue for long and corrective measures should be taken to restore the alternator's excitation or disconnection. The loss of excitation may also lead to a pole of slipping conditions, resulting in voltage reduction for output above the rated load.

9.2.1 Rotor Protection

The circuit in Figure 9.1 shows how to protect the rotor against the earth's fault or open circuit. A small power supply is connected to the pole of the field circuit. A fault at any point on the field circuit will pass a current through the relay. The earth relay is connected to the alarm circuit for indication.

9.2.2 Unbalanced Loading

The relay R current is proportional to the negative generator phase sequence current. Relay R is IDMT and connected to the trip generator circuit breaker. An alarm auxiliary relay shown in Figure 9.2 warns when the maximum continuous permissible negative phase sequence current is exceeded.

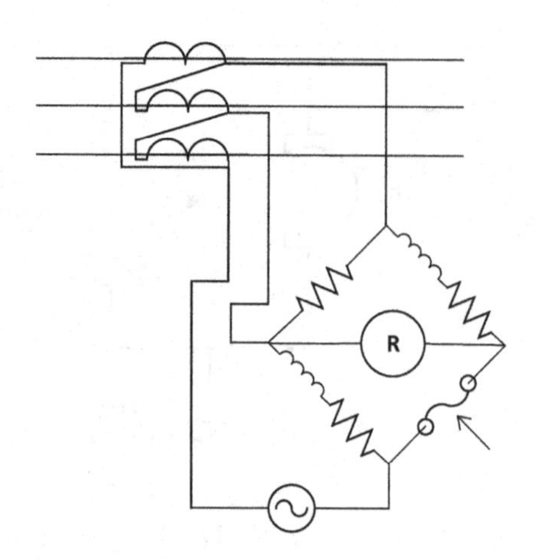

FIGURE 9.2 The negative generator phase sequence current protection.

9.2.3 Overload Protection

The result of overloading is the overheating of the stator windings. An overcurrent relay with time delay may serve the purpose. At the source point regarding overcurrent relay discrimination, such an overcurrent relay will be the slowest in operation and poses a serious problem to the stator windings. Also, consider that temperature rise is due to overloading and depends on the cooling system.

Reliability thermistors or thermocouples are embedded at various points in the stator windings to indicate trip signals for the rating above 50 MW generator. Thermal relays protect lower rating sets.

9.2.4 Overspeed Protection

Such conditions arise when a sudden loss of load occurs. Power relays can detect a reduction in output, which then monitors the steam input to the turbine.

9.2.5 Overvoltage Protection

This form of protection is normally provided for hydro and gas turbine generators. The protection uses an AC overvoltage relay, operating at 110% of the nominal value and operating instantaneously at about 130%–150% of the rated voltage. The relay's operation introduces resistance in the generator or exciter field circuit, and if overvoltage persists, the main generator and field breaker is tripped.

9.2.6 Failure or Prime-Mover

Failure of the prime-mover results in generator motoring. Directional power relays with time delay are used in such protection. Table 9.2 shows a maximum motoring power.

TABLE 9.2
Maximum Motoring Power

Generator Type	Percentage of kW Rating (%)
Steam turbine	3
Water wheel turbine	0.2
Gas turbine	20–30
Diesel engine	35

9.2.7 Loss of Excitation

Large alternators cannot run synchronously for long due to the rapid overheating of the rotor loss of excitation schemes arranged to trip after a certain delay. An offset MHO relay is usually used in such a scheme. The relay setting is so arranged that the relay operates whenever the excitation goes below a certain value. Figure 9.3 shows a generator protection scheme using an MHO relay.

In case of loss of excitation, the generator goes out of synchronism and starts running asynchronously faster than the system, absorbing reactive power from the system. Under these conditions, the stator end regions and part of the rotor get overheated.

This protection will have the following:

- Mho characteristic lying in the third and fourth quadrants of the impedance diagram with adjustable reach and offset.
- An under-voltage and/or overcurrent relay is an additional check.
- A timer with an adjustable range of 1–10 seconds.

9.2.7.1 Recommended Settings

- The diameter of the Mho circle $= X_d$ (synchronous reactance).
- Offset of MHO circuit from the origin $= X_d1/2$(trans.reactance).
- Time delay = 1 seconds.
- Under-voltage relay.

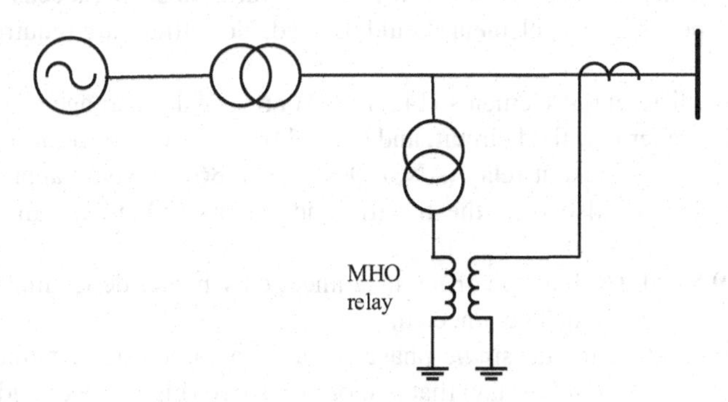

FIGURE 9.3 Generator protection using MHO relay.

9.3 EFFECTS OF GENERATOR BUS FAULTS

For a three-phase fault near the generator, the following characteristics apply:

- *Machine kW and kVAR output:*
  - kVAR out → 5–15 times kVAR normal.
  - kW out → 0; generator cannot transmit kW3φ through the fault.
  - kVA out = ($kVAR_2$ out + kW_2 out)1/2 ≈ kVAR out.
- *Voltage, frequency, power factor, current*:
  - Volts → 0.
  - Frequency → rise to 61–63 Hz.
  - Power Factor → 0.
  - Current → 10–15 times IFLA (a function of X_d'').
- *Machine speed:* Because the fault impedance (Z) is normally very small and the kW out approaches zero, the generator "sees" the fault as an instantaneous drop in load and overspends in a very short time. All of the prime mover kW input goes to accelerating the rotor; if left unchecked, the turbine blades can be seriously damaged (tear out). Speed control by the governor cannot react fast enough, and therefore relays are used to protect the generator.
- *Generator stability:* Faults must be cleared within ~0.3 seconds (18 cycles) to preserve stability. The fault is removed by dropping the generator; load shedding is initiated to prevent large systems' frequency and voltage drops.

9.4 INTERNAL FAULTS

The internal fault types in the generator are as follows.

9.4.1 DIFFERENTIAL PROTECTION (PHASE FAULTS)

The suggested protection for instantaneous and sensitive protection for internal generator faults is differential protection (ANSI Device No. 87G), which is very similar to motor differential protection. A constant percentage, high-sensitivity (e.g., 10%) differential element is recommended. If the CT saturation error exceeds 1%, a lower sensitivity-type (e.g., 25%) element should be used. No settings are required for these elements.

Generator differential elements (Figure 9.4) are usually arranged to simultaneously trip the generator, field circuit, and neutral breakers (if used) through a manually reset auxiliary lockout relay (ANSI Device No. 86). In some applications, the differential element also trips the throttle and admits CO_2 to the fire protection generator.

Figures 9.5 and 9.6 show the different connection schemes depending on whether the generator is wye or delta connected.

As an alternative to the single-phase devices, a three-phase, numerical relay (e.g., SEL-300G or GE 489 relay) that is more sensitive (high-speed) and that offers variable percentage operating characteristics can be used to protect the generator

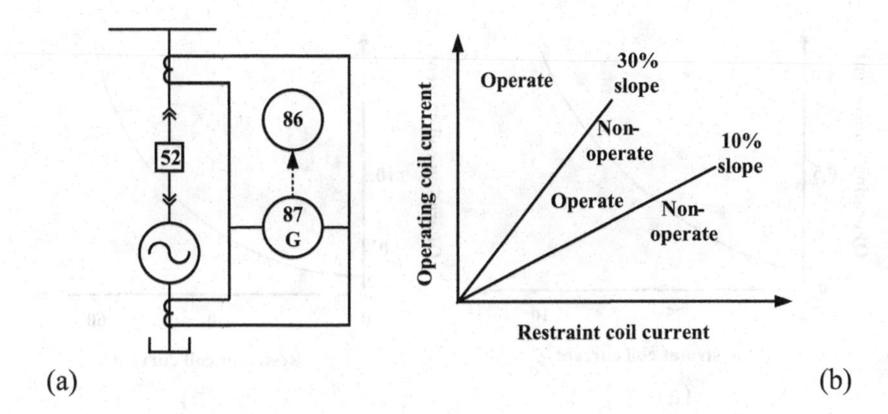

FIGURE 9.4 Fixed slope relay. (a) Circuit diagram and (b) characteristics.

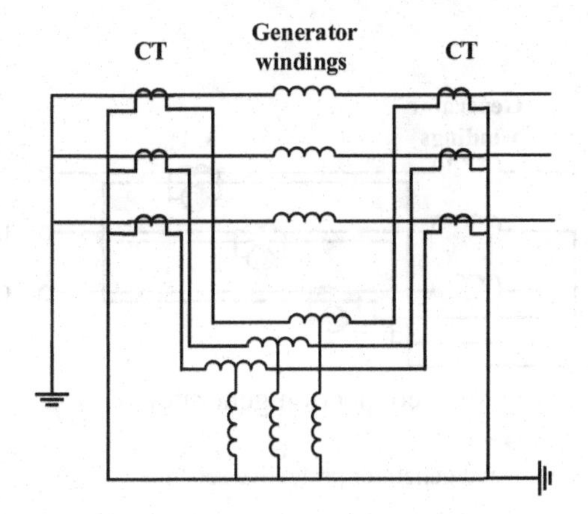

FIGURE 9.5 Differential relay protection for Y-connection generator.

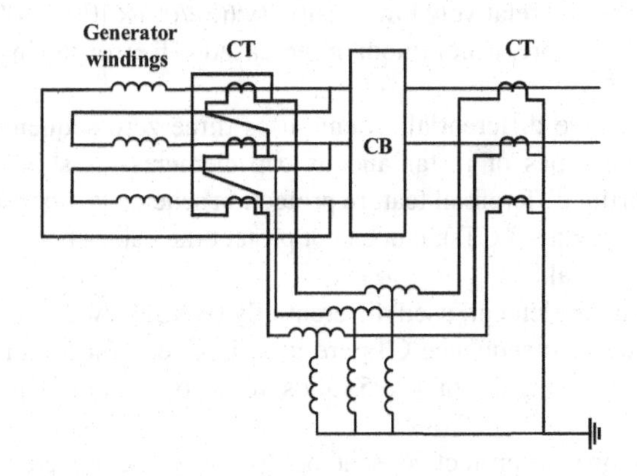

FIGURE 9.6 Differential relay protection for Δ-connection generator.

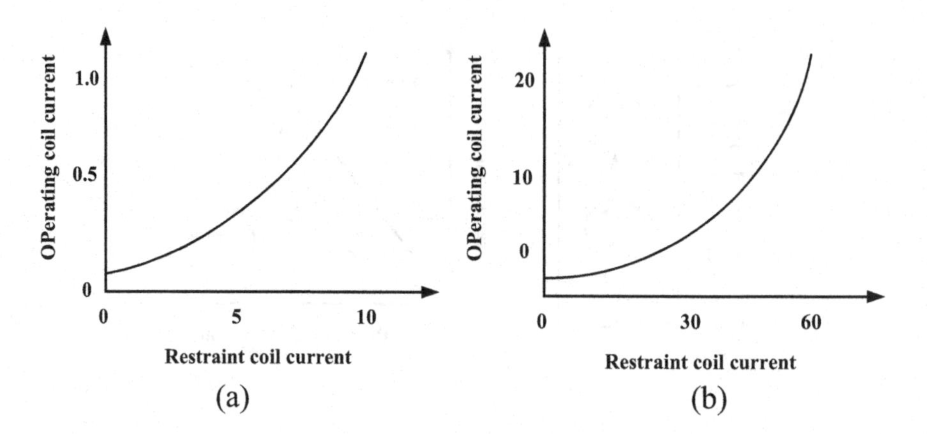

FIGURE 9.7 Typical variable percentage differential relay. (a) With low restraint current and (b) with high restraint current.

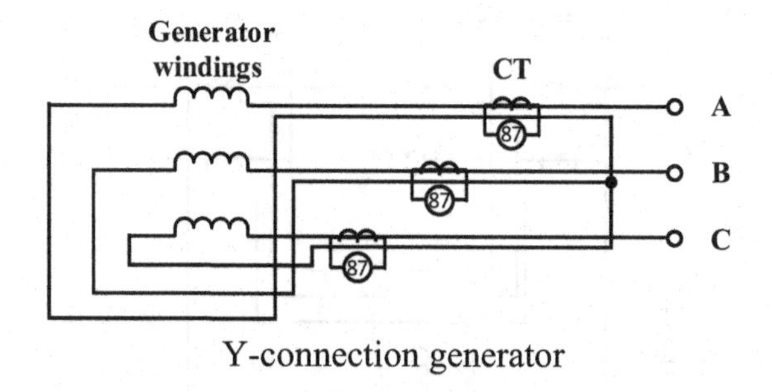

FIGURE 9.8 Wye-connected generator protective scheme.

(Figure 9.7). These relays have very high sensitivity $((0.25/5) \times 100 = 5\%)$ during light internal faults and relatively low sensitivity $((30/60) \times 100 = 50\%)$ during heavy external faults. Therefore, accommodate increased CT error during heavy external faults.

Another alternative differential scheme uses three zero-sequence current transformers and three types of instantaneous trip elements, as shown in Figure 9.8. Although this partial differential feature protection scheme is more sensitive and less costly (e.g., 6 CTs versus 3 CTs), it does not protect the cables between the generator terminals and the breaker.

The instantaneous element, used differentially (ANSI Device No. 87), is typically set at 0.15 A. The zero sequence CTs are usually sized at a 50/5 ratio (most common) with a 4-inch diameter, or 100/5 ratios are also available with 7- and 14-inch diameters.

Figure 9.8 shows a protective scheme for a wye-connected generator, and Figure 9.9 shows a protective scheme for a delta-connected generator.

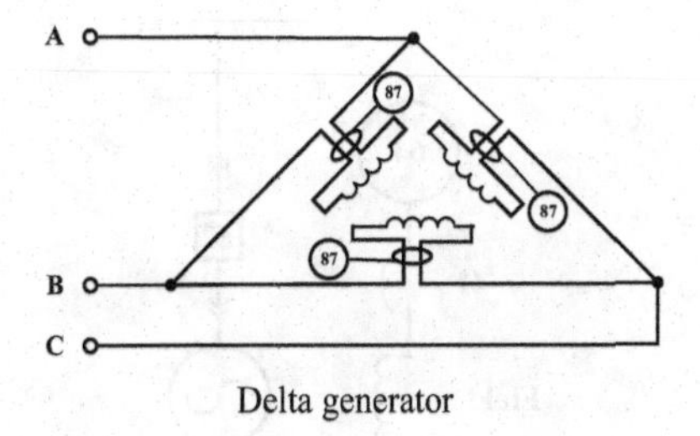

FIGURE 9.9 Delta-connected generator protective scheme.

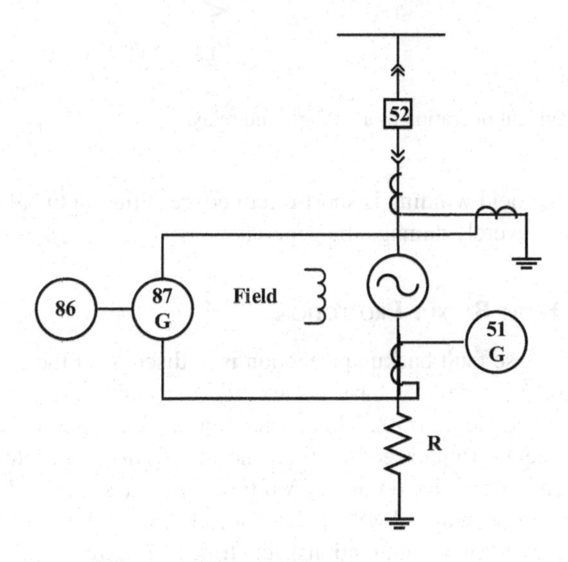

FIGURE 9.10 Typical ground differential scheme.

9.4.2 Differential Protection (Ground Faults)

A separate feature (ANSI Device No. 87G) is essential for generator protection for large generators' internal ground faults. ANSI Device No. 87G supplements (back-ups) the phase differential element that was previously discussed. The element can be set for the minimum time to clear internal faults faster (Figure 9.10). This element operates the lockout relay (ANSI Device No. 86) to trip and lock out the line, field breakers, and prime mover.

9.4.3 Field Grounds

A field ground relay element (ANSI Device No. 64) detects grounds in the generator field circuit (Figure 9.11). This relay uses a very sensitive d'Arsonval movement to measure DC ground currents. The element is used to alarm the first ground occurrence to permit an orderly generator shutdown. If a second ground occurs before the

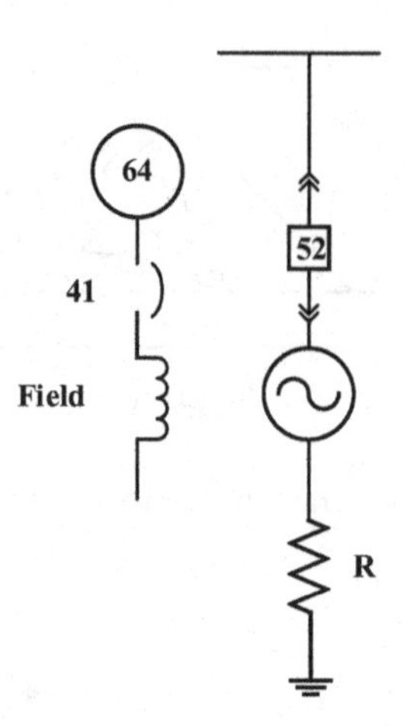

FIGURE 9.11 Typical operation of a field ground relay.

first is cleared, the field winding is short-circuited, resulting in unbalance and vibrations, which may severely damage the generator.

9.4.4 Phase Fault Backup Protection

The function of phase fault backup protection is to disconnect the generator if other downstream protective devices have not cleared the fault (e.g., feeder overcurrent relays, ANSI Device Nos. 51/50). This protection prevents the generator and other auxiliary components from exceeding their thermal limits and protects distribution components against excessive damage. Two types of relays are used to provide this protection. Impedance relays (ANSI Device No. 21) are used to protect unit generators (generator/transformer combinations), and time overcurrent relays (ANSI Device No. 51) are used for non-unit installations typically found in industrial/commercial applications. This tab will restrict discussion to using time overcurrent relays for backup phase fault protection.

9.4.5 The 95% Stator Earth Fault Protection (64G1)

It is an overvoltage relay monitoring the voltage developed across the secondary of the neutral grounding transformer in case of ground faults. It covers the generator, LV winding of the generator transformer, and HV winding of UAT. A pick-up voltage setting of 5% is adopted with a time delay setting of about 1.0 seconds. This will be provided for all machines of ratings 10 MVA and above.

9.4.6 The 100% Stator Earth Fault Protection (64G2)

This is a third harmonic U/V relay. It protects 100% of the stator winding.

There will be a certain third harmonic voltage at the generator's neutral side during the machine running condition. This third harmonic voltage will come down when a stator earth fault occurs, causing this relay to operate. This shall have a voltage check or current check unit to prevent faulty operation of the relay at the generator standstill or during the machine running down period.

9.4.7 Voltage Restrained Overcurrent Protection (51/27 G)

This will operate when the fault current from the generator terminals becomes low due to the excitation system characteristic under voltage criteria. It works as backup protection for system faults with a suitable time delay.

9.4.8 Low Forward Power Relay (37G)

When steam flow through the turbine is interrupted by closing the governor valves in thermal machines, the remaining steam in the turbine generates (low) power. The machine enters motoring conditions drawing power from the system. This protection detects the generator's low forward power conditions and trips the generator breaker after a delay, avoiding generator motoring.

The low forward power relay will be provided with a "turbine trip" interlock in thermal machines. A setting of 0.5% of the rated active power of the generator with a time delay of 2.0 seconds will be adopted.

9.4.9 Reverse Power Relay (32G)

Reverse power protection will be used for all types of generators. When the turbine's input is interrupted, the machine enters into motoring conditions drawing power from the system. The reverse power relay protects the generators from motoring conditions. In thermal machines, reverse power condition appears after low forward power condition.

For the reverse power relay, a setting of 0.5% of the rated active power of the generator with two-stage timer is given subsequently.

- *Stage I:* With turbine trip interlock, a time delay of 2 seconds will be adopted.
- *Stage II:* Without a "turbine trip" interlock, a time delay of about 20 seconds can be adopted to avoid unnecessary unit tripping during system disturbance, causing a sudden rise in frequency or power swing conditions.

9.4.10 Generator Under Frequency Protection (81 G)

The under frequency protection:

- It prevents the steam turbine and generator from exceeding the permissible operating time at reduced frequencies.
- Ensures that the generating unit is separated from the network at a preset frequency value.
- Prevent over-fluxing (*v*/*f*) of the generator (large over-fluxing for short times).

The stator under frequency relay measures the frequency of the stator terminal voltage.

Setting recommendations:

- *For alarm*: 48.0 Hz, 2.0 seconds time delay.
- *For trip*: 47.5 Hz, 1.0 seconds (or)

As recommended by generator manufacturers.

9.4.11 Generator Overvoltage Protection (59 G)

An overvoltage on the generator's terminals can damage the generator's insulation, bus ducting, breakers, generator transformer, and auxiliary equipment. Hence overvoltage protection should be provided for machines of all sizes.

Settings recommendations:

- *Stage I:*
    - Over voltage pickup = 1.15 × Un
    - Time delay = 10 seconds.
- *State II*:
    - Over voltage pickup = 1.3 × Un
    - Time delay = 0.5 seconds.

Example 9.1

Figure 9.12 shows one phase of an alternator winding; the percentage differential relay is used for protection. The relay has a minimum pick-up current of 0.2 A and a percentage slope of 12%.

As shown in Figure 9.12, a high resistance earth fault occurs near the alternator windings' grounded neutral end with the current distribution. Assume a CT ratio of 400/5, and determine where the relay will operate.

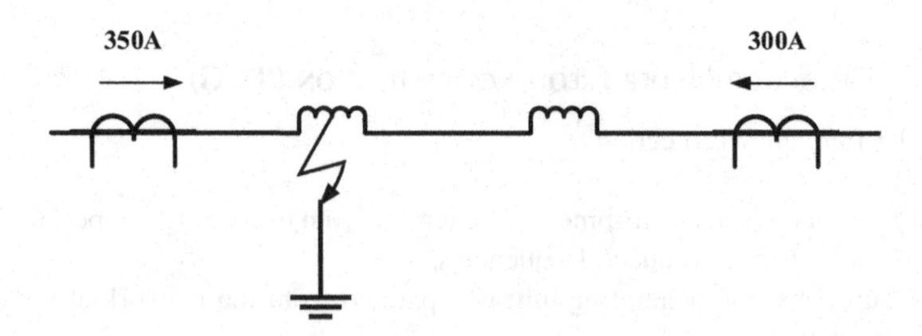

FIGURE 9.12 One phase of an alternator winding of Example 9.1.

Solution

From figure

$$I_1 = 350 \text{ A}$$

$$I_2 = 300 \text{ A}$$

$$I_1 - I_2 = 350 - 300 = 50 \text{ A}$$

$$\text{Operating coil current} = 50 \times \frac{5}{400} = 0.625 \text{ A}$$

$$\text{Average current in restrain coil} = \frac{I_1 + I_2}{2} = \frac{350 + 300}{2} = 325 \text{ A}$$

$$\text{Average current in restrain coil at relay side} = 325 \times \frac{5}{400} = 4.0625 \text{ A}$$

With a 12% slope, $4.0625 \times 12\% = 0.4875$ A.

And with 0.2 A minimum pick-up current $0.4875 + 0.2 = 0.6875$ A.

So, the operating coil current need to operate should be ≥ 0.6875 A.

Since the operating coil current for such a fault condition is 0.625 A (i.e., <0.6875 A), the relay will not operate.

Example 9.2

Figure 9.13 shows a biased percentage differential relay applied to protect synchronous generator windings. The relay has a 0.1 A minimum pick-up current and a slope of 10%.

a. Fault has occurred near the generator's grounded neutral end when the generator is carrying the load. As a result, the currents flowing at each end are shown in the figure. Would the relay operate or not?

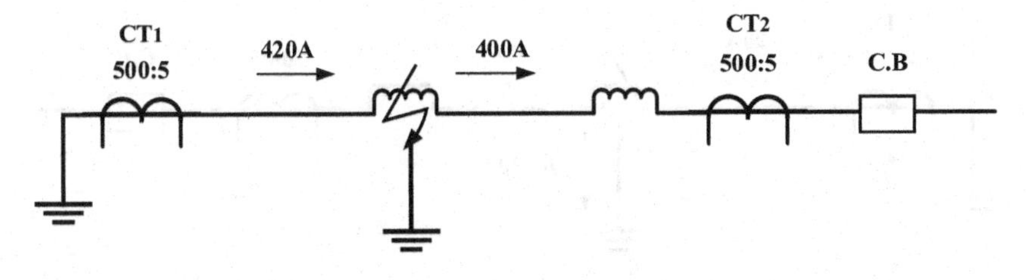

FIGURE 9.13 A biased % differential relay protection of synchronous generator windings of Example 9.2.

b. Will the relay operate at the given fault current value in (a) if the generator was carrying no load?
c. On the same diagram, the relay operating characteristics and the points representing the operating and restraining currents in the relay for the two conditions are shown.

a. Generator under load condition (from Figure 9.13):

$$I_1 = 420\text{ A}$$

$$I_2 = 400\text{ A}$$

$$I_1 - I_2 = 420 - 400 = 20\text{ A}$$

$$\text{Operating coil current} = 20 \times \frac{5}{500} = 0.2\text{ A}.$$

$$\text{Average current in restrain coil} = \frac{I_1 + I_2}{2} = \frac{420 + 400}{2} = 410\text{ A}.$$

$$\text{Average current in restrain coil on relay side} = 410 \times \frac{5}{500} = 4.1\text{ A}.$$

With a 10% slope, $4.1 \times 10\% = 0.41\text{ A}$.
And with 0.2 A minimum pick-up current $0.41 + 0.2 = 0.61\text{ A}$.
So, the operating coil current needed to operate should be ≥0.61 A.
Since the operating coil current for such a fault condition is 0.2 A (i.e., <0.61 A), the relay will not operate.

b. Generator under load condition (see Figure 9.14):

$$I_1 = 420\text{ A}$$

$$I_2 = 400\text{ A}$$

$$I_f = I_1 - I_2 = 420 - 400 = 20\text{ A}$$

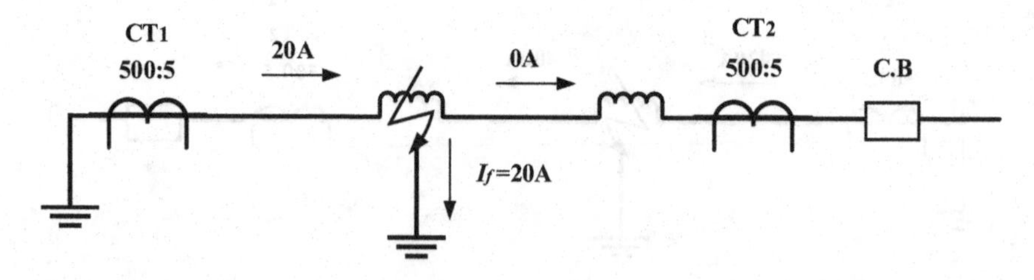

FIGURE 9.14 A biased % differential relay protection of synchronous generator windings of Example 9.2.

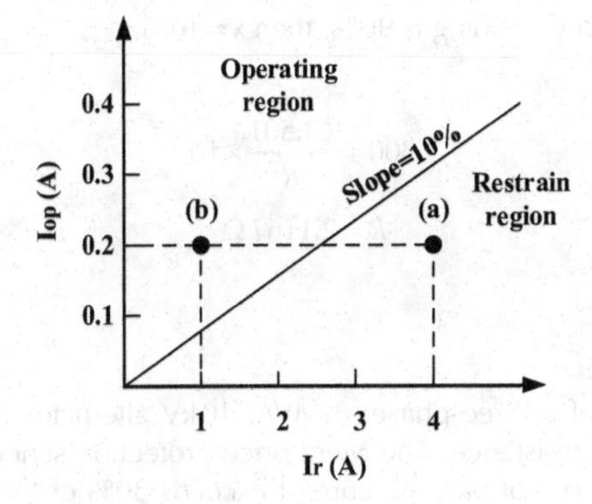

FIGURE 9.15 A biased % differential relay characteristics of Example 9.2.

$$\text{Operating coil current} = 20 \times \frac{5}{500} = 0.2\,\text{A}.$$

$$I_r = \frac{I_{s1} + I_{s2}}{2} = \frac{0.2 + 0}{2} = 0.1\,\text{A}$$

So, the operating coil current $\geq I_r$, the relay will operate (Figure 9.15).

Example 9.3

The neutral point of a three-phase, 20 MVA, 11 kV alternator is earthed through a resistance of 5 Ω. The alternator is protected by a current circulating system set to operate when there is an out-of-balance current of 1.5 A. The CTs have a ratio of 1000/5, determining the % of protected winding. Also, calculate the minimum earthing resistance value required to protect 90% of the winding.

Solution

$$\text{The phase voltage} = \frac{11{,}000}{\sqrt{3}} = 6350\text{ V}$$

Assume $x\%$ is the unprotected portion against an earth fault.

The voltage across this portion is $6350x\% = 63.50x$

The fault current (the primary CT winding current) is $1.5 \times \frac{1000}{5} = 300\text{ A}$

$$\therefore \frac{63.50}{5} x = 300$$

$$x = 23.62$$

$$\therefore \%\text{ of winding protected} = 100 - 23.62 = 76.38\%$$

Since the % protected winding is 90%, then $x = 10\%$

$$300 = \frac{63.501}{R} \times 10$$

$$R = 2.1167\,\Omega$$

Example 9.4

The neutral point of a three-phase, 8 MVA, 10 kV alternator has a reactance of 2 Ω and negligible resistance. The Merz price protection scheme is used, which operates when the out-of-balance current exceeds 30% of the full load current. The alternator is earthed through a resistance of 10 Ω, which determines the % of winding that is protected. Calculate the minimum earthing resistance value required to protect 85% of the winding.

Solution

The alternator's reactance effect can be ignored due to the following reason.

The reactance of winding is proportional to the square of the number of turns

$$X_a \propto N^2$$

Let $x\%$ be the ratio of windings that remain unprotected.

The number of unprotected windings

$$\frac{x}{100} \times N$$

The reactance will be proportional to

$$\frac{x^2}{100^2} \times N^2$$

As 2 Ω reactance, the unprotected reactance will be

$$\frac{x^2}{100^2} \times N^2 \times 2$$

And it is very small, and since it is connected in series with 10 Ω resistance, the effect of this portion can be ignored

$$\text{The phase voltage} = \frac{10000}{\sqrt{3}} = 5773.5\text{ V}$$

Assume $x\%$ be the unprotected portion against earth fault.

The voltage across this portion is $5773.5x\% = 57.735x$.

$$\text{Full load current} = \frac{8\times10^6}{\sqrt{3}\times10\times10^3} = 461.88$$

$$\therefore \frac{57.735}{10}x = 30\% \times 461.88$$

$$x = 24$$

$$\therefore \% \text{ of winding protected} = 100 - 24 = 76\%$$

Since the % protected winding is 85%, then $x = 15\%$

$$461.88 = \frac{57.735}{R}\times 15$$

$$R = 1.875\,\Omega$$

9.5 TYPICAL RELAY SETTINGS

Table 9.3 lists typical relay settings.

9.6 MOTOR PROTECTION

Fuses, thermal overload relays, and contactors have proved themselves effective and economical solutions for up to ~150 hp for small to medium-sized motors. Two basic protections are used for this motor:

i. Thermal overload protection.
ii. Short-circuit (overcurrent) protection.

More sophisticated, flexible, and accurate microprocessor protection relays should be considered on larger, more expensive motors or when maximum motor utilization is required under varying operational conditions. These relays typically include:

i. Thermal overload protection,
ii. Short-circuit protection
iii. Start-up and running stall protection
iv. Phase unbalance protection
v. Single-phasing protection
vi. Earth fault protection
vii. Undercurrent protection

The present-day concept is the use of microprocessor-based numerical relays for both HV and LV motors (say beyond 50 kW), as the relays come with a lot of features that allow them to be interchangeable, ensure site settings, and give valuable feedback on the load details whether a trip occurs or not.

TABLE 9.3
Typical Relay Settings

IEEE No.	Function	Typical Settings and Remarks
24	Overexcitation	PU: 1.1*VNOM/60; TD: 0.3; reset TD: 5 alarm; PU: 1.18*VNOM/60 alarm; delay: 2.5 seconds
25	Synchronism check	Max slip: 6 RPM; Max phase angle error: 10°; Max VMAG error: 2.5% VNOM
32	Reverse power (one stage)	PU: turbine 1% of rated; 15 seconds PU: reciprocating engine: 10% of rated; 5 seconds
32-1	Reverse power nonelectrical trip supervision	PU: same as 32; 3 seconds
40	Loss-of-field (VAR flow approach)	Level 1 PU: 60% VA rating; Delay: 0.2 seconds; Level 2 PU: 100% VA rating: 0.1 seconds
46	Negative sequence overcurrent	I2 PU: 10% irated; $K=10$
49	Stator temperature (RTD)	Lower: 95°C; upper: 105°C
50/87	Differential via flux summation CTs	PU: 10% INOM or less if 1 A relay may be used
50/27 IE	Inadvertent energization overcurrent with 27, 81 supervision	50: 0.5 A (10% INOM); 27: 85% VNOM (81: Similar)
51N	Stator ground over-current (low, med Z Gnd, phase CT residual)	PU: 10% INOM; curve: EI; TD: 4. Inst: none. Higher PU is required to coordinate with the load. No higher than 25% INOM
50/51N	Stator ground over-current (low, med Z Gnd, neutral CT or flux summation CT)	PU: 10% INOM; curve EI, TD4; Inst 100% INOM. Higher PU, if required, to coordinate with the load. No higher than 25% INOM
51GN, 51N	Stator ground over-current (high Z Gnd)	PU: 10% IFAULT at HV term.; curve: VI; TD:4
IEEE No.	Function	Typical settings and remarks
51VC	Voltage controlled overcurrent	PU: 50% INOM; curve: VI; TD: 4. Control voltage: 80% VNOM
51VR	Voltage restrained overcurrent	PU: 175% INOM; curve: VI; TD: 4. Zero restraint voltage: 100% VNOM L-L
59N, 27-3N, 59P	Ground overvoltage	59N: 5% VNEU during HV terminal fault; 27-3N: 25% V3rd during normal operation; TD: 10s 59P: 80% VNOM
67IE	Directional O/C for inadvertent energization	PU: 75%–100% INOM GEN; definite time (0.1–0.25 seconds) Inst: 200% INOM GEN
81	Over/under frequency	Generator protection: 57 and 62 Hz, 0.5 seconds; island detection: 59 and 61 Hz, 0.1 seconds
87G	Generator phase differential	BE1-87G: 0.4 A; BE1-CDS220: Min P.U.: 0.1 * Tap; BE1-CDS220: Min P.U.: 0.1 * Tap
87N	Generator ground differential	BE1-CDS220: Min P.U.: 0.1 times tap; Slope 15%; time delay: 0.1 seconds; choose low tap BE1-67N: current polarization; time: 0.25 A; curve: VI; TD: 2; instantaneous: disconnect
87UD	Unit differential	BE1-87T or CDS220 Min PU: 0.35*tap; tap: INOM; slope 30%

9.6.1 Typical Protective Settings for Motors

a. *Long-time pick-up*
   i. (1.15) times the motor full-load current (FLA) times motor service factor for applications, encountering 90% voltage dip on motor starting.
   ii. (1.25) times motor FLA times motor service factor for applications encountering 80% voltage dip on motor starting.
b. *Long-time delay*
   i. Greater than motor starting time at 100% voltage and the minimum system voltage.
   ii. Less than locked rotor damage time at 100% voltage and the minimum system voltage.
   High-inertia drives are common for the start time to be greater than the locked rotor withstand time. Under these circumstances, set the time to permit the motor to start. Supplemental protection should be added for locked rotor protection. One example is a speed switch set at 25% of the rated speed, tripping through a timer to trip if the desired speed has not been reached in a predetermined time.
c. *Instantaneous pick-up*
   i. Not <1.7 times motor long-time pick-up rated ampere (LRA) for medium-voltage motors.
   ii. Not <2.0 times motor LRA for low-voltage motors.
d. *Earth-fault protection*
   i. Minimum pick-up and minimum time delay for static trip units.
   ii. Core-balance CT and 50 relays set at the minimum for medium-voltage, low-resistance grounded systems.
   iii. Residually connected CT and 50/51 for medium voltage, solidly grounded systems. Minimum tap and time dial equals 1 for 51 relays.
   iv. Minimum tap (not <5 A) for 50 relays.

9.6.2 Motor Protective Device

i. Molded case circuit breaker (MCCB) is used for low-voltage motors of high ratings.
ii. Miniature circuit breakers (MCB) for small motors.
iii. Fuses + contactor + thermal relay for LV motors.
iv. For high-voltage motor: HV circuit breaker and differential protection.

Example 9.5

With a 100 hp (1 hp = 746 W), 480 V, 0.85 lagging power factor, and 85% efficiency, the motor has the starting up to 5.9 of the rated current up to 8 seconds with a voltage dip of 80% during starting. Select the protection means for this motor (Figure 9.16).

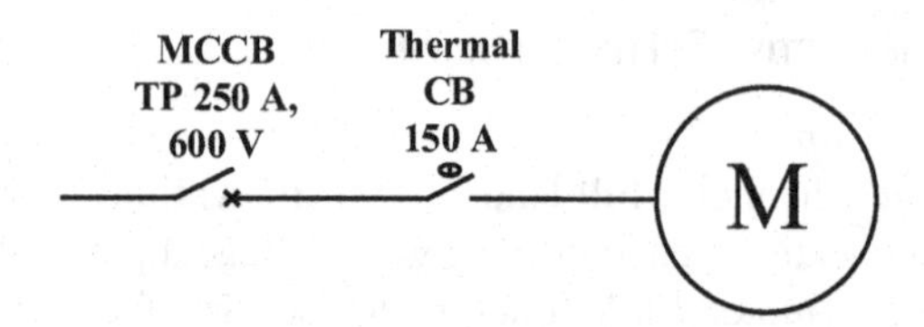

FIGURE 9.16 The motor protection circuit of Example 9.5.

Solution

$$\eta = \frac{P_{out}}{P_{in}}$$

$$P_{in} = \frac{P_{out}}{\eta} = \frac{100 \times 746}{0.85} = 87.764\,\text{kW}$$

$$I_{rated} = \frac{P_{in}}{\sqrt{3} \times V \times p.f} = \frac{87.764\,\text{kW}}{\sqrt{3} \times 480 \times 0.85} = 124\,\text{A}$$

Choose MCCB with both thermal and magnetic trip.
The thermal setting of MCCB:

$$\text{Thermal pick-up setting} \quad (100-120)\%\ I_{rated}$$

Choose 125% I_{rated}

$$\therefore \text{Thermal setting} = 125\% \times 124 = 155\,\text{A}$$

$$\text{Magnetic trip} = 6.75\,I_{rated} = 6.75 \times 124 = 837\,\text{A}$$

$$\text{Circuit breaker rating} = 2 \times I_{rated} = 2 \times 124 = 248\,\text{A}$$

Choose MCCB TP 250 AF/150 AT

9.6.3 Motor Protection by Fuses

In several industrial applications, fuses protect small- and medium-sized motors. To determine the fuse size for a motor, one should refer to Figure 9.17, which shows typical fuse time/current characteristics. These characteristics represent fuse operation where the current is insufficient to operate in the first 1/4-cycle, or 0.005 seconds in a 50 Hz system.

If the motor's starting current is 500 A and the run-up time is 10 seconds, then a 125 A fuse would be required. Examination of the fuse time/current characteristic

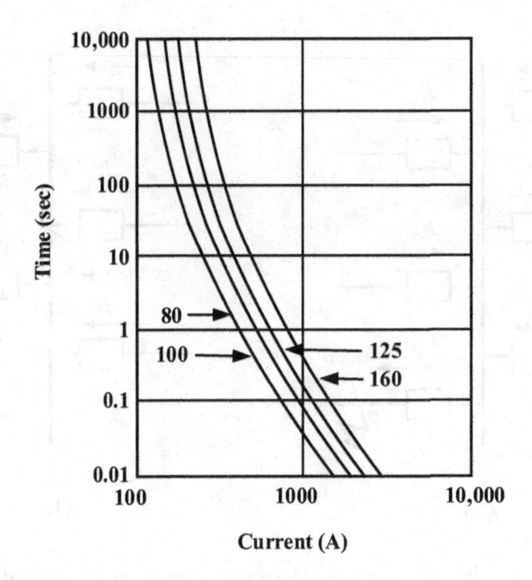

FIGURE 9.17 Current–time fuse characteristics.

shows that at 500 A, the 125 A fuse would operate in 15 seconds. The fuse one size lower, 100 A, would operate in 4 seconds at 500 A and is, therefore, unsuitable. The full-load current of the motor would be 83 A, and although a 100 A fuse would deal with this current, it could not deal with the starting current for the starting time. The fuse does not protect the motor against overload as the fuse's rating is always two to three times the full load current.

To summarize

1. The fuse must be adequately rated to supply normal current to the circuit.
2. The rating must consider any normal healthy overload conditions, e.g., the starting of motors.
3. An allowance must be made if an overload occurs frequently.
4. There must be an adequate margin if discrimination between fuses is required.
5. The fuse must protect any equipment not rated at the full short-circuit rating of the power system, e.g., contactors, cables, switches, and the like.

9.7 BUS BARS PROTECTION

9.7.1 Bus Protection Schemes

Bus protection protects switches, disconnects, instrument transformers, circuit breakers, other bus equipment, and the bus itself.

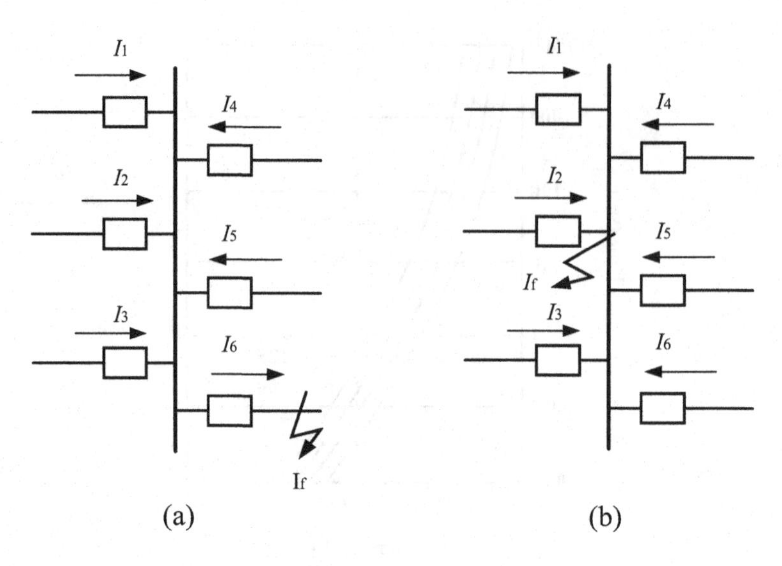

FIGURE 9.18 Simple bus arrangement: (a) external fault and (b) internal fault.

There are several methods of bus protection:

i. Basic differential protection
ii. Differential protection with overcurrent relays
iii. Percentage differential protection
iv. High-impedance voltage differential protection
v. Bus partial differential protection

All these methods are based on KCL, namely, that the sum of all currents entering a node must be zero. Consider the two situations for a simple bus shown in Figure 9.18.

For external fault:

$$I_f = I_6 = I_1 + I_2 + I_3 + I_4 + I_5$$

For internal fault:

$$I_f = I_6 = I_1 + I_2 + I_3 + I_4 + I_5 + I_6$$

9.7.2 Bus Differential Relaying Schemes

9.7.2.1 Basic Differential System

A basic differential system is shown in Figure 9.19. All CTs must have the same ratio and polarity such that the current circulation among them is zero ($I_d = 0$) for all external faults. For internal fault, current $I_d = I_f$ will flow through the relay.

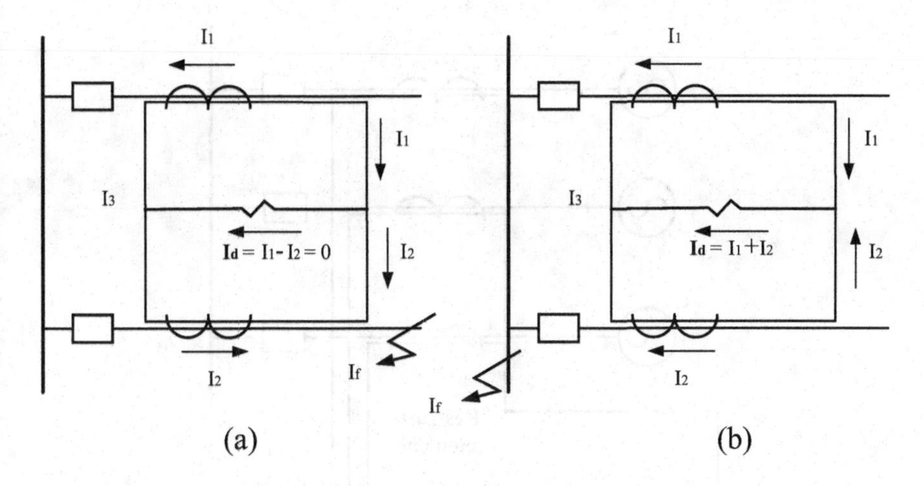

FIGURE 9.19 Basic differential system: (a) external fault and (b) internal fault.

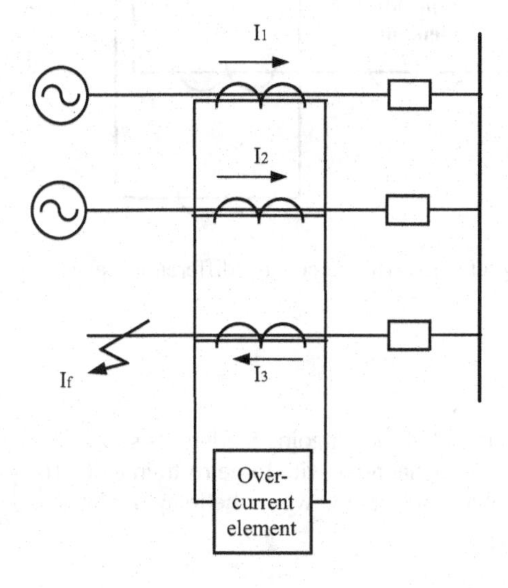

FIGURE 9.20 Bus differential protection with overcurrent relays.

9.7.2.2 Bus Differential Protection with Overcurrent Relays

If the CTs behaved ideally, the differential system shown in Figure 9.19 would be very easy to implement using a simple overcurrent relay, as shown in Figure 9.20.

9.7.2.3 Bus Protection with Percentage Differential Relays

A percentage restrain differential relay takes the fact that there may be an error current in the differential circuit. A simple percentage restrain differential relay is shown in Figure 9.21.

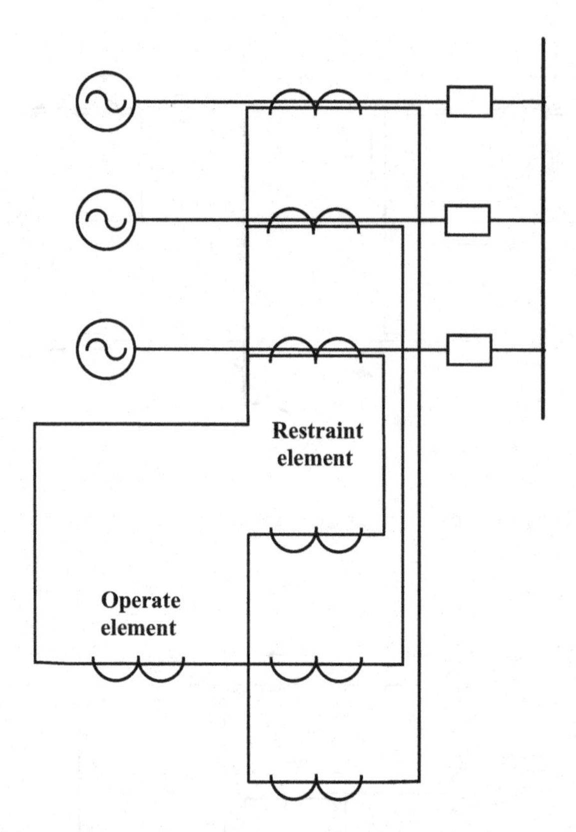

FIGURE 9.21 Bus protection with percentage differential relays.

Example 9.6

Consider a load bus with three outgoing feeders, as shown in Figure 9.22. This bus is protected by a differential relay with three restrain coils. The protection scheme is shown for one phase only. Show when the relay operates and does not operate.

Solution

Case 1: When there is no fault (internal fault):

$$I_1 + I_2 = I_3$$

$$I_1' + I_2' - I_3' = 0 = I_{op} \text{ the relay will not operate.}$$

Case 2: When there is a fault:

$$I_1' + I_2' - I_3' \neq 0 \text{ the relay will operate.}$$

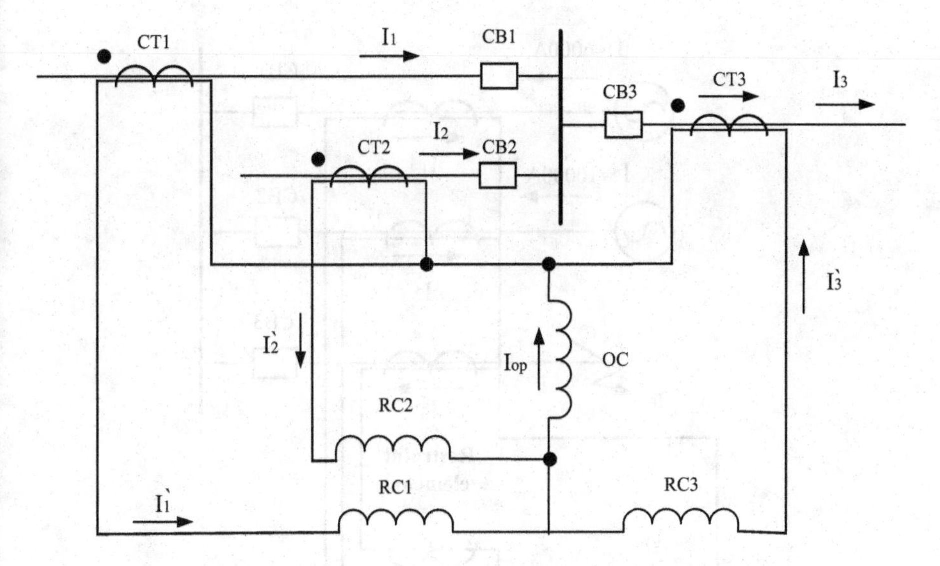

FIGURE 9.22 Bus protection with percentage differential relay of Example 9.6.

Example 9.7

For the system shown in Figure 9.23 (bus protection by differential current relay), an external fault has occurred on feeder no.3; find whether the differential relay will operate. Each CT has a current ratio of 600/5 A.

Solution

$$I_f = I_{F1} + I_{F2} = 6000 + 10,000 = 16,000\ \text{A}$$

$$I_1' = 6000 \times \frac{5}{600} = 50\ \text{A}$$

$$I_2' = 10,000 \times \frac{5}{600} = 83.4\ \text{A}$$

$$I_3' = 16,000 \times \frac{5}{600} = 133.4\ \text{A}$$

$$I_{\text{op}} = I_1' + I_2' + I_3' = 50 + 83.4 - 133.4 = 0\ \text{A}$$

Hence the relay will not operate.

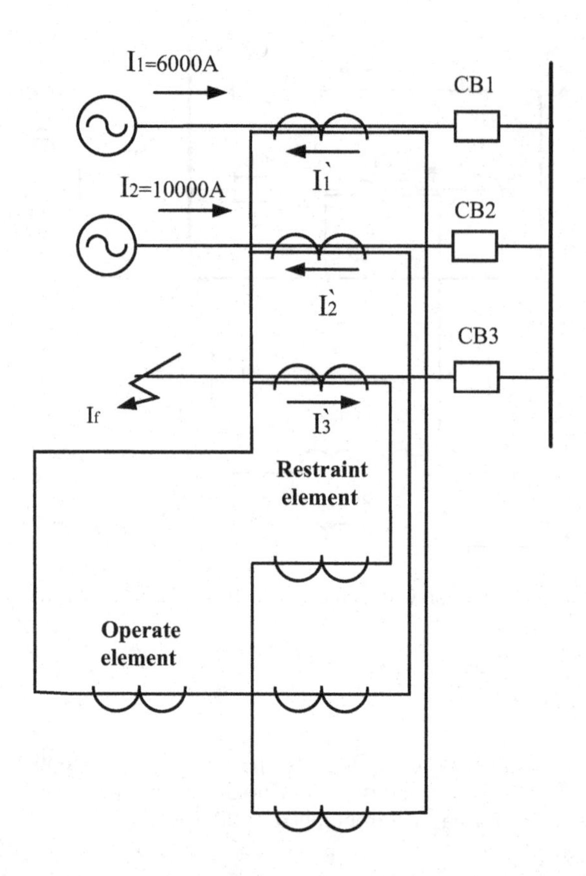

FIGURE 9.23 Bus protection with percentage differential relay (external fault) of Example 9.7.

Example 9.8

For the system shown in Figure 9.24 (bus protection by differential current relay), an internal fault has occurred on feeder no.3, showing whether the differential relay will operate. Each CT has a current ratio of 600/5 A.

Solution

$$I_f = I_{F1} + I_{F2} + I_{F3} = 6000 + 10,000 + 7000 = 23,000\ \text{A}$$

$$I_1' = 6000 \times \frac{5}{600} = 50\ \text{A}$$

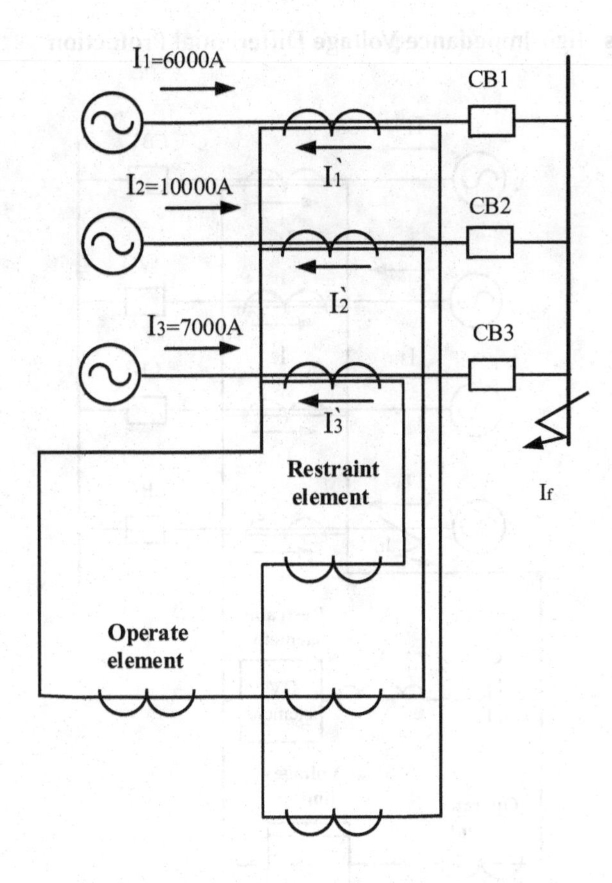

FIGURE 9.24 Bus protection with percentage differential relay (internal fault) of Example 9.8.

$$I_2' = 10000 \times \frac{5}{600} = 83.4\ \text{A}$$

$$I_3' = 7000 \times \frac{5}{600} = 58.3\ \text{A}$$

$$I_{\text{op}} = I_1' + I_2' + I_3' = 50 + 83.4 + 58.3 = 191.7\ \text{A}$$

Hence the relay will operate (Figure 9.25).

9.7.2.4 Bus High-Impedance Voltage Differential Protection

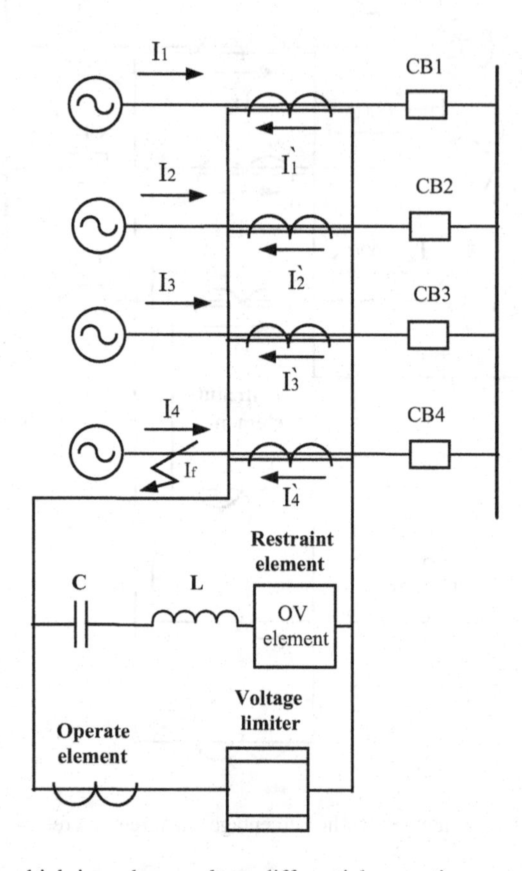

FIGURE 9.25 Bus high-impedance voltage differential protection.

9.7.2.5 Bus Partial Differential Protection

A partial differential scheme is similar to an overcurrent differential scheme, the difference being that all breakers are not monitored. This scheme is used when there are *limited sources and multiple outfeeds.* This scheme is shown in Figure 9.26.

The scheme uses an overcurrent relay fed from paralleled CTs that only monitor the sources to the bus. The overcurrent relay is set to coordinate with the feeder relays.

The advantage of this connection is that it is relatively inexpensive. The drawback being it has a slower clearing time.

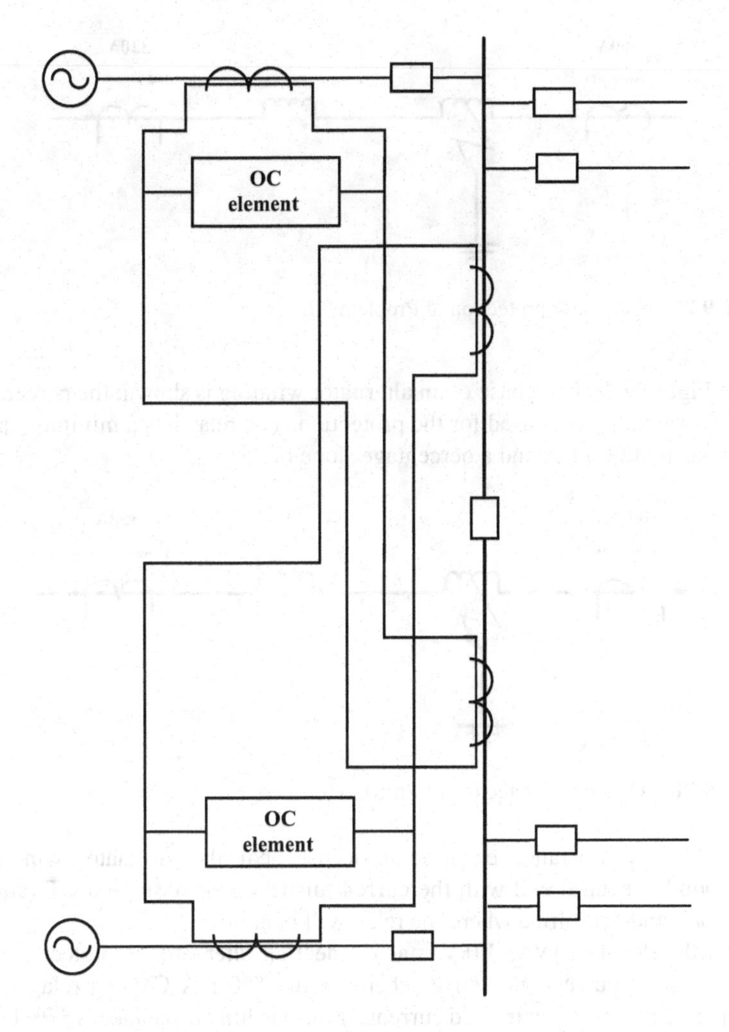

FIGURE 9.26 Bus partial differential protection.

PROBLEMS

9.1. In Figure 9.27, one phase of an alternator winding is shown; the percentage differential relay is used for the protection. The relay has a minimum pick-up current of 0.3 A and a percentage slope of 10%.

A high resistance earth fault occurs near the alternator windings' grounded neutral end with the current distribution. Assume a CT ratio of 400/5, and determine whether the relay will operate.

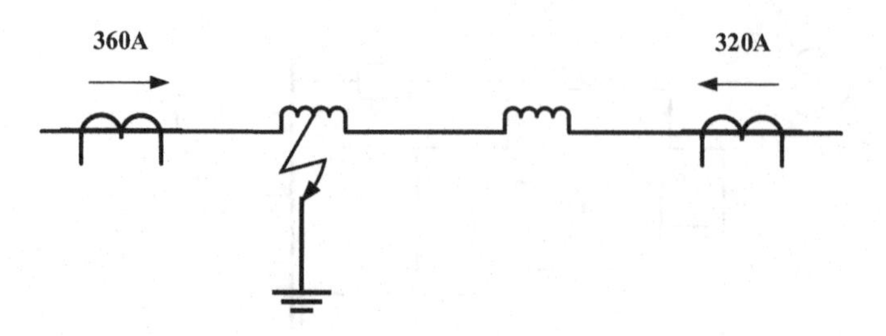

FIGURE 9.27 Generator protection of Problem 9.1.

9.2. In Figure 9.28, one phase of an alternator winding is shown; the percentage differential relay is used for the protection. The relay has a minimum pick-up current of 0.1 A and a percentage slope of 15%.

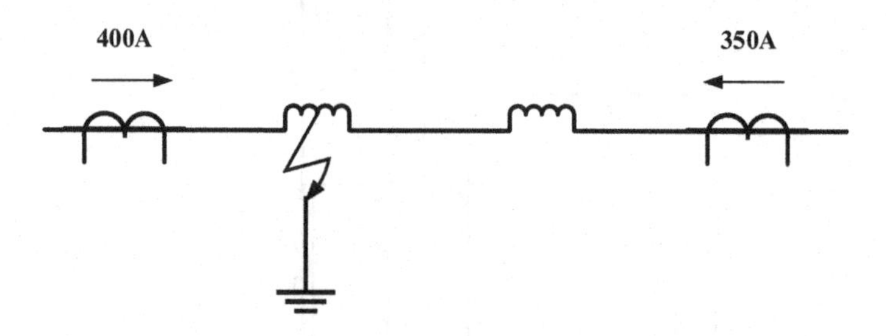

FIGURE 9.28 Generator protection of Problem 9.2.

A high resistance earth fault occurs near the alternator windings' grounded neutral end with the current distribution. Assume a CT ratio of 400/5, and determine where the relay will operate.

9.3. A 40 MW, 44 MVA, 30 kV, star connection alternator is protected by a circulating current protective scheme using 550/1 A CT and relay set to operate at 12% of their rated current. If the earthing resistance is 85% based on the machine rating, calculate the stator winding percentage, which is not protected against the earth's fault.

9.4. A 25 MW, 40 MVA, 30kV, star connection alternator is protected by a circulating current protective scheme using 400/1 A CT and relay set to operate at 10% of their rated current. If the earthing resistance is 90% based on the machine rating, calculate the stator winding percentage, which is not protected against the earth's fault.

9.5. For the system shown in Figure 9.29 (bus protection by differential current relay), an external fault has occurred on feeder no.3. Find if the differential relay will operate. Each CT has a current ratio of 600/5 A.

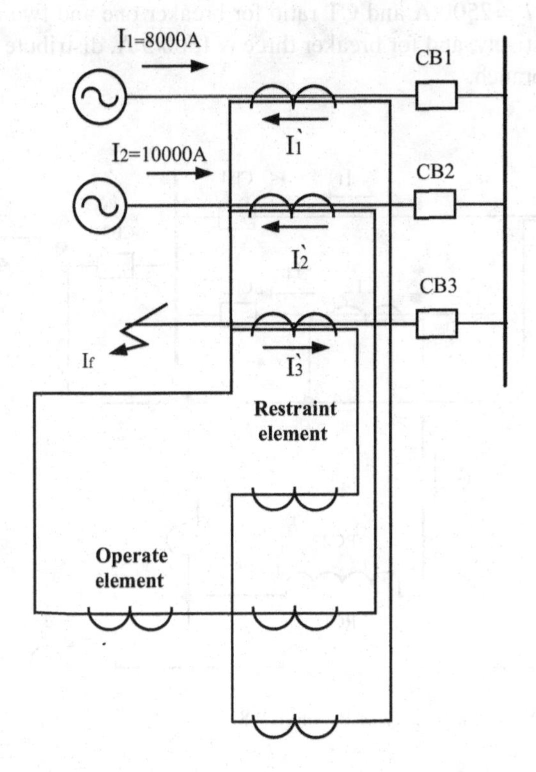

FIGURE 9.29 Bus protection by a differential current relay of Problem 9.5.

9.6. A 80 hp (1 hp = 746 W), 480 V, 0.8 lagging power factor, 80% efficiency, the motor has the starting up to 5.9 of the rated current up to 8 seconds voltage dip of 80% during starting. Select the protection means for this motor.

9.7. A 100 hp, 400 V, 0.7 lagging power factor, 90% efficiency, the motor starts up to 5.9 of the rated current up to 8 seconds with a voltage dip of 80% during starting. Select the protection means for this motor.

9.8. Consider a load bus with three outgoing feeders, as shown in Figure 9.30. This bus is protected by a differential relay with three restrain coils. The protection scheme is shown for one phase only.
   i. Show when the relay operates and does not operate.
   ii. If $I_1 = I_2 = 2500$ A and CT ratio for breaker one and two are 1000/5 A, respectively, and for breaker three is 1500/5 A, distribute the current in each branch.

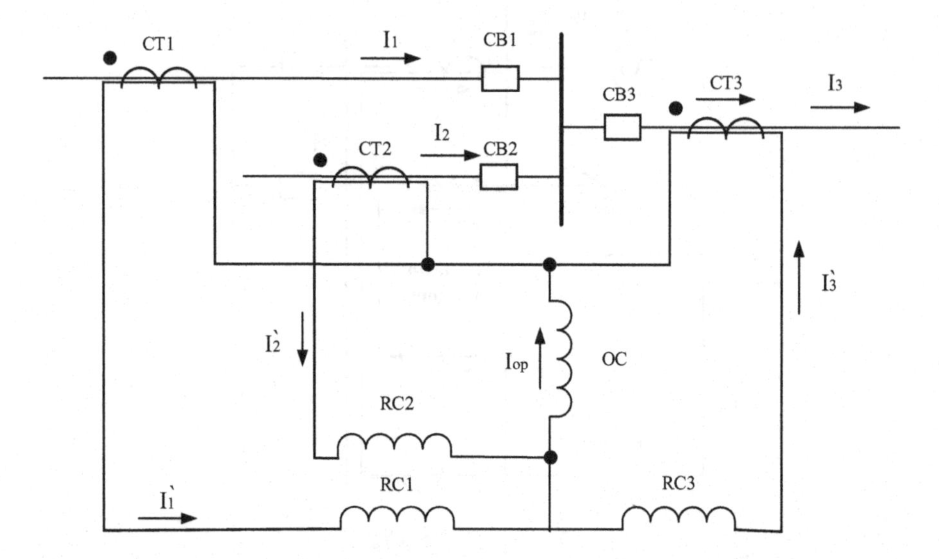

FIGURE 9.30 Differential relay of Problem 9.8.

10 High-Impedance Faults

10.1 INTRODUCTION

Power system distribution feeders are prone to direct contact with neighboring objects, such as tree branches, building walls, or surfaces near them; such a contact constitutes a distribution feeder fault.

This fault condition appears "invisible" to conventional fault detection methods. It is merely a leakage or small current that flows through surrounding objects, which presents high impedance in the current path. High-impedance faults (HIF) are difficult to recognize and detect by traditional monitoring equipment because their presence results in a slight increase in load current; thus, they can be confused with a normal increase in load. The failure of HIF's detection leads to potential hazards to human beings and potential fire hazards. Moreover, such conditions also constitute a loss of energy to the power companies as not all the produced electrical power is delivered to their appointed loads.

This chapter focuses on the technique based on Fourier transform, and an artificial neural network (ANN) is presented, simulated, and used. Primarily, this method detects and recognizes HIFs in electric distribution power systems. Moreover, it is used to detect and recognize system state changes.

10.2 CHARACTERISTICS OF HIFs

Figure 10.1 shows a HIF due to a tree branch contact. This situation is aggravated under wet weather conditions. Table 10.1 furnishes the current values of HIF for various fault situations.

FIGURE 10.1 HIF due to tree contact.

DOI: 10.1201/9781003394389-10

TABLE 10.1
Typical HIF's Current Values for Various Materials

Material	Current (A)
Dry asphalt	Very low ≈ 0
Concrete(non-reinforced)	Very low ≈ 0
Dry sand	Very low ≈ 0
Wet sand	15
Dry sod	20
Dry grass	25
Wet sod	40
Wet grass	50
Concrete reinforced	75

The distribution feeder under the HIF condition undergoes an awkward behavior. The weird nature of the HIF current and voltage causes such behavior.

The characteristics of the HIF are

- The fault current is low and much less than the feeder load current.
- A random arc can cause it.
- Consists of high-frequency harmonic components and inter-harmonics.
- Nonlinear behavior due to arcing conditions.
- Asymmetry of the current waveform.

10.3 HIF's DETECTION

All HIF detection schemes' objective is to identify special features in the voltage and current patterns associated with HIF.

In general, identification and detection techniques comprise two basic steps, these are:

i. Feature extraction.
ii. Pattern recognition (classification).

10.3.1 Feature Extraction

Researchers and protection engineers have proposed feature extractors based on factual techniques, digital signal processing, crest factor, and wavelet transform in high-frequency noise patterns and dominant harmonic vectors.

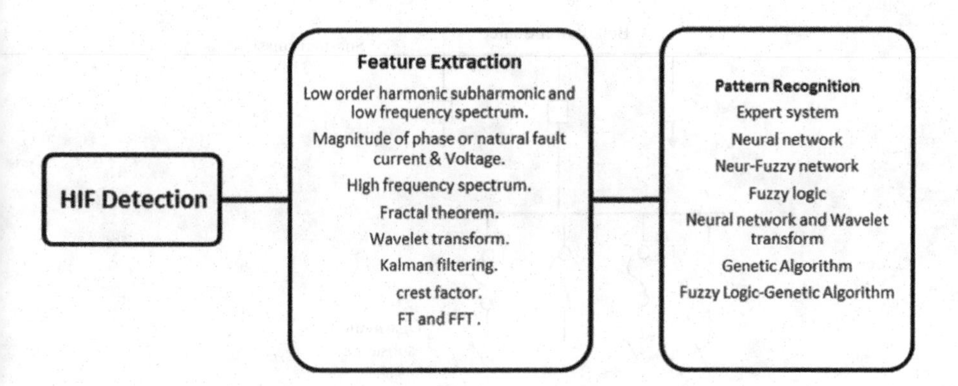

FIGURE 10.2 Methods used to detect HIF.

10.3.2 Pattern Recognition (Classification)

Pattern recognition methods, including expert systems and ANNs, are used to classify the fault based on the extracted features. Pattern recognition methods aim at detecting characteristic voltage and current distortions caused by arcing faults. Figure 10.2 shows more details of the technique used to detect HIFs.

To detect HIFs and recognize them from normal switching operations and load increase, nonlinear load, induction motor start, and capacitor switching comprise two stages.

In the first stage (feature extraction), the feeder's current and voltage signals at the source side are obtained and analyzed using Fourier transform (FT) to get the spectrum of the harmonic content. In the second stage (classification), the analyzed signals are classified using an ANN to detect, recognize, and categorize it as a fault, normal load switching, and increases, the nonlinear, induction motor or capacitor switching event.

This chapter has established a model to detect HIFs (L-G and L-L) in a medium voltage distribution system and distinguish it from other normal operation events, such as; normal load operation, induction motor starting, capacitor switching, and nonlinear load.

10.4 POWER DISTRIBUTION NETWORK

Distribution circuits come in different configurations and circuit lengths. Most share many common characteristics. Figure 10.3 shows a "typical" distribution circuit, and Table 10.2 shows the typical parameters of a distribution network. The main feeder is the three-phase backbone of the circuit, often called the main or mainline. Utilities often design the main feeder for 400 A and often allow an emergency rating of 600 A. Branching from the main are one or more laterals, called taps, lateral taps, branches, or branch lines. The laterals normally are fused to separate them from the mainline if they are faulted.

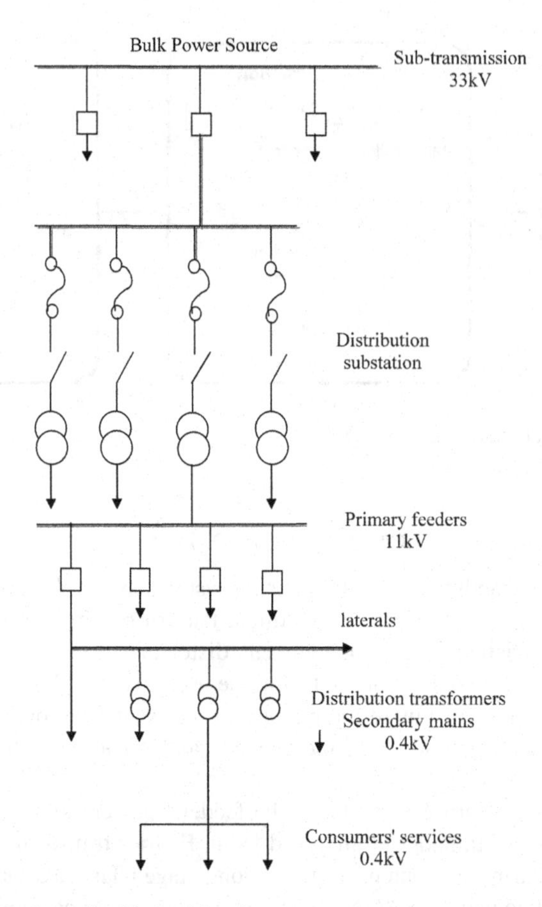

FIGURE 10.3 Typical distribution circuit.

TABLE 10.2
Typical Distribution Network Parameters

Characteristics	Value
Substation Characteristics	
Voltage	3.3–34.5 kV
Transformer size	5–60 MVA
Number of feeders per bus	1–7
Feeder Characteristics	
Peak current	100–500 A
Peak load	1–15 MVA
Power factor	0.8 lagging to 0.95 leading
Number of customers	50–5000
Length of feeder mains	3.3–25 km
Length including laterals	6.6–40 km
Area covered	0.5–500 km^2

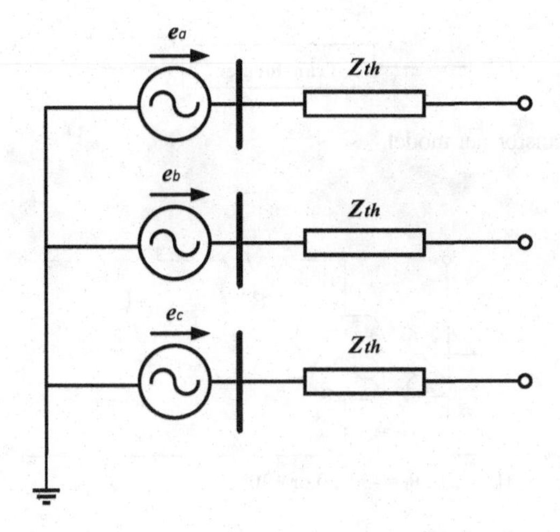

FIGURE 10.4 Source model.

10.5 SOURCE MODEL

A realistic representation of the source can be realized using the short-circuit level of the source bus. This is done by calculating the Thevenin's impedance (Z_{th}), simply using

$$Z_{th(PU)} = \frac{S_b}{SCL} \quad (10.1)$$

The source's balanced model can be obtained by adding the ideal voltage source behind the Z_{th} calculated and as shown in Figure 10.4.

An ideal source model is considered to reduce the complication arising while modeling the source side due to the vast variants of configurations in practice. In other words, an infinite bus source of zero Thevenin's impedance is considered.

10.6 POWER TRANSFORMER MODEL

Power transformers are essential components in any power system. At the distribution end of the system, transformers reduce the voltage to values suitable for utilization. Transformers can be modeled by a four-terminal network admittance equation, symbolically shown in Figure 10.5. The basic admittance is given as follows:

$$\begin{bmatrix} I_i \\ I_j \end{bmatrix} = \begin{bmatrix} Y^{pp} & Y^{ps} \\ Y^{sp} & Y^{ss} \end{bmatrix} \begin{bmatrix} V_i \\ V_j \end{bmatrix} \quad (10.2)$$

where Y^{pp}, Y^{ps}, Y^{sp}, and Y^{ss} are submatrices of the bus admittance matrix.

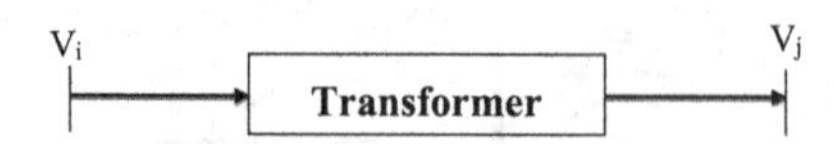

FIGURE 10.5 Transformer model.

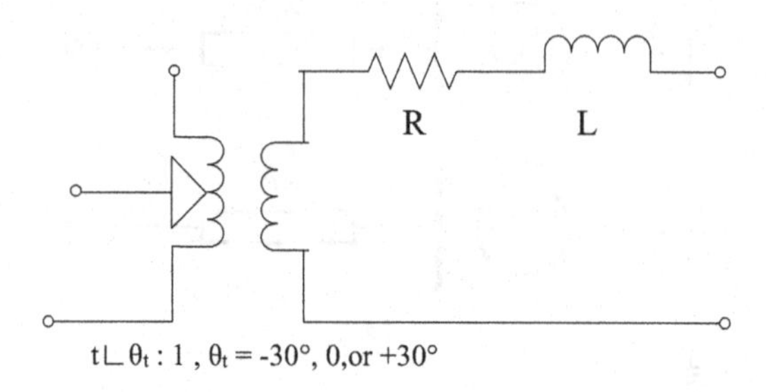

FIGURE 10.6 Transformer harmonic flow model.

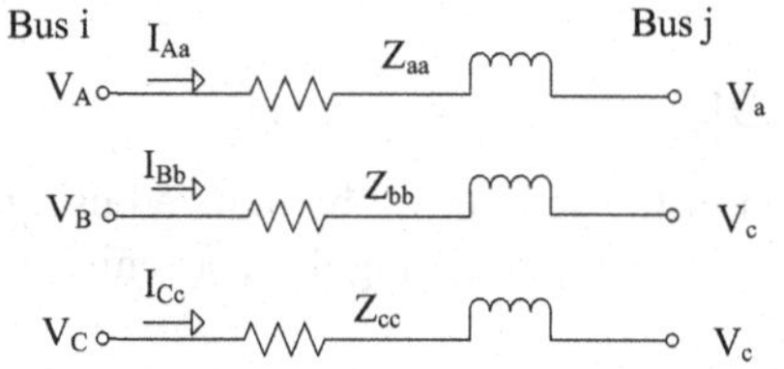

FIGURE 10.7 Short transmission line model.

A simple model is produced by simplifying the transformer standard equivalent circuit by ignoring the shunt branch and assuming constant leakage inductance. This model is depicted in Figure 10.6. The phase shift shown in Figure 10.6 represents the different winding connections.

10.7 LINE MODEL

Along streets, alleys, woods, and residential sites, many distribution lines that feed customers are overhead structures. If a distribution feeder of 80km in length or less, the shunt capacitance can be neglected entirely without much accuracy loss. A line can be modeled with only series *R*–*X* losses, called short-line representation.

A short transmission line model is shown in Figure 10.7, where Equation 10.3 is the impedance matrix for this model.

$$[Z_{abc}] = \begin{bmatrix} Z_{aa} & Z_{ab} & Z_{ac} \\ Z_{ba} & Z_{bb} & Z_{bc} \\ Z_{ca} & Z_{cb} & Z_{cc} \end{bmatrix} \tag{10.3}$$

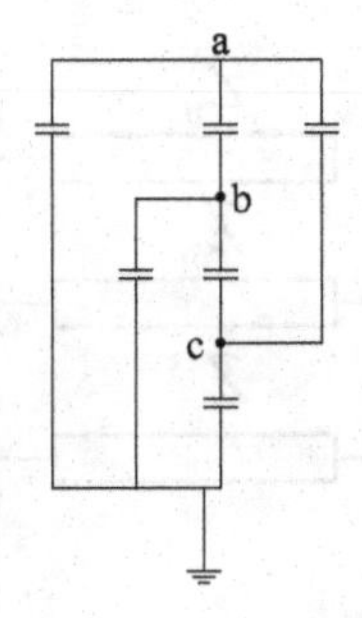

FIGURE 10.8 Shunt branch.

The relations between the bus voltages and branch currents in Figure 10.5 can be expressed as follows:

$$\begin{bmatrix} V_a \\ V_b \\ V_c \end{bmatrix} = \begin{bmatrix} V_A \\ V_B \\ V_C \end{bmatrix} - \begin{bmatrix} Z_{aa} & Z_{ab} & Z_{ac} \\ Z_{ba} & Z_{bb} & Z_{bc} \\ Z_{ca} & Z_{cb} & Z_{cc} \end{bmatrix} \times \begin{bmatrix} I_{Aa} \\ I_{Bb} \\ I_{Cc} \end{bmatrix} \tag{10.4}$$

Underground distribution has been used for decades, practically in crowded urban areas. Cables, connectors, and installation equipment have advanced considerably in the past decades, making underground distribution installations faster and less expensive.

For underground feeders, the shunt branch is shown in Figure 10.8; similar considerations as for the series impedance matrix lead to

$$[Y] = \begin{bmatrix} Y_{aa} & Y_{ab} & Y_{ac} \\ Y_{ba} & Y_{bb} & Y_{bc} \\ Y_{ca} & Y_{cb} & Y_{cc} \end{bmatrix} \tag{10.5}$$

The series impedance (series branch) and shunt admittance (shunt branch) lumped π-model representation of the three-phase line are shown in Figure 10.9.

10.8 LOAD MODEL

Generally, loads can be connected at a bus (spot load) or assumed to be uniformly distributed along a line (distributed load). Loads can be three-phase (balanced and unbalanced) or single-phase. Three-phase loads can be connected in wye or delta, while single-phase loads can be connected line-to-neutral or line-to-line. All loads can be modeled as constant kW and KVAr, constant impedance (Z), or constant current (I).

All the load data are given in three-phase kW and KVAr at a rated voltage (1.0 per unit).

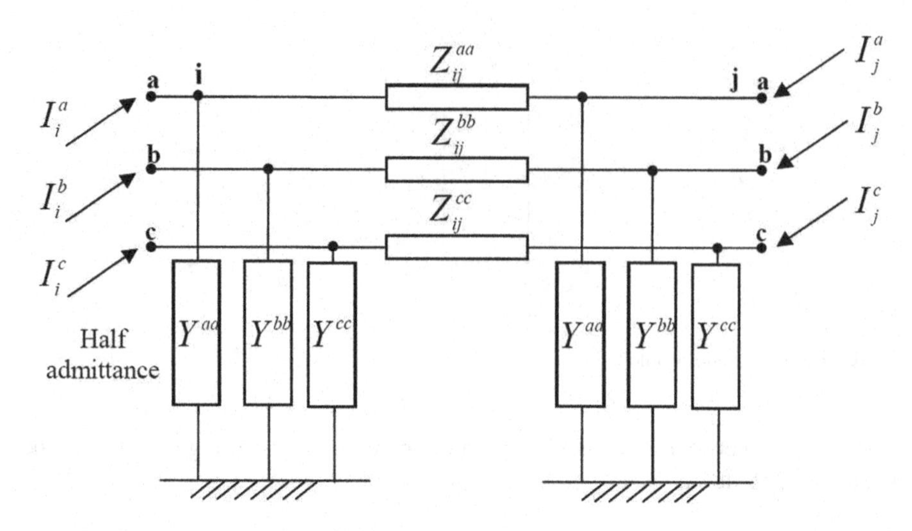

FIGURE 10.9 Full circuit representation.

10.9 SHUNT CAPACITOR MODEL

Most industrial loads operate at moderately low-power factors. Around 60% of the utility load consists of motors, so the power system's overall power factor is low. Depending on the level of the load, these motors are inherently low-power factor devices. The motors' power factor varies from 0.30 to 0.95, depending on the motor's size and other operating conditions. Therefore, the power factor level is always a concern for industrial power systems, utilities, and users. The system performance can be improved by correcting the power factor. The system power factor is given by

$$\text{Power factor} = \frac{P}{S} \tag{10.6}$$

where P and S are the real and apparent power, respectively.

The relation between the power factor and the Q/P ratio is shown in Table 10.3, from which even at 90% power factor, the reactive power requirement is 48% of the real power. At low-power factors, the reactive power demand is much higher. Therefore, some form of power factor correction is required in all industrial facilities.

The power factor correction capacitors can be installed at the high voltage bus, the medium voltage distribution bus, or the load bus. The power factor correction capacitors can be installed for a group of loads, at the branch location, or for a local load. The benefits of the power factor correction for the utility are a release in system generation capacity, savings in transformer capacity, reduction in line loss, and improved voltage profile. The benefits of power factor correction to the customer are reduced rate associated with power factor improvement, reduced loss causing lower peak demand, reduced energy consumption, and increased short-circuit rating. Figure 10.10 shows the representation of the shunt capacitor bank.

TABLE 10.3
Power Factor and *Q/P* Ratio

Power Factor %	Angle Degree	*Q/P* Ratio
100	0	0.00
95	11.4	0.20
90	26.8	0.48
805	31.8	0.62
80	36.8	0.75
70.7	45	1.00
60	53.1	1.33
50	60.0	1.73

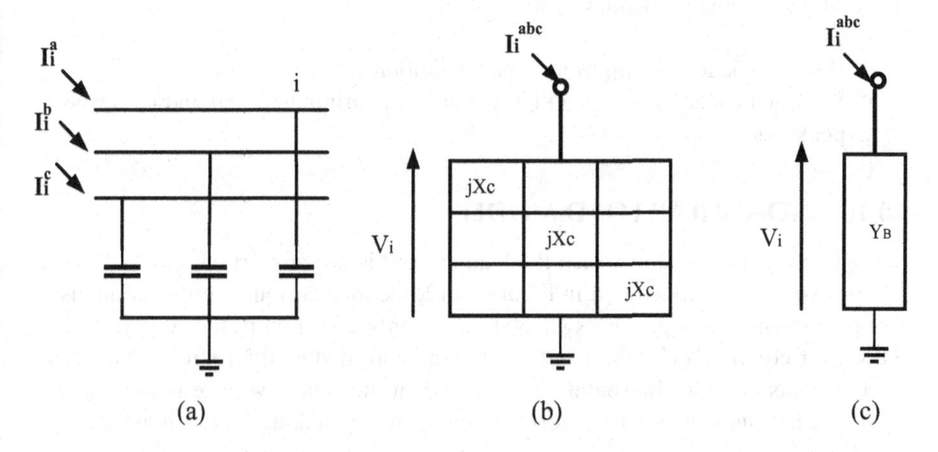

FIGURE 10.10 Representation of shunt capacitor bank. (a) Y-capacitor bank; (b) matrix equivalent; and (c) admittance representation.

The admittance matrix for shunt elements is usually diagonal, as there is normally no coupling between each phase's components. This matrix is then incorporated directly into the shunt branch's admittance matrix, contributing only to the particular node's self-admittance or used as a reactive power source (i.e., incorporated into the load of the particular node).

Shunt capacitors applied to distribution systems are generally located on the distribution lines or substations. The distribution capacitors may be in pole-mounted racks, pad-mounted banks, or submersible installations. The distribution banks often include three to nine capacitor units connected in three-phase grounded wye, ungrounded wye, or a delta configuration; Figure 10.11 shows these three types.

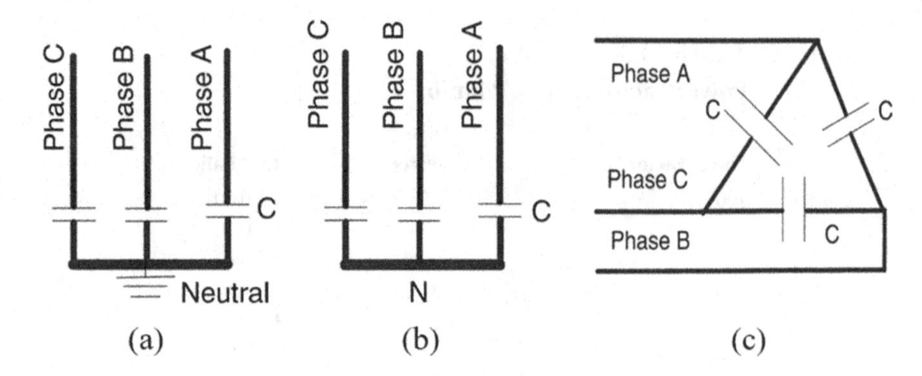

FIGURE 10.11 Capacitors bank. (a) Y grounded; (b) Y ungrounded; and (c) delta.

The distribution capacitors can be fixed or switched depending on the load conditions. The following guidelines apply:

- Fixed capacitors for minimum load conditions.
- Switched capacitors for lead levels above the minimum load and up to the peak load.

10.10 NONLINEAR LOAD MODEL

A nonlinear load is one in which the load current is not proportional to the instantaneous voltage. As illustrated in Figure 10.12, the load current is not continuous in this case. Nonlinear loads are switched on for only a portion of the AC cycle as in a thyristor-controlled circuit or pulsed as in a controlled rectifier circuit. Nonlinear load currents are non-sinusoidal, and even when the source voltage is a clean sine wave, the nonlinear load will distort the voltage wave, making it non-sinusoidal.

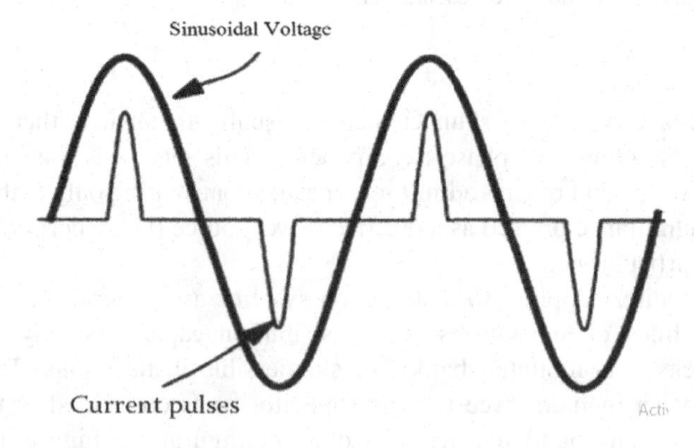

FIGURE 10.12 Typical waveforms of nonlinear load current.

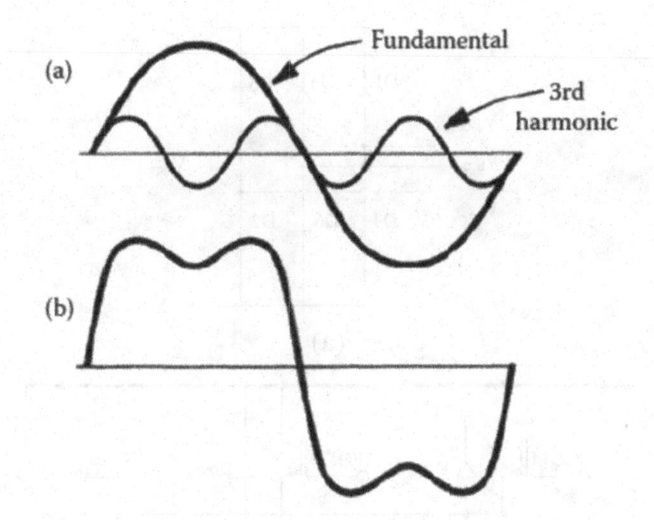

FIGURE 10.13 The addition of 50 Hz fundamental and third harmonic wave shapes: (a) individual waveforms and (b) resulting in a distorted wave shape.

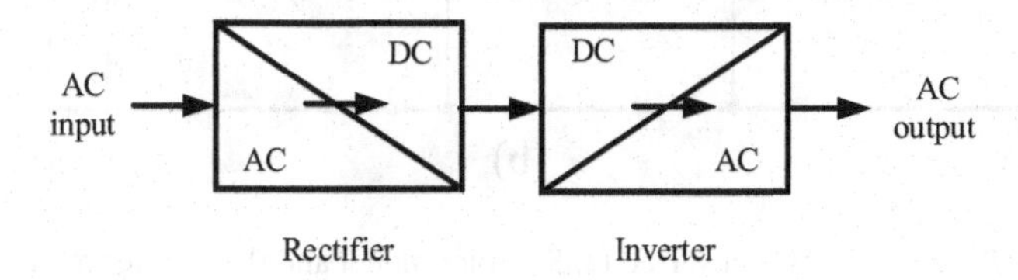

FIGURE 10.14 Nonlinear load (NLL).

Although traditional linear loads allow voltages and currents of the fundamental frequency to appear in the power system with little or no harmonic currents, nonlinear loads can introduce significant harmonics levels. Harmonic energies combine with the fundamentals to form distorted waveforms of the type shown in Figure 10.13. The amount of distortion is determined by the frequency and amplitude of the harmonic currents.

One of the most used nonlinear loads is an AC/DC/AC converter, and a sample nonlinear load is shown in Figure 10.14. This model consists of a rectifier and an IGBT inverter. Figure 10.15 shows the rectifier and inverter model, where (V_a, V_b, and V_c) is the input voltage to the full-wave rectifier and V_{dc} is the output DC voltage of the rectifier, and the V_{dc} is given by

$$\begin{aligned} V_{dc} &= \frac{6}{\pi}\int_{0}^{\frac{\pi}{6}} \sqrt{3}V_m \cos\omega t \, d(t) \\ &= \frac{3\sqrt{3}}{\pi}V_m = 1.654\,V_m \end{aligned} \tag{10.7}$$

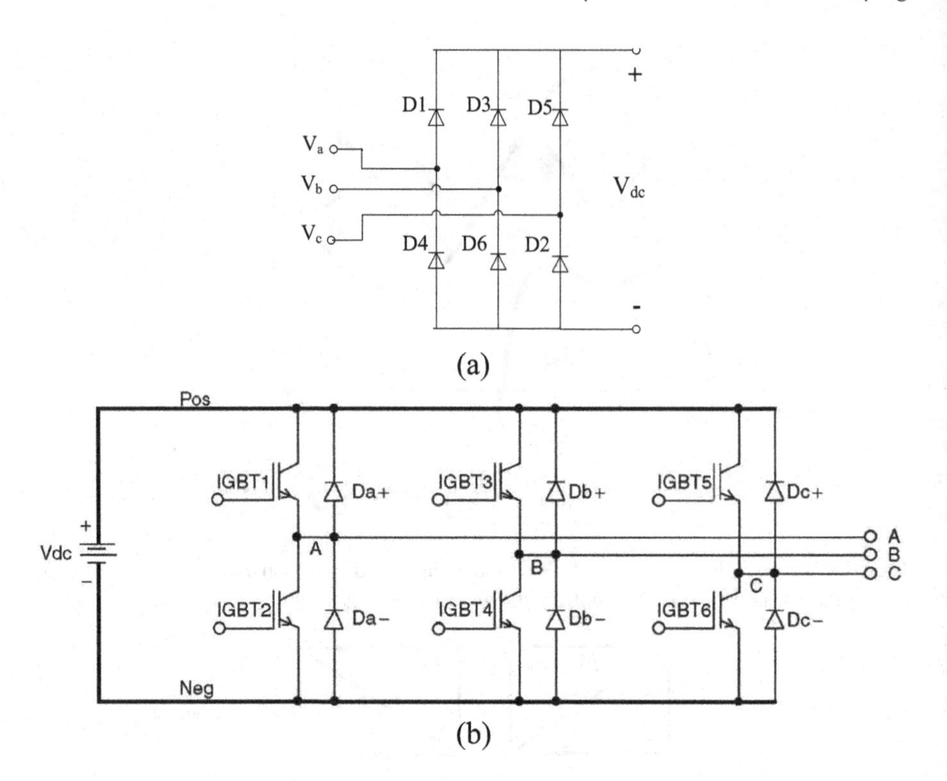

FIGURE 10.15 AC/DC/AC converter. (a) Rectifier model and (b) inverter model.

V_m: peak value of the phase voltage, and for the inverter, the line-to-line voltage can be found from

$$V_L = \frac{1}{2\pi}\int_0^{\pi} V_{dc}^2 \, d(\omega t) = \sqrt{\frac{3}{2}}\, V_{dc} = 0.8165 V_{dc} \tag{10.8}$$

and for the nth component of the line, voltage is

$$V_{Ln} = \frac{4V_s}{\sqrt{2}}\frac{1}{n\pi}\sin\frac{n\pi}{3} \tag{10.9}$$

where:

V_{dc}: DC input to the inverter.
V_{Ln}: nth component of the line output voltage

10.11 INDUCTION MOTOR MODEL

Electric motors represent an important fraction of the residential, commercial, and industrial loads; motors of some kind consume about 50%–60% of the electric energy in the world. Motor loads comprise fans, pumps of all kinds, including refrigerators and air-conditioners, power tools from hand drills to lawnmowers, and even electric streetcars – anything electric that moves.

Therefore, it is important to model a motor load in the distribution feeder. An induction motor is modeled because it has large starting currents that alter the current characteristics of the line momentarily; Figure 10.16 shows a sample induction motor starting current. Induction-starting synchronous motors are treated similarly.

An induction machine is generally modeled by specifying its VA rating, line voltage, and percent impedance. The per-phase *d*- and *q*-axes equivalent circuit is shown in Figure 10.17.

From the equivalent circuit, the *d*- and *q*-axes representation of the motor is

$$V_{qs} = Rs\ i_{qs} + \frac{d}{dt}\varphi_{qs} + \omega\varphi_{ds} \tag{10.10}$$

$$V_{ds} = Rs\, i_{ds} + \frac{d}{dt}\varphi_{ds} - \omega\varphi_{ds} \tag{10.11}$$

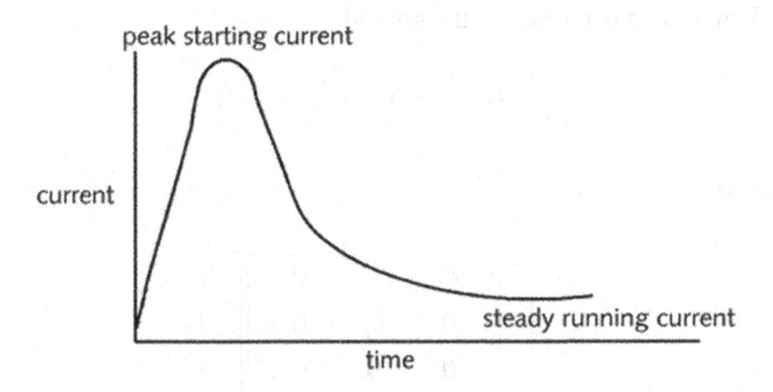

FIGURE 10.16 Induction machine starting current curve.

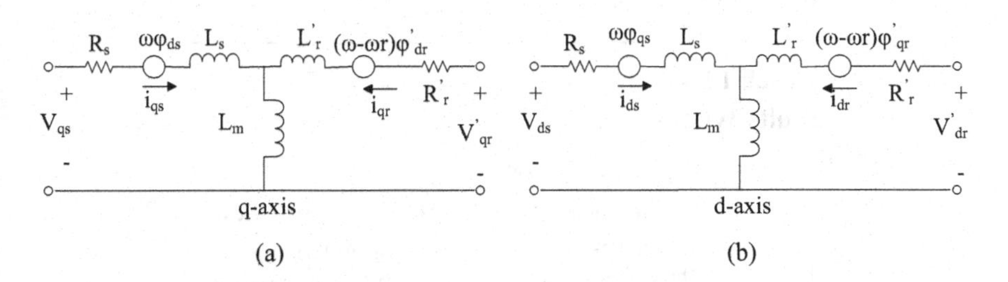

FIGURE 10.17 Per-phase equivalent circuit of induction motor: (a) quadrature axis equivalent circuit. (b) Direct axis equivalent axis.

$$V'_{qr} = R'_r i'_{qr} + \frac{d}{dt}\varphi'_{qr} + (\omega - \omega_r)\ \varphi'_{qr} \tag{10.12}$$

$$V'_{dr} = R'_r i'_{dr} + \frac{d}{dt}\varphi'_{dr} + (\omega - \omega_r)\ \varphi'_{qr} \tag{10.13}$$

10.12 FAULT MODEL

A fault in a circuit is any failure that interferes with the normal flow of electric power. Faults occur in symmetrical and unsymmetrical faults, as classified in Table 10.4, excluding open circuit faults.

Unsymmetrical short-circuit faults are more common. The most common type is a line-to-ground fault. Approximately 70% of the faults in power systems are single line-to-ground faults.

Since unsymmetrical faults are the most common, the unsymmetrical component analysis of faults will be considered.

10.12.1 Symmetrical Fault Model

The three phases are short-circuited through equal fault impedances Z_f, as shown in Figure 10.18. The vector sum of fault currents is zero, as the symmetrical fault is considered, and there is no path to the ground.

$$I_a + I_b + I_c = 0, \quad \text{and} \quad I_o = 0 \tag{10.14}$$

As the fault is symmetrical:

$$\begin{bmatrix} V_{fa} \\ V_{fb} \\ V_{fc} \end{bmatrix} = \begin{bmatrix} Z_f & 0 & 0 \\ 0 & Z_f & 0 \\ 0 & 0 & Z_f \end{bmatrix} \begin{bmatrix} I_{fa} \\ I_{fb} \\ I_{fc} \end{bmatrix} \tag{10.15}$$

TABLE 10.4
Faults Types

	Faults	
	Symmetrical	**Unsymmetrical**
1	Three-phase	Single line-to-ground
2	Three-phase-to-ground	Double line-to-ground
3	-	Line-to-line

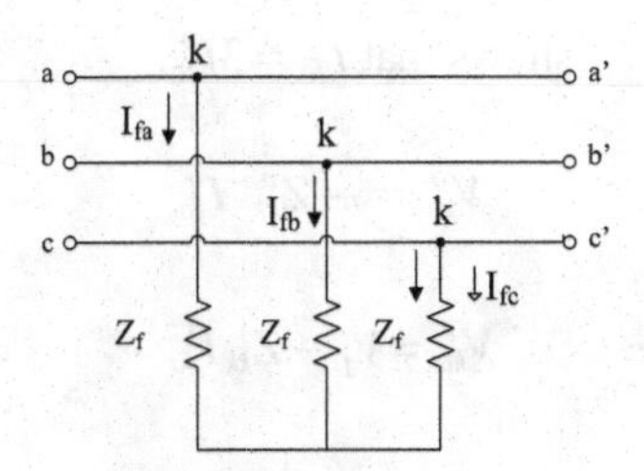

FIGURE 10.18 Three-phase symmetrical fault.

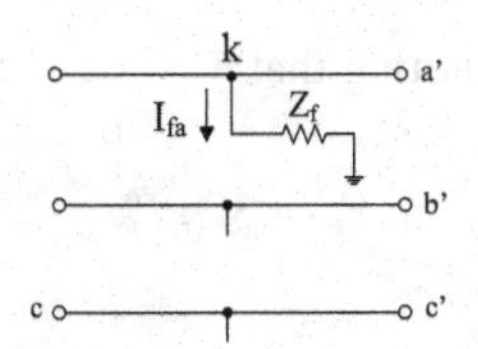

FIGURE 10.19 Line-to-ground fault in a three-phase system.

where

$$I_f = \frac{V_f}{Z_f} \tag{10.16}$$

10.12.2 Line-to-Ground Fault Model

A single line-to-ground fault through impedance Z_f, the hypothetical stubs on the three lines, is connected, as shown in Figure 10.19, where phase a is the one on which the fault occurs. The following equations express the conditions of the faulty bus.

$$I_{fb} = 0, \quad I_{fc} = 0, \quad V_{ka} = Z_f I_{fa} \tag{10.17}$$

The symmetrical components of the stub current are given by

$$\begin{bmatrix} I_{fa}^o \\ I_{fa}^1 \\ I_{fa}^2 \end{bmatrix} = \frac{1}{3}\begin{bmatrix} 1 & 1 & 1 \\ 1 & a & a^2 \\ 1 & a^2 & a \end{bmatrix}\begin{bmatrix} I_{fa} \\ 0 \\ 0 \end{bmatrix} = \frac{1}{3}\begin{bmatrix} I_{fa} \\ I_{fa} \\ I_{fa} \end{bmatrix} \tag{10.18}$$

and performing the multiplication yields

$$I_{fa}^o = I_{fa}^1 = I_{fa}^2 = \frac{1}{3} I_{fa} \tag{10.19}$$

Substituting I_{fa}^{o} for I_{fa}^{1} and I_{fa}^{2} shows that $I_{fa} = 3I_{fa}^{o}$,

$$V_{ka}^{o} = -\ Z_{kk}^{o}\ I_{fa}^{o} \tag{10.20}$$

$$V_{ka}^{1} = V_f - Z_{kk}^{1} I_{fa}^{1} \tag{10.21}$$

$$V_{ka}^{2} = -Z_{kk}^{2} I_{fa}^{2} \tag{10.22}$$

Summing these equations and noting that

$$V_{ka} = 3Z_f\, I_{fa}^{o} \tag{10.23}$$

$$V_{ka} = V_{ka}^{o} + V_{ka}^{1} + V_{ka}^{2} = V_f - \left(\ Z_{kk}^{o} + Z_{kk}^{1} + Z_{kk}^{2}\right) I_{fa}^{o} = 3Z_f\, I_{fa}^{o} \tag{10.24}$$

Solving for I_{fa}^{o} and combining the result with Equation 10.19 gives

$$I_{fa}^{o} = I_{fa}^{1} = I_{fa}^{2} = \frac{V_f}{Z_o + Z_1 + Z_2 + 3Z_f} \tag{10.25}$$

Equation 10.25 is the fault current equations particular to the single line-to-ground fault through impedance Z_f. They are used with symmetrical component relations to determine the voltages and currents at the fault point. The Thevenin's equivalent circuits of the system's three sequence networks are connected in series, as shown in Figure 10.20.

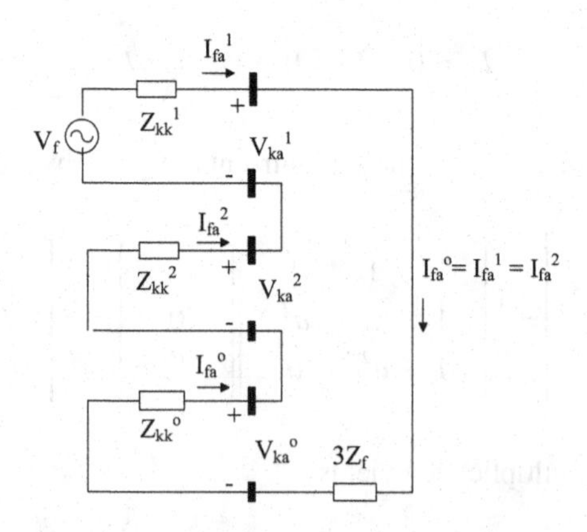

FIGURE 10.20 The Thevenin equivalent of the sequence networks simulates a single line-to-ground fault on phase an at bus *k*.

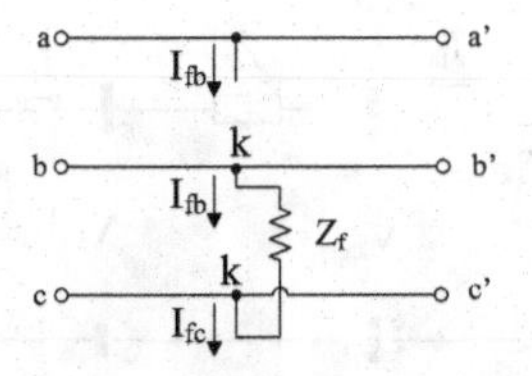

FIGURE 10.21 Line-to-line fault in a three-phase system.

10.12.3 Line-to-Line Fault Model

The hypothetical stubs on the three lines at fault are connected to represent a line-to-line fault through impedance Z_f, as shown in Figure 10.21. Bus k is the fault point, and the line-to-line fault is in phases b and c. The following relations must be satisfied at the fault point

$$I_{fa} = 0, I_{fb} = -I_{fc}, V_{kb} - V_{kc} = I_{fb} Z_f \tag{10.26}$$

Since $I_{fb} = -I_{fc}$ and $I_{fa} = 0$, the symmetrical components of the current are

$$\begin{bmatrix} I_{fa}^o \\ I_{fa}^1 \\ I_{fa}^2 \end{bmatrix} = \frac{1}{3} \begin{bmatrix} 1 & 1 & 1 \\ 1 & a & a^2 \\ 1 & a^2 & a \end{bmatrix} \begin{bmatrix} 0 \\ I_{fb} \\ -I_{fb} \end{bmatrix} \tag{10.27}$$

$$I_{fa}^o = 0 \tag{10.28}$$

$$I_{fa}^1 = -I_{fa}^2 \tag{10.29}$$

To satisfy the requirement that $I_{fa}^1 = -I_{fa}^2$. Let's connect the Thevenin equivalent of the positive- and negative-sequence network in parallel, as shown in Figure 10.22.

To show that this connection of the networks also satisfies the voltage equation $V_{kb} - V_{kc} = I_{fb} Z_f$, each side of that equation can be expanded as follows:

$$\begin{aligned} V_{kb} - V_{kc} &= \left(V_{kb}^1 + V_{kb}^2\right) - \left(V_{kc}^1 + V_{kc}^2\right) = \left(V_{kb}^1 - V_{kc}^1\right) + \left(V_{kb}^2 - V_{kc}^2\right) \\ &= \left(a^2 - a\right)V_{ka}^1 + \left(a - a^2\right)V_{ka}^2 = \left(a^2 - a\right)\left(V_{ka}^1 - V_{ka}^2\right) \end{aligned}$$

$$I_{fb} Z_f = \left(I_{fb}^1 + I_{fb}^2\right) Z_f = \left(a^2 I_{fb}^1 + a I_{fb}^2\right) Z_f$$

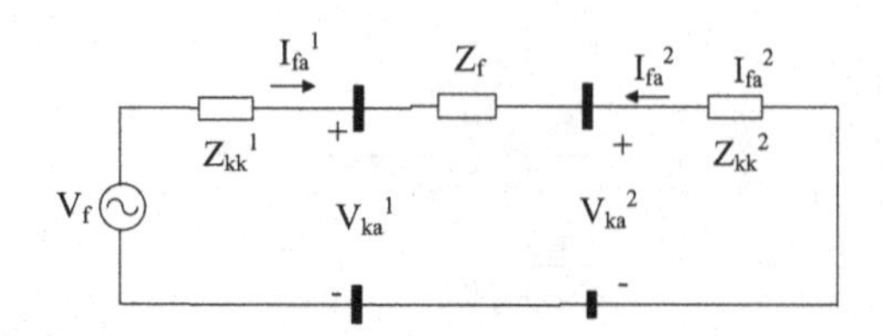

FIGURE 10.22 Connection of the Thevenin equivalents of the positive- and negative-sequence networks for a line-to-line fault between phases *b* and *c* at bus *k*.

Equating both terms and setting $I_{fa}^2 = -I_{fa}^1$ as in Figure 10.22, they yield

$$\left(a^2 - a\right)\left(V_{ka}^1 - V_{ka}^2\right) = \left(a^2 + a\right) I_{fa}^1 Z_f$$

or

$$\left(V_{ka}^1 - V_{ka}^2\right) = I_{fa}^1 Z_f \quad (10.30)$$

The equation for the positive-sequence current in the fault can be determined directly from Figure 10.22, so that

$$I_{fa}^1 = -I_{fa}^2 = \frac{f}{Z_{kk}^1 + Z_{kk}^2 + Z_f} \quad (10.31)$$

For a bolted line-to-line fault, $Z_f = 0$.

10.13 PROCEDURAL EVENTS MODELING AND TECHNIQUES

In three-phase distribution feeder system, the events are to be considered normal load, load switching, and induction motor starting. Nonlinear load application.

i. Capacitor switching.
ii. Fault occurrence [HIF(L-G and L-L)], three-phase low-impedance fault (LIF).

Before actually simulating any of the events mentioned earlier, the system's normal operating response is obtained. That is, getting the source bus voltage and current waveform. These time domain waveforms are then transformed into the frequency domain using the FT technique. A variety of loading states are treated likewise, and a neural network is trained using the data obtained. As such, the starting ground for the whole study is furnished, and the source bus equipment is ready to detect any abnormality if it occurs.

10.14 THE FOURIER TRANSFORM

The voltage and current signal of each case, assumed in this chapter, are recorded and converted to the frequency domain, arranged in a matrix form of both voltage and current signal.

The idea of Fourier transforms a natural extension of the idea of the Fourier series. A function, $F(x)$, with periodicity λ, in the sense $F(x+\lambda)=F(x)$ is represented by the series

$$F(x)=\frac{1}{\sqrt{2\lambda}}a_0+\sum_{n=1}^{\infty}\left(a_n\sqrt{\frac{2}{\lambda}}\cos\frac{2n\pi x}{\lambda}+b_n\sin\frac{2n\pi x}{\lambda}\right) \tag{10.32}$$

where

$$a_n=\sqrt{\frac{2}{\lambda}}\int_{\frac{-\lambda}{2}}^{\frac{\lambda}{2}}F_{(x)}\cos\frac{2\pi nt}{\lambda}\,dt \tag{10.33}$$

$$b_n=\sqrt{\frac{2}{\lambda}}\int_{\frac{-\lambda}{2}}^{\frac{\lambda}{2}}F_{(x)}\sin\frac{2\pi nt}{\lambda}\,dt \tag{10.34}$$

Inserting Equations 10.33 and 10.34 into Equation 10.32 yields

$$F(x)=\frac{1}{\lambda}\int_{\frac{-\lambda}{2}}^{\frac{\lambda}{2}}F_{(t)}dt+\frac{1}{\lambda}\sum_{n=1}^{\infty}F_{(t)}\cos\frac{2\pi n(x-t)}{\lambda}\,dt \tag{10.35}$$

If $F_{(t)}$ has compact support and putting $k=\frac{2\pi n}{\lambda}$ and identifying $\delta_k=\frac{2\pi}{\lambda}$ and letting $\lambda\to\infty$, Equation 10.35 becomes

$$F(x)=\frac{1}{\pi}\int_{0}^{\infty}dk\int_{-\infty}^{8}F_{(t)}\cos(k(x-t))\,dt \tag{10.36}$$

by expanding the cos in Equation 10.36 to yield:

$$F(x)=\frac{1}{2\pi}\int_{-\infty}^{\infty}dk\int_{-\infty}^{8}F(t)e^{ik(t-x)}\,dt \tag{10.37}$$

From the exponential form of the Fourier integral Equation 10.37, we obtain the Fourier transformation relations

$$F(X) = \frac{1}{\sqrt{2\pi}} \int_{-\infty}^{\infty} f(t)e^{-ikt}\, dt \quad (10.38)$$

$$f(t) = \frac{1}{\sqrt{2\pi}} \int_{-\infty}^{\infty} F(k)e^{ikt}\, dk \quad (10.39)$$

PROBLEMS

10.1. Explain the characteristics of a high-impedance fault.
10.2. Give the difference between HIF and LIF?
10.3. Which fault occurs as a high-impedance fault? How does the high-impedance fault occur in a system? What is the difference between high-impedance and low-impedance REF protection?
10.4. What is a high-impedance relay?
10.5. What are the methods of detection of HIF?

11 Grounding of Power System

11.1 INTRODUCTION

Power systems must be grounded for several technical and safety reasons; grounding is achieved by embedding metallic structures (conductors) into the earth and electrically connecting them to the power system's neutral. In this manner, the low impedance is provided between the power system neutral and the vast soil, which guarantees that the neutral concerning the earth's voltage will be down under all conditions. Grounding is necessary for several reasons

i. To support the voltage during transitive conditions.
ii. Minimize the likelihood of flashover during transients.
iii. To dissipate lightning strokes.
iv. To ensure the correct procedure of electrical devices.
v. To supply safety during normal or fault conditions.

This chapter contributes to power system protection and is dedicated to modeling various systems and configurations used in substation grounding. Different available theoretical approaches will be analyzed and compared for this specific purpose. The model includes all influence factors, such as short-circuit level, soil resistivity, electrode type, size, materials, and configuration. This chapter analyzes and discusses the various ways of soil resistivity measurements.

11.2 THE CONCEPT OF GROUNDING

The earth-embedded metal structures will produce a grounded system and provide a conducting course of electricity to the ground; hence the purpose of grounding is to provide a low-impedance power contact between the neutral of the electrical power system and the earth. Ideally, the potential of the simple three-phase system needs to be the same as that of earth. In this instance, individuals are safe when they touch metallic buildings linked to the system neutrals. However, the abnormal procedure includes highly unbalanced working or fault conditions and public safety. Depending on the potential difference between earth points and grounded structures, a hazardous condition may be produced for humans. This may result from three specific possibilities

i. A person holding a grounded structure with a different potential because of the point of the ground at which the person is standing. In this case, the person is exposed to a voltage that will generate an electric current through the body (called touch voltage).

DOI: 10.1201/9781003394389-11

ii. A person walking in the area will experience a voltage between their feet. This kind of voltage will generate an electric body current, called (step voltage).
iii. A person is touching the substation fence (in a remote area) where the shock ac electricity may be approaching (or equal to) the full ground potential rise (GPR) of a ground electrode.

Grounding systems should be designed so that an operator or bystander's possible electric body current does not exceed a specific limit under any foreseeable adverse conditions. The analysis procedure to determine the safety of power installations will be partitioned into two parts. The first analysis problem addressed is determining the maximum voltage level of grounded structures [potential ground rise (GPR)] under all not-far-off adverse conditions. For this purpose, the fault condition resulting in the greatest GPR must be determined and examined. The second analysis problem details the maximum body current computation in a person found in the ground field, given the earth's potential rise. Based on available experimental data and accepted safety factors, the ANSI/IEEE standard-80 suggests that the electric body current should be below $0.116/\sqrt{t}$.

Due to the increasing complexities of modern power system networks, improving the existent protection functions and developing new ones have recently attracted much attention. The goal is to enhance the overall power system performance, and fault location estimation can be considered the first attempt to realize this aim; fault location estimation requires more accurate and sophisticated computation routines. Thus, the need for fault location algorithms is obvious.

11.3 PURPOSES OF SYSTEM GROUNDING

There are several important reasons why a grounding system should be installed:

1. Prevent hazardous voltages due to system faults,
2. Prevent disturbances to electronics,
3. Minimize danger and damage due to lightning strikes,
4. Ensure the fast operation of protection devices,
5. Controlling the voltage as for the earth, or ground, inside unsurprising points of confinement, and
6. Providing a stream of current that will permit discovering an undesirable association between system conductors and ground. Such identification may then start using programmed gadgets to expel the voltage source from these conductors with undesired associations with the ground (Figures 11.1 and 11.2).

Electric circuits are linked to the ground (earth) for several reasons. In main run equipment, exposed metal parts are linked to containment surfaces to prevent user contact with dangerous voltage when electrical insulation fails. A protective ground is an essential part of the safety grounding system in electrical power distribution systems.

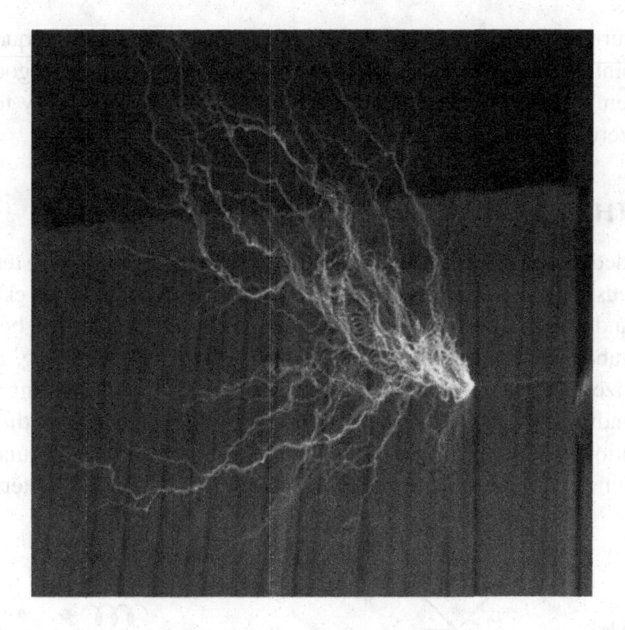

FIGURE 11.1 Hazardous voltages (http://www.iaacblog.com/programs/final-research-phase-i-on-the-issue-of-waves/).

FIGURE 11.2 Lightning strikes (https://www.pinterest.com/explore/lightning-strikes/).

For measurement purposes, our planet serves as a (reasonably) frequent potential reference point against which other possibilities can be measured. A good electrical ground system should have an appropriate current-carrying capacity to serve as a conduit for zero-voltage reference level.

11.4 METHODS OF SYSTEM-NEUTRAL GROUNDING

Most grounded systems utilize some technique for grounding the system neutral at least one focuses. These strategies can be separated into two general classes: strong grounding and impedance grounding. Impedance grounding might be partitioned into a few subcategories: reactance grounding, resistance grounding, and ground-fault neutralizer grounding. Figure 11.3 shows cases of these grounding methods.

The grounding system's design could differ greatly depending on the site's situation, application, and geography. For instance, if the system to be grounded features high AC electricity, then low-impedance grounding could lead to potential hazards

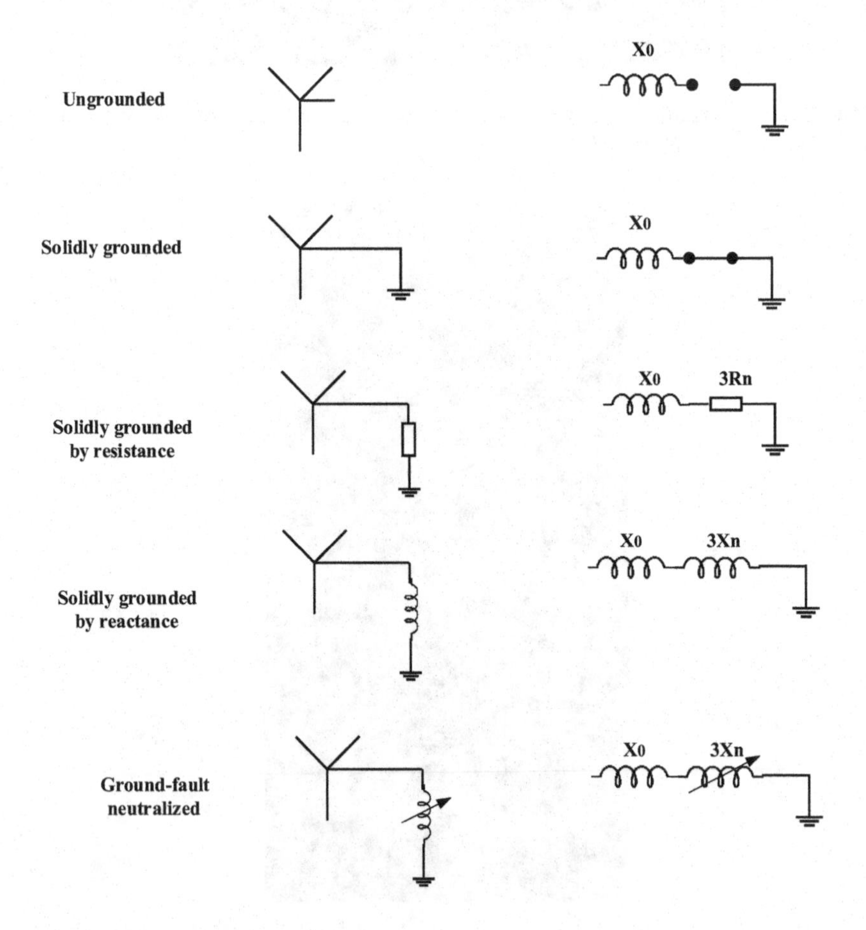

FIGURE 11.3 System-neutral circuit and equivalent diagram.

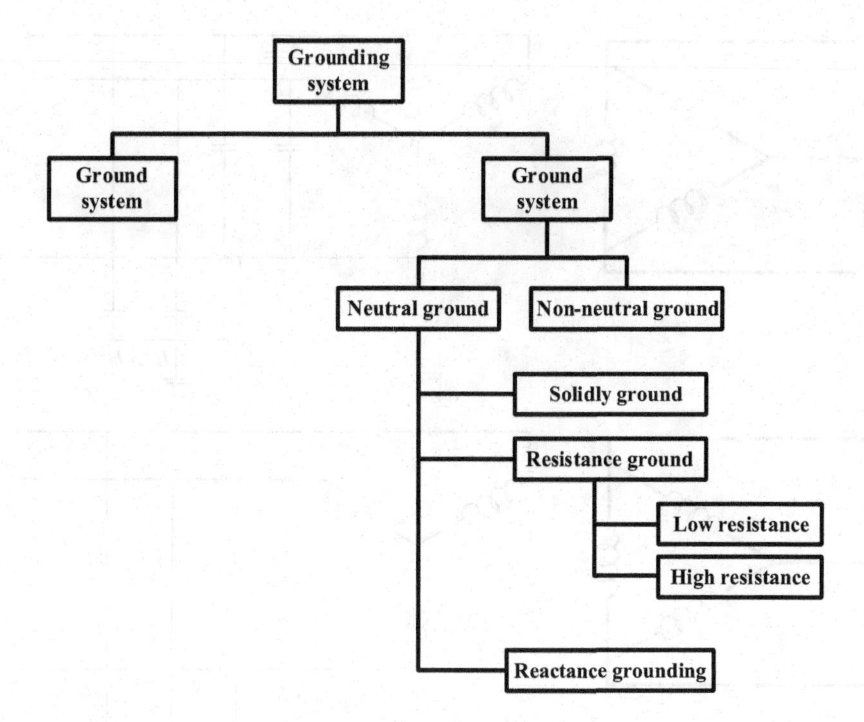

FIGURE 11.4 Classification of grounding systems.

relating to fire hazards. The site's geography is considered because the soil resistivity could customize the impedance of the grounding system. The classification of grounding systems is given in Figure 11.4.

11.4.1 Ungrounded System

In an ungrounded system, there is no intentional connection between the system conductor's ground. However, as shown in Figure 11.5a, there always exists a capacitive coupling between one system conductor and another and between system conductors and ground. Consequently, the so-called ungrounded system is a capacitance grounded system. The distributed capacitance from the system conductors to the ground, since the capacitance between phases has little effect on the system's grounding characteristics, will be disregarded. For simplicity, the distributed capacitive reactance to ground, X_{co}, is assumed to be balanced. In a non-faulted condition, with balanced three-phase voltages applied to the lines, the capacitive charging current, I_{co}, in phase will be equal and displaced 120° from one another. The phase voltages to the ground will also be equal and displaced 120° from one another.

Since the neutral of the distributed capacitances is at earth potential, the transformer's neutral is also at earth potential, held there by the capacitance to ground.

If one of the system conductors, phase C, faults to ground, current flow through that capacitance to the ground will cease since no potential difference exists across

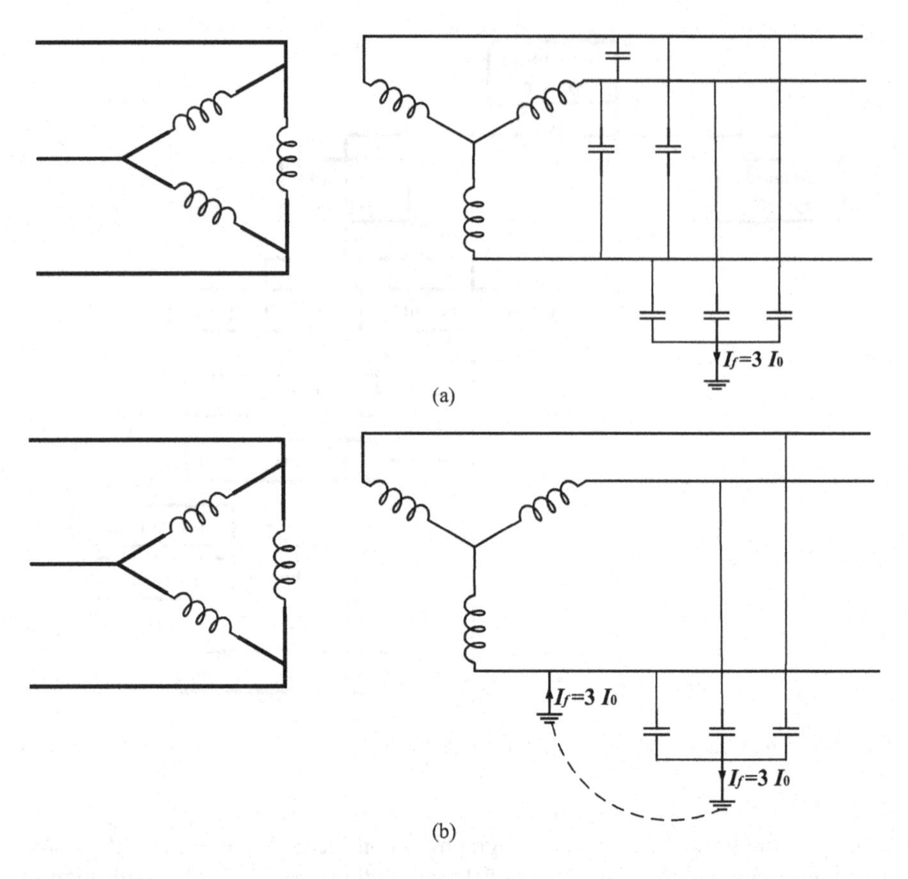

FIGURE 11.5 Ungrounded system: (a) circuit configuration, (b) single line to the ground on an underground system circuit configuration.

it. However, the voltage across the remaining two distributed capacitors to the ground will increase from line to neutral to line to line. The capacitive charging current, I_{co}, in the two non-faulted phases will increase by the square root of 3.

Hence, the vector sum of the capacitive charging current to ground is no longer zero but is $3I_{co}$ or three times the original charging current per phase. If flowing from the faulted conductor to the ground, the fault current leads the original line-to-neutral voltage ($V_{nc} = -V_{an}$) by ~90°.

In an ungrounded system, destructive transient overvoltage can occur throughout the system during restricting ground faults. This overvoltage, which can be several times normal in magnitude, result from a resonant condition being established between the inductive reactance of the system and the distributed capacitance to the ground.

11.4.2 Methods of System-Neutral Grounding

In a resistance-grounded system, the transformer or generator's nonpartisan is associated with the ground through at least one resistor. A resistance-grounded neutral

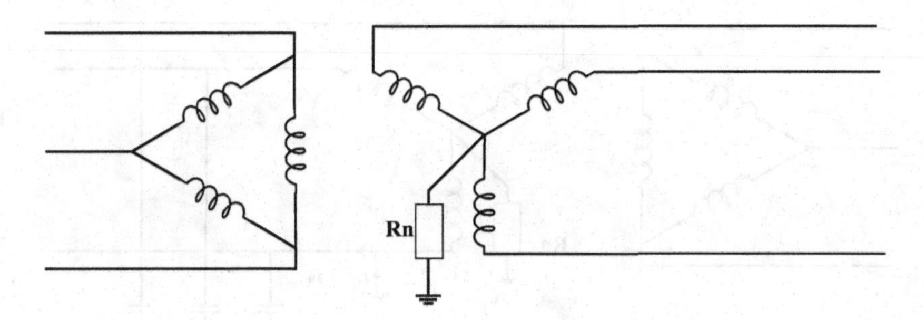

FIGURE 11.6 Resistance ground system.

system appears in Figure 11.6. As regularly introduced, the resistance has an extensively greater ohmic extent than the resistor area's system reactance. The line-to-ground voltages that exist amid are about the same as those for an ungrounded system.

The purposes behind restricting the current by resistance grounding incorporate the accompanying:

a. To lessen consuming impacts in fault electric, for example, switchgear, transformers, l, and machines.
b. To lessen mechanical worries in circuits and device conveying issue streams.
c. To decrease the current impact of the work power, which may have unintentionally brought on or happened to be close to the ground fault.
d. To lessen the transient line-voltage plunge caused by the event and clearing of a ground fault.
e. To secure control of transient overvoltage while, in the meantime, staying away from the shutdown of a fault circuit in the event of the principal ground fault.

Resistance grounding might be of two classes, either high or low, recognized by the size of the ground-fault current allowed to stream.

High-resistance grounding utilizes a neutral resistor of high ohmic value. High-resistance grounding has the accompanying choices:

a. Service congruity is kept up. The main ground fault does not require prepared gear to be closed.
b. Transitive overvoltage because of reducing ground flaws is reduced (to 250% of typical).
c. A sign pursuing or heartbeat system will encourage finding a floor fault.
d. The requirement for and cost of facilitated ground-fault handing-off is wiped out.

Figure 11.7 shows a circuit diagram for a single line to the ground on a high-resistance grounded system, and Figure 11.8 shows a scheme for detecting a single line to the ground on a high-resistance grounded system

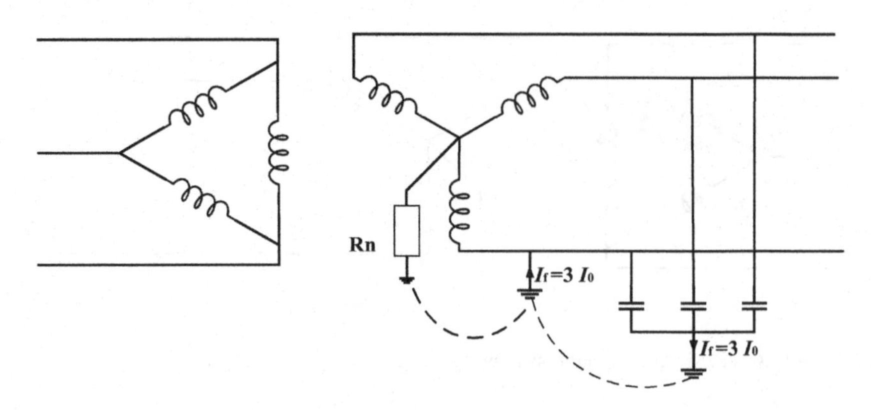

FIGURE 11.7 Circuit diagram for single line to the ground on a high-resistance grounded system.

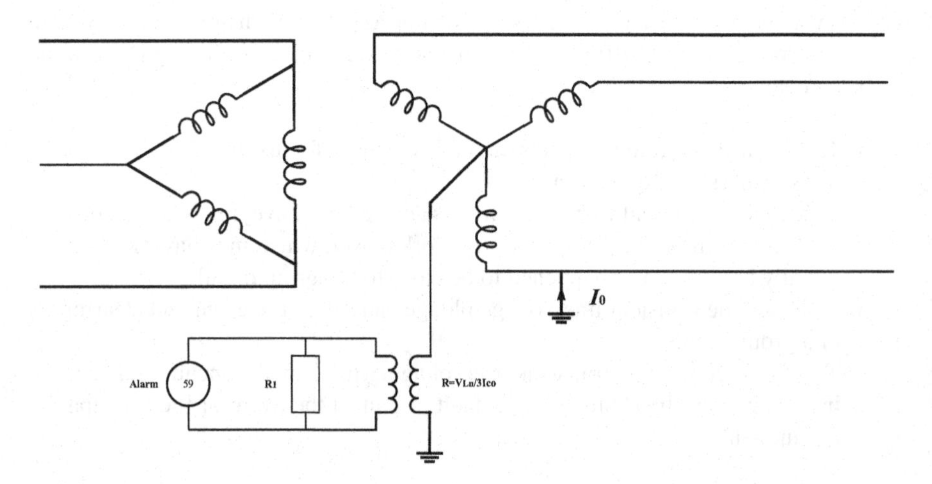

FIGURE 11.8 Scheme for detecting single line to the ground on a high-resistance grounded system.

Low-resistance grounding is when resistance is added over the grounding to keep faults power within limits. Limiting the fault current here aims to protect the grounding caudillo's efficiency. Low-resistance grounding gets a well-grounded circuit. This requires the minimum ground-fault current to be large enough to positively actuate the applied ground-fault relay. This method is presented in Figure 11.9.

11.4.3 Reactance Grounding

The term reactance grounding portrays the case in which a reactor is associated with the system unbiased and grounded, as appeared in Figure 11.10. Since the ground fault that may stream in a reactance-grounded system is an element of unbiased reactance, the extent of the ground-fault current is regularly utilized as a paradigm for

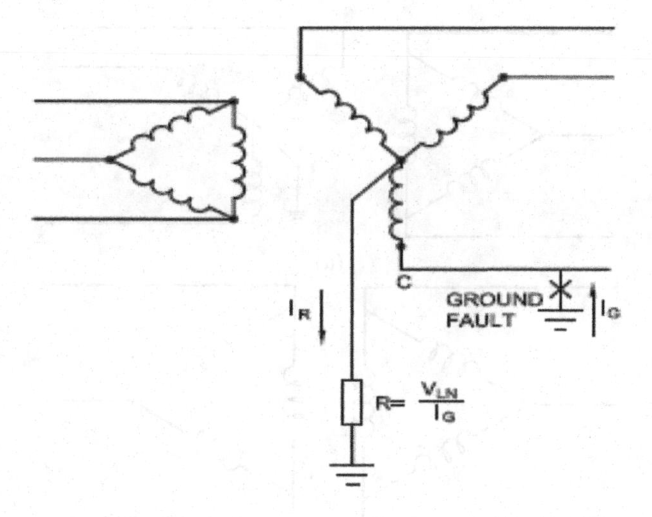

FIGURE 11.9 Low-resistance ground system.

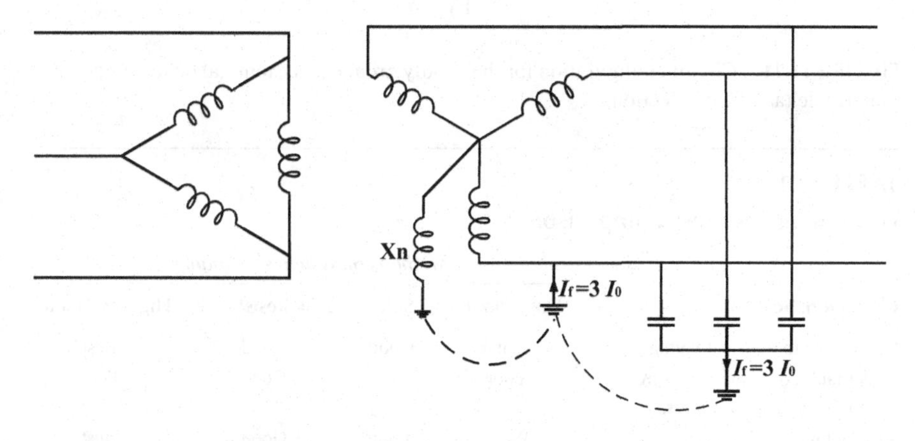

FIGURE 11.10 Circuit configuration for single line to the ground on a low-resistance grounded system.

portraying the level of grounding. In a reactance-grounded system, the accessible ground-fault current should be no <25% ($X_0 = 10X_1$) and ideally 60% ($X_0 = 3X_1$) of the three-stage fault current to counteract genuine transient overvoltage. The term X_0, as utilized, is the aggregate of the source zero-succession reactance, X_0, in addition to three times the grounding reactance, $3X_n$, ($X_0 = X_0$ source $+ 3X_n$).

Two cases of emphatically grounded systems appear in Figure 11.11. In contrast, due to the reactance of the grounded generator or transformer in arrangement with the neutral circuit, a very good ground relationship does not provide a zero-impedance impartial circuit.

Table 11.1 condenses the previously mentioned grounding plans, which would give a better understanding

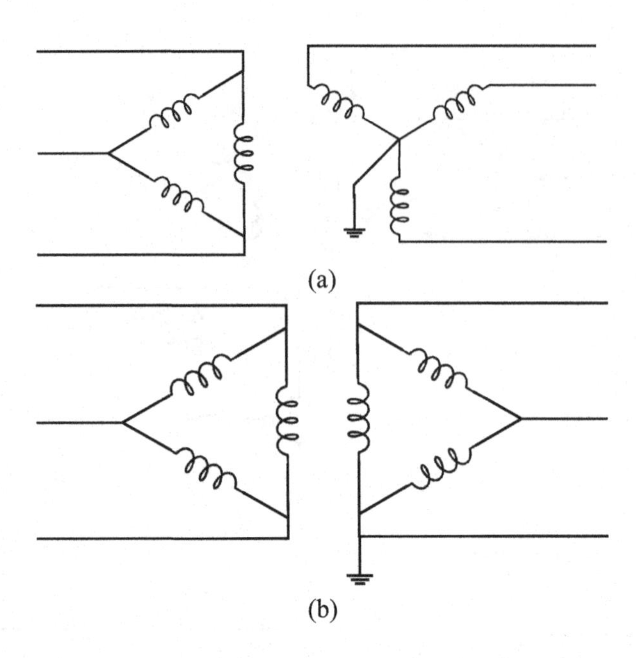

FIGURE 11.11 Circuit configuration for the solidly grounded system: (a) delta/star connection; (b) delta/delta connection.

TABLE 11.1
Grounding Systems: Comparison

	Methods of System Grounding			
Characteristics	**Ungrounded**	**Solid**	**Low Resistance**	**High Resistance**
Susceptible to transient voltage	Worst	Good	Good	Best
Under fault conditions (line-ground), an increase in voltage stress	Poor	Best	Good	Poor
Arc fault damage	Worst	Poor	Good	Best
Personnel safety	Worst	Poor	Good	Best
Reliability	Worst	Good	Better	Best
Economics – maintenance	Worst	Poor	Poor	Best
Plant continues to operate under single line-to-ground fault	Fair	Poor	Poor	Best
Ease of locating ground faults (least time)	Worst	Good	Better	Best
System coordination	Not possible	Good	Better	Best
Upgrade of the ground system	Worst	Good	Better	Best
Two voltage levels on the same system	Not possible	Possible	Not possible	Best
Reduction in the number of faults	Worst	Better	Good	Not possible
Initial fault current into the ground	Best	Worst	Good	Better
System potential flashover to ground	Poor	Worst	Good	Best

11.5 EQUIVALENT-CIRCUIT REPRESENTATION OF GROUNDING SYSTEMS

The analysis of grounding systems is often better understood using equivalent circuits. Specifically, it can represent a general grounding system with an equivalent circuit. The conceptual basis for such an equivalent circuit is illustrated in Figure 11.12. The figure illustrates three-conductor segments buried in the earth. Assume that each conductor segment is connected to a thin wire brought outside the soil.

Further, assume that the thin wires are insulated from the soil so electric current may not flow from the thin wire's surface into the earth. Under these conditions, the thin wires' presence does not affect the electric current flow or the soil's voltage distribution. On the other hand, the entire system appears as a system with three terminals. It is well known from the theory that given any linear system with terminals, no matter how complex, it can be represented with a circuit with the same input/output relationships as the actual system. Thus, the grounding system of Figure 11.12a can

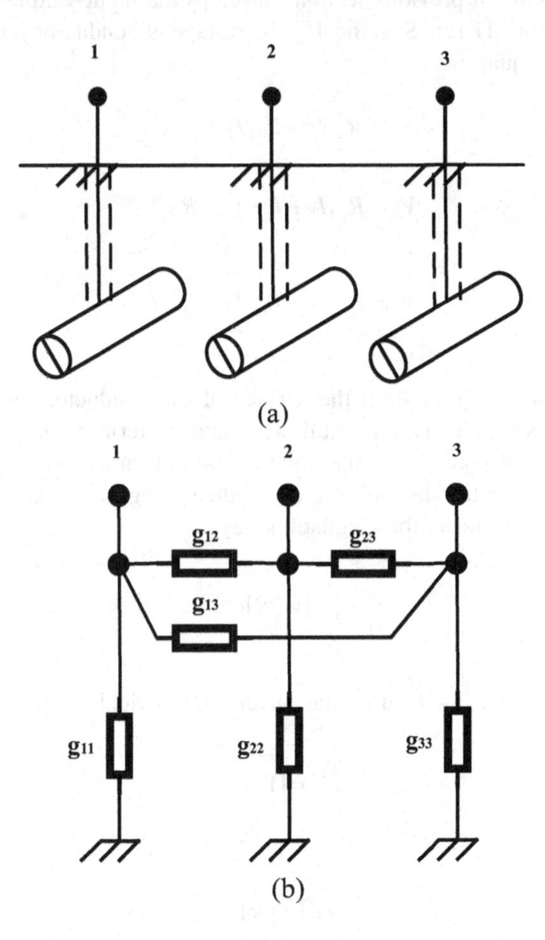

FIGURE 11.12 Representation of the soil surrounding a grounding system by an equivalent circuit. (a) Earth-embedded conductors and (b) equivalent circuit of the surrounding soil.

be represented with the equivalent circuit of Figure 11.12b. The equivalent circuit is purely resistive because it assumes that the grounding system is energized with low-frequency currents and voltages.

The equivalent circuit parameters are computed from the requirement that the input/output relationship of the systems of Figure 11.12a and b should be identical. The input/output relationship of the system of Figure 11.12b is expressed in terms of the admittance matrix as follows:

$$\begin{bmatrix} I_1 \\ I_2 \\ I_3 \end{bmatrix} = \begin{bmatrix} g_{11}+g_{12}+g_{13} & -g_{12} & -g_{13} \\ -g_{12} & g_{12}+g_{22}+g_{23} & -g_{23} \\ -g_{13} & -g_{23} & g_{13}+g_{23}+g_{33} \end{bmatrix} \begin{bmatrix} V_1 \\ V_2 \\ V_3 \end{bmatrix} \quad (11.1)$$

Applying the results of previous sections develops the input–output relationship for the system of Figure 11.12b. Specifically, the voltage of conductor segments 1, 2, and 3 is given by the equations

$$V_1 = R_{t11}I_1 + R_{t12}I_2 + R_{t13}I_3 \quad (11.2)$$

$$V_2 = R_{t21}I_1 + R_{t22}I_2 + R_{t23}I_3 \quad (11.3)$$

$$V_3 = R_{t31}I_1 + R_{t32}I_2 + R_{t33}I_3 \quad (11.4)$$

where:

Ii: electric current flows from the surface of the conductor segment *i* into the earth, the same as the electric current flowing into the terminal *i*.

V_i: The voltage of segment *i* is the same as that of terminal *i*.

R_{tij}: voltage distribution factor between conductor segments *i* and *j*.

In compact matrix form, these equations read

$$[V] = [R][I] \quad (11.5)$$

where:

The equation above is solved for the currents [*I*] to yield

$$[I] = [Y][V] \quad (11.6)$$

where

$$[Y] = [R]^{-1} \quad (11.7)$$

Equation 11.5 represents the input/output relationship of the system in Figure 11.18a.

$$g_{11} + g_{12} + g_{13} = Y_{11}, \quad -g_{12} = Y_{12}, \quad \text{etc.,}$$

where:

Y_{ij} is the (i, j) entry of the matrix $[Y]$, Equation 11.7.

Upon solution for the unknown conductance of the equivalent circuit will be:

$$g_{12} = -y_{12},\ g_{13} = -y_{13},\ g_{11} = y_{11} + y_{12} + y_{13},$$

$$g_{23} = -y_{23},\ g_{22} = y_{12} + y_{22} + y_{32}, \quad g_{33} = y_{13} + y_{23} + y_{33}$$

The conductor segments can be part of the same or different conductors, not electrically connected. Assume that the grounding system is divided into n segments. Writing one equation for the voltage of each segment i, Equation 11.5 in a compact matrix form is obtained.

$$[V] = \begin{bmatrix} V_1 \\ V_2 \\ V_3 \end{bmatrix} \quad \text{and} \quad [I] = \begin{bmatrix} I_1 \\ I_2 \\ I_3 \end{bmatrix} \tag{11.8}$$

where:

V_i: voltage of the outside surface of conductor segment i.

I_i: total current emanating from the surface of conductor segment i.

$[R]$: symmetric $n \times n$ matrix.

The solution of Equation 11.5 for the currents $[I]$ yields

$$[I] = [Y][V] \tag{11.9}$$

where

$$[Y] = [R]^{-1} \tag{11.10}$$

The parameters of the equivalent circuit are obtained from $[Y]$ as follows

1. The negative value of the entry Y_{ij}, $i \neq j$ of the matrix $[Y]$ equals the conductance of an element connected between conductor segments i and j.
2. $\sum_{j=1}^{n} Y_{ij}$ Equals the conductance of an equivalent-circuit element connected between the conductor's segment i and remote earth.

Note that the equivalent circuit represents the soil surrounding the grounding system. Since a grounding system is typically connected to a power system, the equivalent circuit can represent the power network's grounding system. The equivalent-circuit approach is useful in analyzing systems with multiple grounds.

11.6 TOUCH AND STEP VOLTAGES

The effect of an electric current passing through the essential elements of an individual body relies upon the length, magnitude, and frequency of this current. The most dangerous consequence of such exposure could be ventricular fibrillation. Humans are incredibly vulnerable to the effect of electric current at frequencies of 50 and 70 Hz. Currents about 0.1 A can eventually be lethal. The following equation describes the related energy absorbed by the body

$$I_b^2 \times t = k \tag{11.11}$$

where:

I_b: is (RMS) the magnitude of the current flowing through the body,

t: is the duration of this current flow,

k: is an empirical constant related to electric shock energy tolerated by $x\%$ of a given population. Figures 11.13 and 11.14 illustrate human beings in the vicinity of a substation ground mat subjected to step and touch voltages, respectively.

Step voltage: Step potential is the step voltage between the feet of a person standing near an energized grounded object. It is equal to the difference in voltage, given by the voltage distribution curve, between two points at different distances from the electrode. A person could be at risk of injury during a fault simply by standing near the grounding point.

Touch voltage: Touch potential is the touch voltage between the energized object and the person's feet in contact with the object. It is equal to the difference in voltage between the object and a point some distance away. The touch potential or touch voltage could be nearly the full voltage across the grounded object if that object is grounded at a point remote from where the person is in contact. For example, a crane grounded to the system neutral and contacted an energized line would expose any person in contact with the crane or its uninsulated load line to a touch potential nearly equal to the full fault voltage.

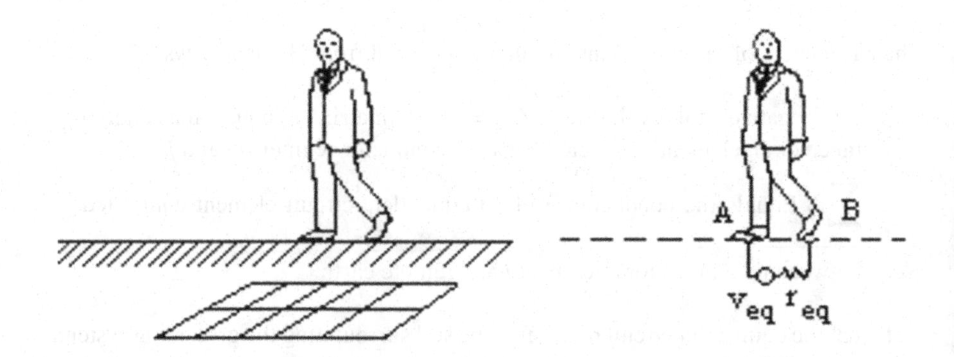

FIGURE 11.13 Definition of an equivalent circuit for the computation of body currents due to step voltage.

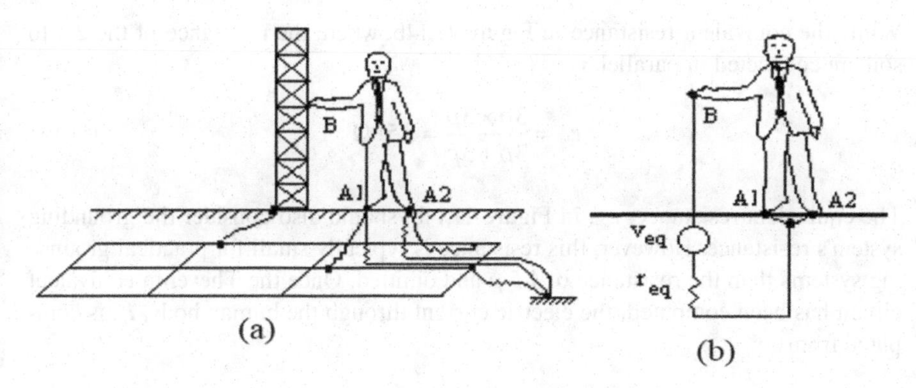

FIGURE 11.14 Definition of an equivalent circuit for computing body currents due to touch voltage. (a) Body effected by touch voltage. (b) Equivalent circuit for computing body currents due to touch voltage.

The human foot can be modeled for fast but approximate computation as a plate touching the earth's surface. The resistance of the plate to remote earth is approximately

$$R = \frac{\rho}{4b} \tag{11.12}$$

where:

ρ: is the resistivity of the earth,
b: the radius of the plate.

The human foot is not a circular plate. However, it has been observed with scale models and numerical studies that the area of the foot in touch with the earth is the determining variable. For this reason, b can be approximated with the following:

$$b = \sqrt{\frac{A}{\pi}} \tag{11.13}$$

where

A: is the area of the foot in touch with the earth. For an adult with large feet, the area of the person's feet is ~200 cm^2.

Thus, the value of b is computed to be $b \approx 0.08$ m. Hence the resistance of one foot touching the earth is

$$R = \frac{\rho}{4 \times 0.08} = 3\rho\Omega \tag{11.14}$$

ρ is in Ωm.

Thus, approximately the equivalent resistance in Figure 11.13 is

$$R = 3\rho + 3\rho = 6\ \rho \tag{11.15}$$

While the equivalent resistance in Figure 11.14b, where the resistance of the 2 ft to soil are connected in parallel, is

$$r_{eq} = \frac{3\rho \times 3\rho}{3\rho + 3\rho} = 1.5\rho\Omega \tag{11.16}$$

The equivalent resistance, r_{eq}, in Figure 11.14b, should also consider the grounding system's resistance. However, this resistance is typically small for practical grounding systems than the resistance of 1.5 ρ and omitted. Once the Thevenin equivalent circuit has been computed, the electric current through the human body, I_b, is computed from

$$I_b = \frac{v_{eq}}{r_{eq} + r_b} \tag{11.17}$$

where

r_b: is the human body's resistance between the points of contact. The human body will depend on many factors, such as size, skin condition, contact pressure, and voltage equivalent. The electric body current provides the basis for the safety assessment of grounding systems. Based on available experimental data, the ANSI/IEEE standard-80 shows that the electric body currents below $\frac{0.116\text{ A}}{\sqrt{t}}$ can be tolerated by the average person. Thus, according to this standard, the maximum allowable body current is

$$I_b = \frac{0.116}{\sqrt{t}} \tag{11.18}$$

where

t: the duration of the electric current in seconds. On the other hand, the electric body current is

$$I_b = \frac{v_{eq}}{r_{eq} + r_b} \tag{11.19}$$

V_{eq}: is the Thevenin equivalent voltage, which equals the step or touch voltage. Thus, combining the two preceding equations, the maximum allowable step or touch voltage is computed:

$$v_{eq.\,\text{allowable}} = \left(r_{eq} + r_b\right)\frac{0.116}{\sqrt{t}} \tag{11.20}$$

To obtain the maximum allowable step voltage, r_{eq}, should be replaced with 6ρ. To obtain the maximum allowable touch voltage, r_{eq}, should be replaced with 1.5ρ. For body resistance, r_{eq} the value of 1000 Ω is suggested, yielding

$$v_{\text{touch allowable}} = (1.5\rho + 1000)\frac{0.116}{\sqrt{t}} \tag{11.21}$$

And

$$v_{\text{step allowable}} = (6\rho + 1000)\frac{0.116}{\sqrt{t}} \tag{11.22}$$

Hence, the maximum touch and step voltage should not exceed this ideal for a safe grounding system. It is clear from the equations above that safety can be evaluated in touch and step attention conditions rather than body power. Safety assessment refers to how the actual maximum touch and step voltages are computed and compared to the tolerable (safe) touch and step voltages.

11.7 TYPICAL INSPECTION

The proper installation of bonding and grounding devices is important in the protection of personnel and equipment. At the installation time, a resistance test is needed to confirm electrical continuity to the ground. Also, an effective inspection and maintenance program is needed to ensure the continued adequacy of the system.

In evaluating maintenance requirements, the bonding and grounding system can be divided into three categories:

- The point-type clamps equipped with flexible leads are used for the temporary bonding of portable containers to the building grounding system;
- The fixed grounding cables and bus bars are used to connect the flexible leads and fixed equipment to the ground;
- The grounding electrode itself.

The flexible leads are subjected to mechanical damage, wear, corrosion, and general deterioration. For this reason, they should be inspected frequently. This inspection should evaluate the cleanliness and sharpness of the clamp points, stiffness of the clamp springs, evidence of broken strands in the cables, and solidity of cable attachments. A more thorough inspection should be made regularly, using an intrinsically safe ohmmeter to test ohmic resistance and continuity.

One lead of the ohmmeter is connected to a clean spot on the container, and the other is connected to the paint grounding bus, metallic piping, or other fixed equipment. The measured resistance should be <25 Ω and will usually be about 1 Ω. The fixed leads and the bus bars are not usually related to injury or wear as the temporary connectors. These should be checked with an ohmmeter on an annual basis. One lead of the ohmmeter should be connected to the fixed lead or bus bar, and the other should be connected to the plant grounding electrode or the building's structural steel. The measured resistance should be <1 Ω.

Conductive hoses should be checked regularly after repair or replacement for electrical continuity and resistance. The conductive segments may break and not be repaired properly, thus rendering the hoses nonconductive or with abnormally high resistance. Nonconductive hoses having an internal spiral conductor should be installed to contact adjacent metallic fittings.

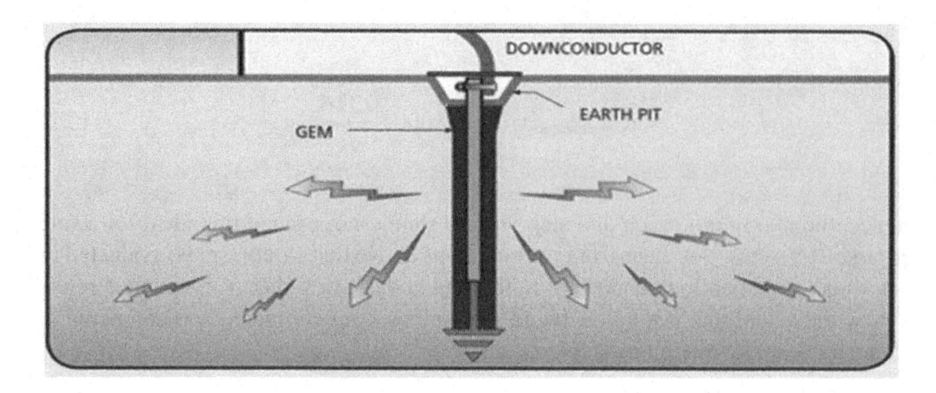

FIGURE 11.15 Characteristics of a good grounding system. (Characteristics of a Good Grounding System, August 15, 2015, by Phil.)

The grounding inspection aims to prevent accidental electrical shock injuries caused by using damaged or improper electrical equipment on the job and to conduct maintenance and inspection procedures for electrical equipment (Figure 11.15).

11.8 GROUNDING ELECTRODES

The bonding/grounding system's final component is the "grounding electrode," which passes static charges into the soil. This may be a device installed solely for grounding purposes, such as a driven rod (copper clad) or buried plate, or an underground metal water pipe. If the building has a steel structure frame grounded for lightning protection or is otherwise effectively grounded, it is adequate for static grounding; no separate static grounding electrode is needed.

Underground piping equipped with cathodic protection is not appropriate ground. Underground piping made of cement, asbestos, or plastic would not be as satisfactory as ground. It is also possible for metal piping to have sections of plastic or cement asbestos, which would make it unsatisfactory. Water meters should have jumper cables permanently installed around them to provide a continuous electrical path. When underground piping is utilized as a ground, any disconnections for alterations or repair may make the grounding system ineffective. Sprinkler piping and electrical conduit should be avoided because of the increased resistance to joints and connectors' ground. A break in continuity can also result when piping and conduit are removed for repair or alterations (Figure 11.16).

11.9 GROUNDING VERIFICATION CONTROL SYSTEM

Properly labeled "the invisible enemy," static electricity cannot be seen but poses extreme risks if not properly attended to. Engineers rely on visual means to confirm that a ground clamp and lead are in place for proper grounding or bonding, with periodic confirmation via resistance meters. Newer "electronic verification systems" now take the guesswork out of proper grounding techniques. These verification systems

FIGURE 11.16 Grounding electrode.

FIGURE 11.17 Grounding verification control system installation.

offer a continuous visual/electronic confirmation of ground to a high-integrity ground point (ground bus). Through "interlock" functions, they can control pumps, valves, motors, and the like, or interface with a PLC or DCS control system to ensure that nothing happens until a good ground is achieved. Engineers may also initiate sound alarms if required. Further information on the various systems can be found in this booklet (Figures 11.17 and 11.18).

FIGURE 11.18 Soil measurements.

11.10 SOIL MEASUREMENTS

The soil model can be established through a volume of field tests. The Wenner and driven fishing rod methods are the most widely used. Both methods are incredibly simple to implement. Right now, there are several commercial courses for performing these measurements. Then the soil may get ready in a two-layer model. This model can be done by several methods, such as weighted at a minimum square and steepest descent.

11.10.1 THE SOIL MODEL

Introducing a statistical model to symbolize soil's electrical properties can be a formidable process due to the earth's widely nonstandard characteristics. The variables ρ_1 and ρ_2 are generally determined by interpreting the apparent resistivity values measured using the Wenner (or four probes) arrays.

Unlike most design problems, the interpretation of soil resistivity measurements is an "inverse" problem, i.e., from the electrical response to impressed current at specific locations on the entire world surface, the electrical properties of the conducting media (earth) are to be determined. Conventional electrostatic problems determine the electrical response or excitation current sources, depending on the conducting material's known properties. They are known as the Laplace and Dirichlet problems. The "inverse" problem, where the physical constants of the material are unidentified, presents more difficulties than the patient's problems, where the material's physical constants are known functions of the position.

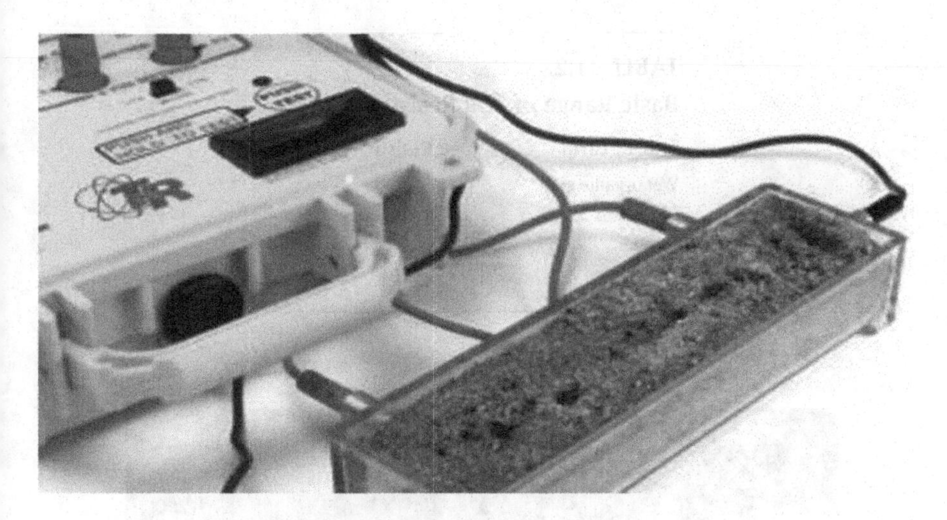

FIGURE 11.19 The soil model inspection.

Moreover, the number of parameters required to represent a model of the earth system is usually so great it is difficult to choose initial values for these parameters and have some type of computer algorithm converge to an acceptable solution within a practical time. Consequently, the selection of initial ideals becomes a fundamental job in the interpretation of the measurement (Figure 11.19).

11.10.2 Soil Characteristics

The world's soil can be an immaculate resistance; along these lines is the last area where a fault current is scattered. Soil resistance can contain a current up to a basic sum, which differs depending upon the soil, and now, circular electrical segments can create on the surface of the soil that can charge protests at first glance, for example, a man. The current stream can influence a soil's resistivity by being warm, making the soil dry out and more resistant. Wet soil has considerably less resistance than dry soil, so the establishing framework and bars should be situated on clammy earth in a perfect world. Ordinarily, soil resistance rapidly increments when its dampness content is under 15% of the soil weight, and the resistance scarcely changes once the dampness content is no <22%. Table 11.2 demonstrates a fundamental accumulation of soil resistivity, relying upon the dampness and sort.

Table 11.2 demonstrates that wet or even clammy soil has little resistance, so it is useful to keep the establishing soil as soggy as could be allowed. A typical practice to cause a finish is to utilize a surface material layer, such as rock. Not exclusively does a surface material enormously diminish the measure of soil vanishing. However, it ordinarily has a high resistance, decreasing the sizes and odds of stun streams. Soil attributes and the surface layer's sort to be utilized shift contingent upon the region on the planet in which the substation is found and required by the establishing framework.

TABLE 11.2
Basic Range of Soil Resistivity

Type of Earth	Average Resistivity (Ω·m)
Wet organic soil	10
Moist soil	10^2
Dry soil	10^3
Bedrock	10^4

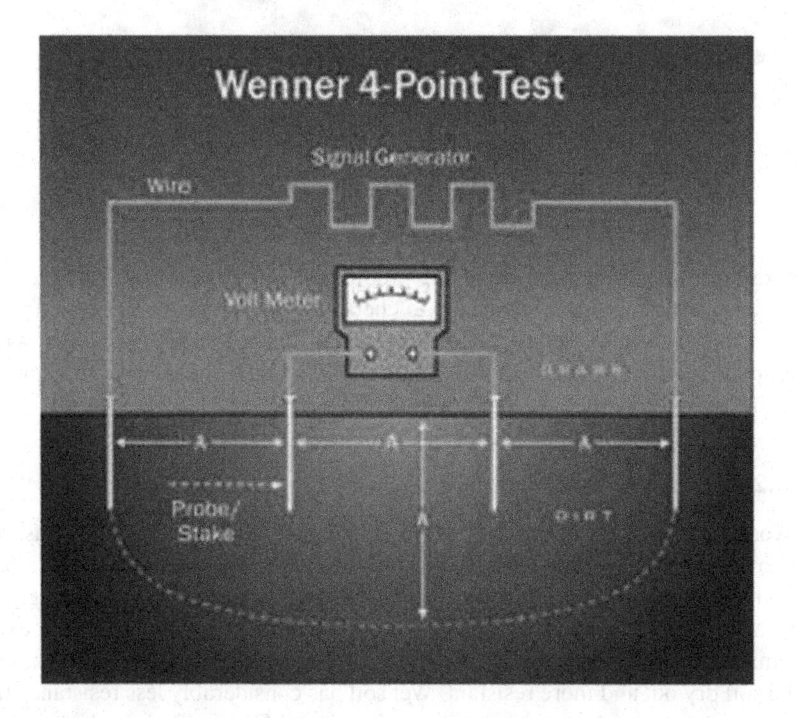

FIGURE 11.20 Wenner method.

11.10.3 Wenner Method

The Wenner strategy is the most generally used in practice. That involves positioning four small pins into the planet in a straight line, as illustrated in Figure 11.20. A source linked between outer pins produces a power current, injected into the globe from one pin and collected at the other pin. The flow of this electric current in the entire world generates a potential distribution in the ground. As a result, the location of both the inner pins is nonzero. The voltage between the two inner hooks is measured with a voltmeter. The injected current I and the scored voltage V are related to the resistivity of the soil.

Assume that the span of the pin is very small when compared to the parting distance between them. In cases like this, the two outer hooks can be viewed as point current types of current I and $-I$, respectively, located on the earth's surface. The voltage at a point along the pins line, located far away x from the PIN injecting current I in the soil (see Figure 11.20). The following equation gives

$$V(x) = \frac{\rho \times I}{2\pi x} - \frac{\rho \times I}{2\pi(3a - x)} \tag{11.23}$$

The two inner pins' voltage is $V(a)$ and $V(2a)$, respectively. Thus, the voltage V between the two inner pins is

$$V(x) = V(a) - V(2a) = \frac{\rho \times I}{2\pi\, a} \tag{11.24}$$

Solving for the soil resistivity ρ gives

$$\rho = \frac{2\pi a V}{I} \tag{11.25}$$

where:

ρ: resistivity of soil Ωm,
a: probe spacing m,
V: voltmeter reading volts,
I: ammeter reading amperes.

In uniform soil, the four-pin arrangement should provide the same soil resistivity irrespective of separation distance a; when the soil is not uniform, which is the most common cause, the method will provide the "apparent soil resistivity," which depends on the separation distance a. From this information, it is possible to determine a nonuniform soil model (Figure 11.21).

11.10.4 Driven Rod Technique

The driven rod method contains inserting an earth rod into the soil. Every time a length l of the earth rod is driven in the soil, the ground rod's resistance concerning the remote floor is measured. For this specific purpose, a source is employed, linked between the powered rod and auxiliary electrode (current probe) located a distance D away from the driven rod, as illustrated in Figure 11.22b. This connection triggers electric current I to be injected into the planet from the driven fly fishing rod collected at the auxiliary electrode. One other electrode (voltage probe) is put away from the driven rod and away from the current to minimize the distraction.

The driven rod level of resistance concerning distant earth is around $R = V/I$. This resistance relates to soil resistivity. Roughly expression of a surface rod resistance to remote control earth is:

$$R = \frac{\rho}{2\pi l} \ln \frac{2l}{r} \tag{11.26}$$

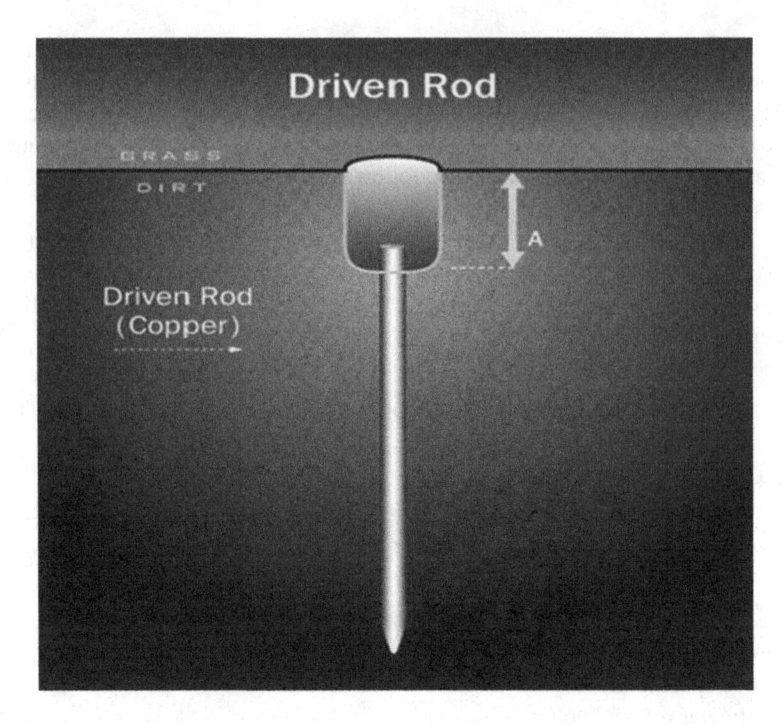

FIGURE 11.21 Driven rod technique.

Upon solving Equation 11.26 for the soil resistivity

$$\rho = \frac{2\pi l R}{\ln \frac{2l}{r}} \tag{11.27}$$

where:

l: rod length,

$$R = V / I,$$

r: radius of the rod,

Suppose the soil is uniform (constant resistivity throughout). In that case, the powered rod method should supply the same soil resistivity despite the length l of the driven rod in contact with the soil. When the driven rod method is applied to non-uniform soil, it will provide an "apparent garden soil resistivity," which will fluctuate with the ground rod's length in contact with the ground.

Inside the application of the driven rod method, the current and voltage probe position is important. Specifically, the current probe should be located away from the powered rod so that the current probe's occurrence does not damage the electric field around the ground rod. Likewise, the voltage probe should be put at approximately no voltage, whose voltage is not affected by the ground rod's presence and the current probe.

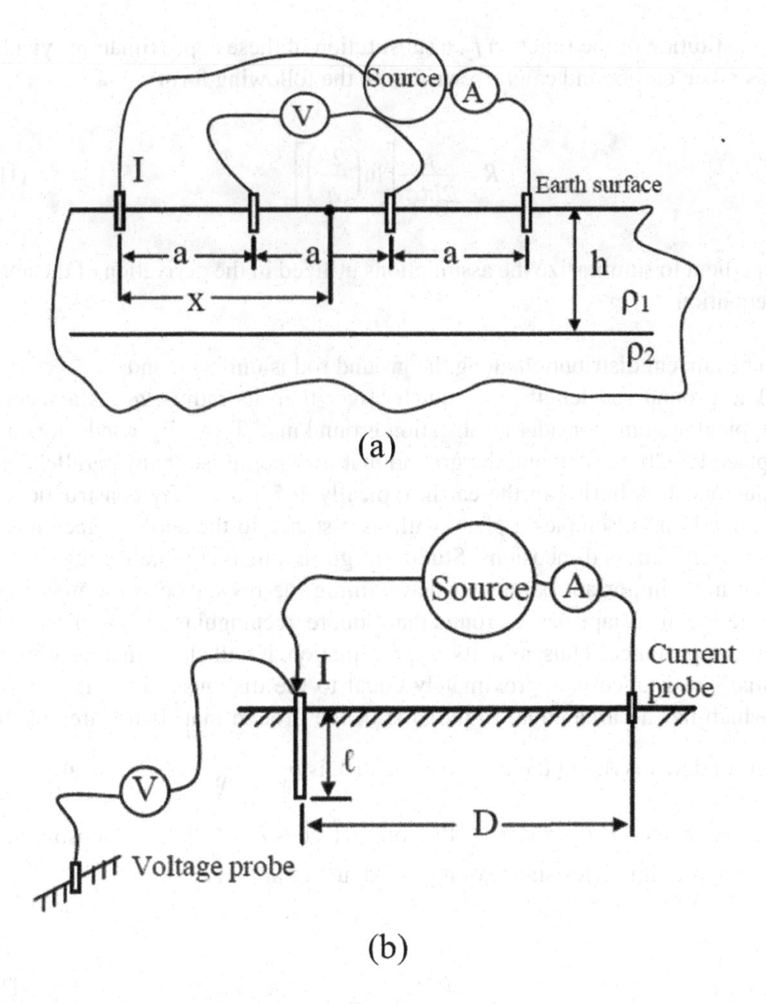

FIGURE 11.22 Soil resistivity measurement system arrangements. (a) Wenner method, (b) driven rod method.

11.11 RESISTANCE OF GROUNDING SYSTEMS

This section will examine the simplified equations for the resistance of two typical cases: a ground rod and a substation ground mat. Consider a ground rod in Figure 11.22a. Typically, a ground rod is ~8 ft long and of diameter on the order of <1 inch thus $l \gg a$. Assume that the ground rod is buried so that the rod's top point is near the earth's surface. In this case, the z coordinate of the center of the ground rod is $|z_1| = \ell/2$. Assuming uniform current distribution along the ground rod, the ground rod's resistance is given as follows, assuming that $\ell = 2L$ and $|z_1| = \ell/2 = L$. Substitution is given as

$$R = \frac{1}{16L^2 \pi \sigma}\left[f_2(4L,a) - f_2(2L,a) + f_2(-2L,a) - 3a\right] \tag{11.28}$$

Upon substitution of the function f_2, a substitution of these approximations yields the result as given earlier and can be rewritten in the following form

$$R = \frac{\rho}{2l\pi\sigma}\left[\ln\left(\frac{2l}{a}\right)\right] \tag{11.29}$$

It is expedient to summarize the assumptions utilized in the derivation of the approximate equation:

i. The current distribution along the ground rod is uniform, and
ii. The ground rod length, ℓ, is much larger than its radius, a. As a second typical system, consider a substation ground mat. Typically, conductors are placed 5–20 ft apart, and the ground mat may comprise many parallel conductors. It is buried in the earth, typically 1–5 ft deep. By construction, a ground mat resembles a plate, with its distance to the soil's surface much smaller than its dimensions. Studies of ground mats' resistance reveal that the most important parameter determining the resistance is the mat area. The specific shape of the ground mat (square, rectangular, etc.) is of secondary importance. Thus, as a first approximation, it will claim that the ground mat's resistance is approximately equal to the disk near the soil surface, which has an area equal to the area of the ground mat. If the area of the ground mat is A and the radius of the disk is b, $b = \sqrt{\frac{A}{\pi}}$. Now, recall that the resistance of the disk near the soil surface is $R = \frac{\rho}{4}b$. Upon substitution, the approximate resistance of a ground mat of area A

$$R = \frac{\rho}{4}\sqrt{\frac{A}{\pi}} \tag{11.30}$$

Note that since the earlier equation is approximate, alternative formulae are possible.

11.12 TYPES OF THE ELECTRODE GROUNDING SYSTEM

Grounding systems vary considerably from a simple vertical rod to a substation mat with vertical and horizontal components of different lengths and angles. The design of grounding systems of substations and electrical systems, in general, has the primary purpose of ensuring the safety and well-being of personnel, anyone who may come close to conductive media electrically coupled to grounding mats during unbalanced fault conditions. In general, an unbalanced fault will cause a potential rise of the system neutral and a conductive medium electrically connected to the neutral. During a fault, hazardous transfer voltages may be generated on these elements.

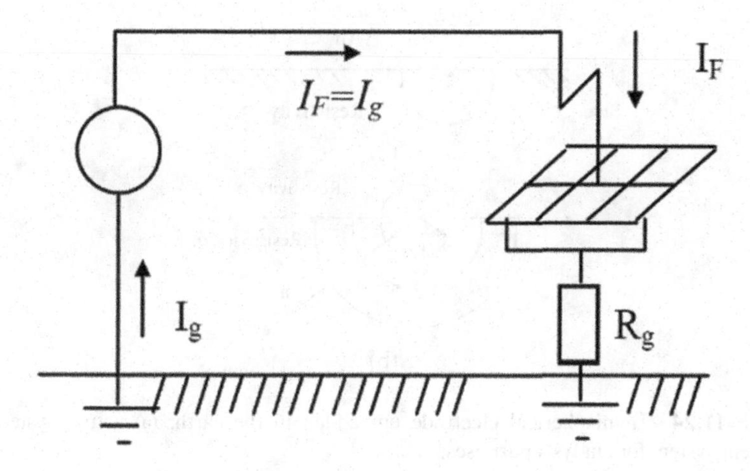

FIGURE 11.23 Schematic of earthling system.

A safe grounding design has two objectives:

1. Provide means to carry and dissipate electric currents into the ground under normal and fault conditions without exceeding operating and equipment limits or adversely affecting service continuity.
2. Assure such a degree of human safety that a person working or walking near grounding facilities is not exposed to the danger of critical electric shock.

For instance, if a substation is supplied from an overhead line, a low grid resistance is important because a substantial part of the total ground-fault current enters the earth, causing an often-steep rise of the local ground potential as shown in Figure 11.23.

A simple grounding system will be explained in this section. The analysis of these systems is simple and offers the basic ideas underlying the appearance of grounding systems.

11.12.1 Hemispherical Electrode Hidden in Globe

The easiest grounding system is a hemispherical electrode stuck in the earth of resistivity ρ, as shown in Figure 11.23; the center of the hemispherical electrode is located on the top of the earth. Assume that the hemisphere's potential is v; in cases like this electrode, current will flow from the electrode's surface in the earth.

As a result of symmetry, the flow of the electric current in the semi-infinite earth will be identical to in the system of shape provided in Figure 11.24a, which shows a sphere embedded within an infinite medium of resistivity ρ. In other words, the flow of the current will be so that the equipotential surfaces made will be concentric circular surfaces. If a total current I flow from the top of the hemisphere into earth (Figure 11.24a), total current $2I$ will flow from the ball into earth (Figure 11.24b).

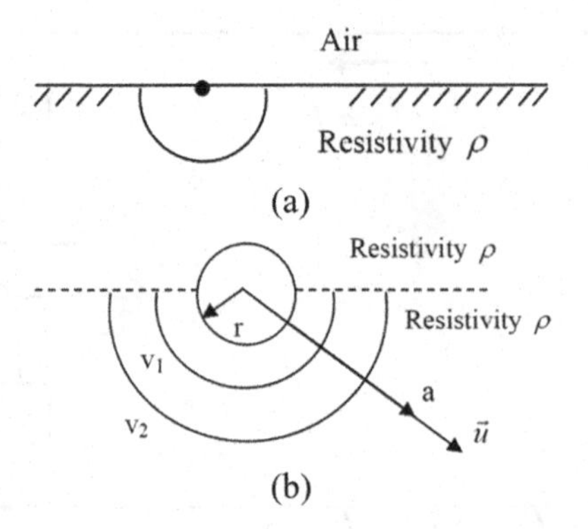

FIGURE 11.24 Hemispherical electrode embedded in the earth: (a) actual system, (b) equivalent system for analysis purposes.

The existing density *J*(*a*) at an area located at a distance from the middle of the electrode will be

$$J(a) = \frac{2I}{4\pi r^2}\vec{u}\, r \geq u \left(\frac{\text{A}}{\text{m}^2}\right) \tag{11.31}$$

where *r* is the radius of the hemisphere and $\vec{u}$ is a unit vector in the radial direction.

By Ohm's low, the electric field intensity at a point located at a distance from the center of the hemisphere will be

$$\vec{E}(a) = \rho \times J(a)\vec{u} \;\; a \geq r \tag{11.32}$$

The equation will give the potential of the hemisphere concerning a point *x* located at a distance $a = a_1$ from the center of the hemisphere

$$v(a_1) = \int_{a=r}^{a1} J(a) \times \rho\, da \tag{11.33}$$

Upon the substitution and evaluation of the integral

$$v(a_1) = \frac{\rho \times I}{2\pi}\left(\frac{1}{r} - \frac{1}{a_1}\right) \tag{11.34}$$

The potential of the sphere concerning remote earth, v_∞, is obtained by letting $a_1 \rightarrow \infty$.

$$v_\infty = \frac{\rho \times I}{2\pi r} \tag{11.35}$$

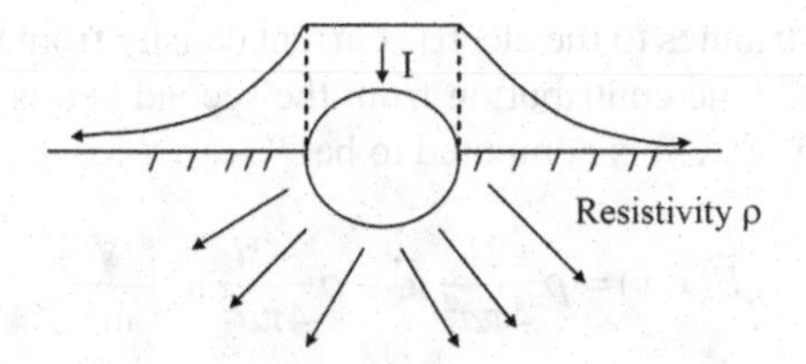

FIGURE 11.25 A hemisphere generates potential distribution on the surface of the earth.

The potential on the earth's surface along a line passing through the center of the hemisphere is illustrated in Figure 11.25. The resistance of the hemisphere to remote earth is:

$$R = \frac{v}{I} = \frac{\rho}{2\pi r} \tag{11.36}$$

11.12.2 Two Hemispheres Inserted in Earth

The current source is linked between the two hemispheres, causing total electric current to flow through the globe. Figure 11.26a supposes the space between the two hemispheres is much larger than their radii. In cases like this, the results can be used directly. The solution for this case is obtained by superposition. The electric current density *J*(*x*, *y*) at a point (*x*, *y*), illustrated in Figure 11.26a, is:

$$J(x,y) = \frac{2I}{4\pi a_1^2}\vec{u_1} - \frac{-2I}{4\pi a_2^2}\vec{u_2}\ \text{A/m}^2 \tag{11.37}$$

where a_1, a_2 are distances illustrated in Figure 11.26a and $\vec{u_1}$ and $\vec{u_2}$ are unit vectors.

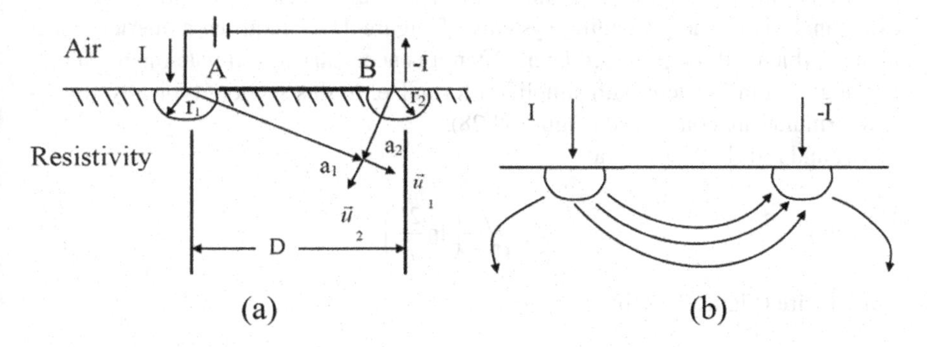

FIGURE 11.26 Two hemispherical electrodes: (a) configuration and (b) lines of current flow.

The first term contributes to the electric current density from the first hemisphere, and the second term is the contribution from the second hemisphere. Similarly, the electric field intensity $E(x, y)$ is computed to be

$$\vec{E}(x,y) = \rho \frac{2I}{4\pi a_1^2} \vec{u_1} - \rho \frac{-2I}{4\pi a_2^2} \vec{u_2} \quad \frac{\text{V}}{\text{m}^2} \tag{11.38}$$

The voltage between the electrodes is computed from

$$v = \int \vec{E}(x,y)\, du$$

Selecting an integration path along the line AB and carrying out the integration yield

$$v = \rho \frac{I}{2\pi}\left(\frac{1}{r_1} - \frac{1}{D - r_2} + \frac{1}{r_2} - \frac{1}{D - a_2}\right) \tag{11.39}$$

If both hemispheres are identical (i.e., $r_1 = r_2$), then

$$v = \rho \frac{I}{\pi}\left(\frac{1}{r} - \frac{1}{D - r}\right) \tag{11.40}$$

The resistance between the two hemispheres is

$$R = \frac{V}{I} = \rho \frac{1}{\pi}\left(\frac{1}{r} - \frac{1}{D - r}\right) \tag{11.41}$$

The lines of electric current are illustrated in Figure 11.26b.

11.12.3 Other Simple Grounding Systems

The practical grounding system contains ground rods, pieces, rings, disks, and pads. Several of the simplest useful grounding electrodes are specified in Figure 11.27. The exact analysis of the grounding systems of Figure 11.27 requires numerical techniques, which will be provided later. Often it is necessary to estimate the resistance of the grounding system with simplified formulae. In this section, typically simplified formulae are considered (Figure 11.28).

Ground rod (Figure 11.27a)

$$R = \frac{\rho}{2\pi l}\left(\ln \frac{2l}{r}\right) \tag{11.42}$$

Buried wire (Figure 11.27b)

$$R = \frac{\rho}{2\pi l}\left(\ln \frac{2l}{r} + \ln \frac{2}{2z}\right) \quad z \geq 6r \tag{11.43}$$

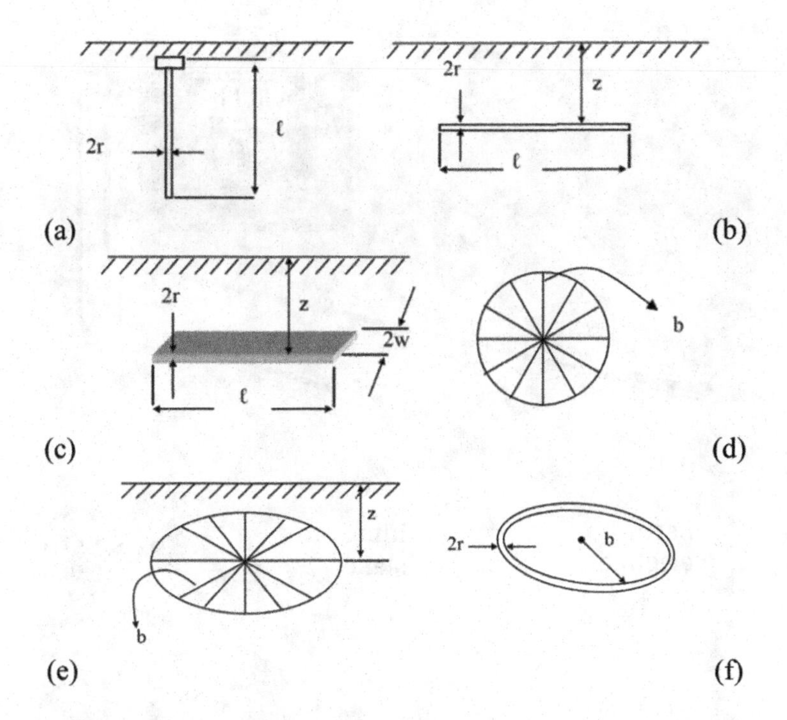

FIGURE 11.27 Simple grounding systems. (a) Ground rod, (b) buried wire, (c) buried strip, (d) thin plate in an infinite medium, (e) thin plate near the soil surface, and (f) ring in an infinite medium.

Buried strip (Figure 11.27c)

$$R = \frac{\rho}{2\pi l}\left(\ln\frac{2l}{w} + \ln\frac{2}{2z}\right) \quad z \geq 3w \tag{11.44}$$

Disk in infinite medium (Figure 11.27d)

$$R = \frac{\rho}{8b} \tag{11.45}$$

Disk near the soil (Figure 11.27e)

$$R = \frac{\rho}{4b} \tag{11.46}$$

Ring in infinite medium (Figure 11.27f)

$$R = \frac{\rho}{4\pi^2 b}\left(\ln\frac{8b}{r}\right) \tag{11.47}$$

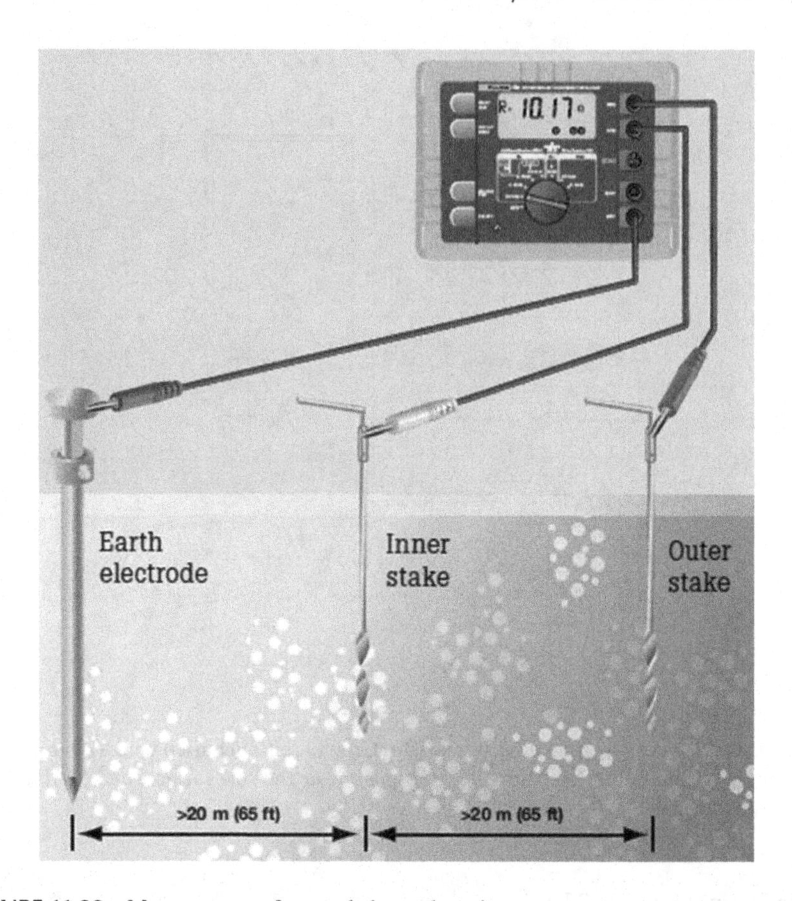

FIGURE 11.28 Measurement of ground electrode resistance.

11.13 MEASUREMENT OF GROUND ELECTRODE RESISTANCE

The resistance of grounding systems; what are you examining? It will be better to give some information about measurement methods or principal features of grounding resistance for measurement purposes.

Several methods are available for measuring the resistance of an installed ground electrode. Of these, the three-electrode and the fall-of-potential methods are the most common.

A lot of methods are available for measuring the resistance of an installed ground electrode. Of these, the three-electrode and the fall-of-potential methods are the most common (Figure 11.29).

11.13.1 Three-Electrode Method

Think about a ground electrode E whose resistance is to be measured experimentally. This requires that two other electrodes, P and R, be temporarily hidden in the ground, as determined in Figure 11.30. The electrodes are shown as hemispheric

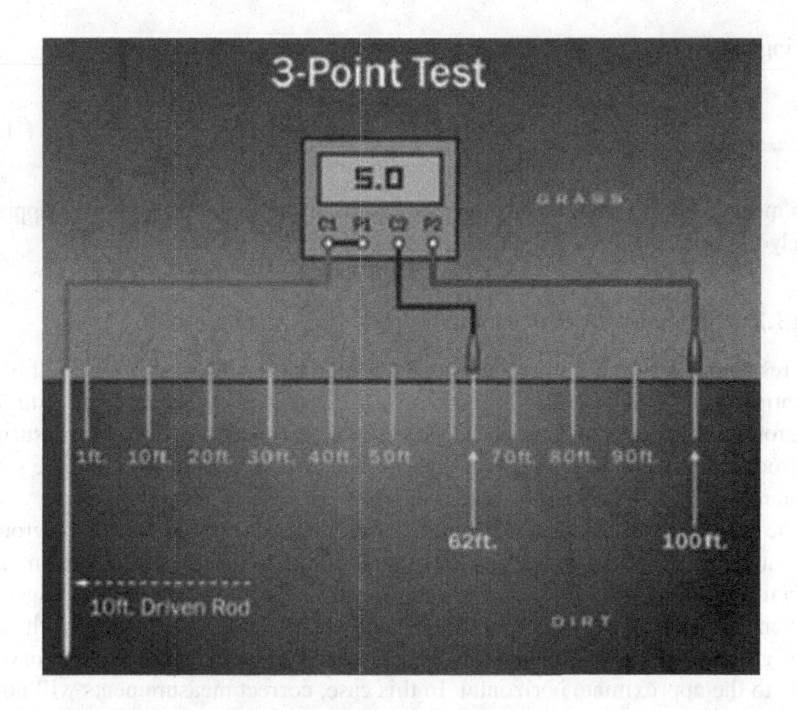

FIGURE 11.29 Three-electrode method. (*Grounding Compendium for PV Systems*, by Rebekah Hren, Brian Mehalic Contents (Aug/Sep 13: Issue 6.5).)

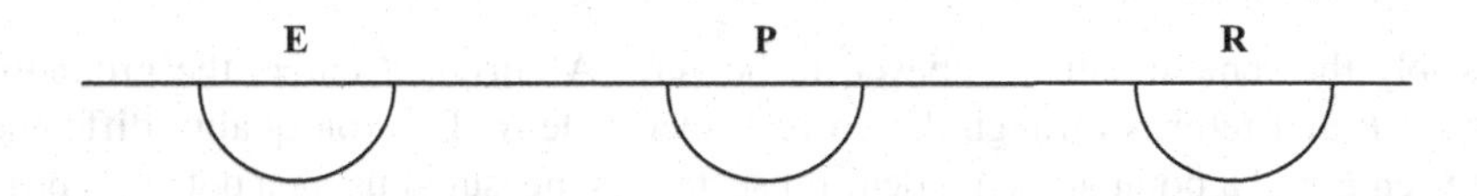

FIGURE 11.30 Three ground electrodes provide a system for measuring ground electrode resistance.

electrodes. They might be driven ground rods or other shapes. Building the soil in two layers using the measured least square method necessary so that the electrode intervals are large compared to their radii; otherwise, problems are introduced. The other measuring outlet linked with one airport terminal on electrode E and one on electrode S procedures:

$$R_x = R_E + R_P \tag{11.48}$$

Similarly, R_y is measured between electrodes P and R and R_z between electrodes R and E, with the result

$$R_y = R_P + R_R \tag{11.49}$$

$$R_z = R_R + R_E \tag{11.50}$$

Solving these three equations simultaneously gives the value for R_E as:

$$R_E = \frac{R_x + R_y + R_z}{2} \tag{11.51}$$

This measurement method gives the best results if the three resistors have approximately the same resistance magnitude.

11.13.2 Show Up of Potential Method

The resistance of ground interconnection might also be measured by the fall of the potential method presented in Figure 11.31. It involves growing a current I through the grounding system, another electrode, and another electrode called the returning electrode. This current's passage produces X from the voltage drops V_X in the soil at a distance X; V_X is measured by potential probe P.

The zone V_X/I is an evident resistance, giving the true resistance R_E of the grounding system under certain conditions. The simplest form of the fall-of-potential method is obtained when P and R are on the same brand. The most widely used arrangement is when S is situated between E and R. If the distance D is large enough (with value to the grounding system dimensions), the center part of the fall-of-potential curves is likely to the approximate horizontal. In this case, correct measurements will not be obtained unless this trade is already a good idea of the complete probe P position.

11.13.3 Theory of the Fall of Potential

Possibly the remote soil is believed to be zero. A current I enters the grounding system E and returns through the go-back electrode R. The volt quality difference between E and a point at the garden soil surface is measured using a potential probe P (Figure 11.31).

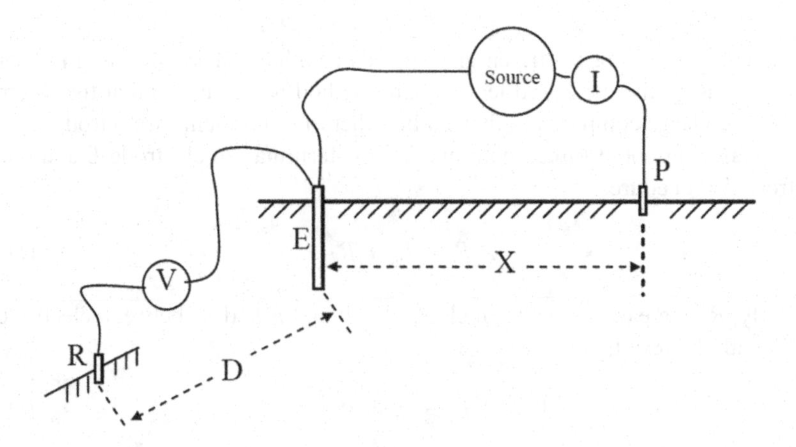

FIGURE 11.31 The fall of the potential method of measuring ground resistance.

Let GM sixth v be the potential of electrode G ($G = E$ or R) at point M ($M = P$ or E). The assumption is that electrode G has a current of 1 A. Therefore

GM v is in V/A. The following equations can be written:

$$U_p = v_P^E \cdot I + v_P^R \cdot (-I) \tag{11.52}$$

$$U_E = v_E^E \cdot I + v_E^R \cdot (-I) \tag{11.53}$$

where U_P and U_E are the potentials of electrodes P and E, respectively.

The voltage v measured by the fall-of-potential method is

$$V = U_E - U_P$$

Thus,

$$V = I\left(v_E^E - v_E^R - v_P^E + v_P^R\right) \tag{11.54}$$

Where v_E^E is the potential rise of electrode E, assuming a current of 1 A. This is the resistance R_E (or impedance) of electrode E. Therefore, Equation 11.29 can be written as:

$$R = \frac{V}{I} = \left(v_E^E - v_E^R - v_P^E + v_P^R\right) \tag{11.55}$$

where v_P^R, v_E^R, and v_P^E are functions of the distance between the electrodes, the configuration of the electrodes, and soil characteristics.

Let us define the following functions η, ϕ, and ψ concerning the coordinate system shown in Figure 11.32:

$$v_E^R = \eta(D) \tag{11.56}$$

$$v_P^R = \phi(D - X) \tag{11.57}$$

$$v_R^E = \psi(X) \tag{11.58}$$

According to Equation (11.55), the measured resistance will be equal to the true resistance if

$$\left(v_P^R - v_E^R - v_P^E\right) = 0$$

$$\phi(D - X) - \eta(D) - \psi(X) = 0 \tag{11.59}$$

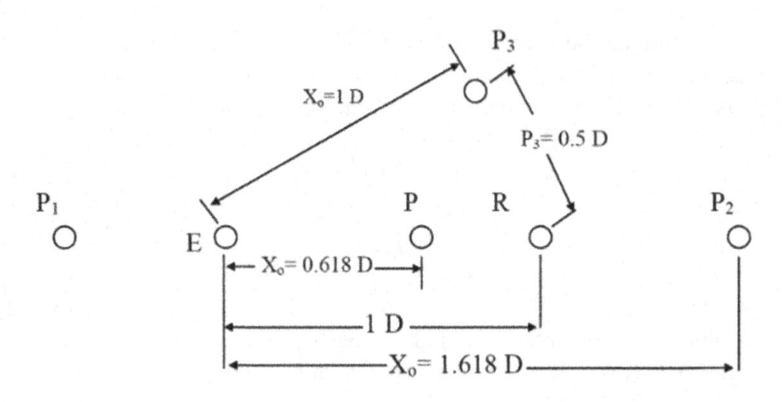

FIGURE 11.32 Electrode arrangement, which minimizes error in the fall-of-potential method.

For identical electrode and large spacing (electrodes E and R are identical), then $\phi = \psi$ and if D is large enough that

$$v_E^R = \eta(D) \approx 0$$

Then, the earlier condition becomes

$$\phi(D - X) - \psi(X) = 0$$

thus

$$X_o = D/2$$

11.13.4 Hemispherical Electrodes

If electrodes E and R are hemispheres and their radius is small compared to X and D, and if the soil is uniform, then the potential functions ϕ, η, and ψ are inversely proportional to the distance following the hemisphere center. In the event the source of the axis is at the center of hemisphere E, then, Equation 11.59 will be proportional to the following equation:

$$\frac{1}{D - X} - \frac{1}{D} - \frac{1}{X} = 0 \tag{11.60}$$

The positive root of Equation (11.60) is the exact potential probe location $X_o = 0.618D$. If the potential probe L is at location P_1 (E side; Figure 11.32), $D - X$ should be replaced by $D + X$ in Equation (11.60). In this case, the equation has a complex root only. If P is at location P_2 (R side), then $D - X$ should be changed by $X - D$. The good root is

$$X_o = 1.618\,D$$

FIGURE 11.33 Substation-measured earth resistance. (G. F. Tagg, "Measurement of the resistance of an earth electrode system covering large area." *Proceeding IEE Transaction*, vol. 116, no. 3, 1969, pp. 475–480.)

11.13.4.1 General Case

If the soil is not uniform and E and R have complex configurations, then the functions ϕ, η, and ψ are not easy to calculate. In such cases, computer solutions are generally required.

11.13.5 Electrical Center Method

If its equivalent hemisphere replaces the earth electrode and the fall-of-potential users, the measured resistance is given by the expression (Figure 11.33)

$$\frac{R}{R_\infty} = 1 - \frac{1}{\frac{ac}{r}} - \frac{1}{\frac{ap}{r}} + \frac{1}{\frac{ac}{r} - \frac{ap}{r}} \tag{11.61}$$

where:

ac: distance from the arbitrary starting point to the current electrode.
ap: distance from the arbitrary starting point to the potential electrode.
r: radius of the equivalent hemisphere.
$R = V/I$ is the measured resistance.
R_∞: true resistance of an earth system.

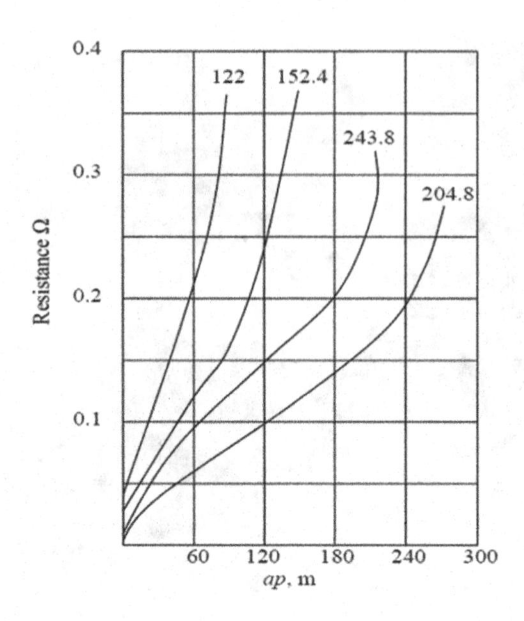

FIGURE 11.34 Substation-measured earth resistance curves. (G. F. Tagg, "Measurement of the resistance of an earth electrode system covering large area." *Proceeding IEEE Transaction*, vol. 116, no. 3, 1969, pp. 475–480.)

And if the true resistance is obtained when $R = R_\infty$ and $ap = 0.618\,ac$. This will likely give very appropriate results. Yet two conditions are necessary. First, it ought to be possible to consider our planet's electrode system as a hemisphere. Several measurements demonstrate that the resistance figure of most electrode systems closely approximates that of the hemisphere, except at points very near the system. The second issue is to know at what point to start the measurements of the distances to the current and potential electrode; i.e., where is the center of the equivalent hemisphere? It is very challenging to decide where this is in a complicated system of rods, tapes, and so forth. So away is required, in which it is not essential to know this exact center.

A test of the measurement was made at the large substation, its area, and the grounding system involves many earth plates and fishing rods joined together by copper mineral cables. The testing range was run out from approximately midway along one side, and the latest electrode was put at distances of 122, 152.4, 243.8, and 204.8 m from the starting point.

Figure 11.34 shows the substation-measured earth resistance curves.

PROBLEMS

11.1. What are the various methods of neutral grounding? Compare their performance concerning (i) protective relaying, (ii) fault levels, (iii) stability, and (iv) voltage levels of power systems.

11.2. Explain the phenomenon of "Arcing grounds" and suggest the method to minimize this phenomenon's effect.

11.3. Discuss the advantages of (i) grounding the system's neutral and (ii) keeping the neutral isolated.
11.4. A transmission line has a capacitance of 0.1 μF per phase. Determine the inductance of the Peterson coil to neutralize the effect of the capacitance of (i) the complete length of the line, (ii) 97% of the line, and (iii) 90% length of the line. The supply frequency is 50 Hz.
11.5. A 132 kV, 50 Hz, three-phase, 100 km long transmission line has a capacitance of 0.012 μF per km per phase. Determine the arc suppression coil's inductive reactance and kVA rating suitable for the line to eliminate arcing ground phenomena.
11.6. A 132 kV, three-phase, 50 Hz overhead line of 100 km length has a capacitance to the earth of each line of 0.01 μF per km. Determine the inductance and kVA rating of the arc suppression suitable for this line.

Appendix A
Relay and Circuit Breaker Applications

Appendix A briefly explains some relays and their characteristics.

A.1 EIGHT-POLE INDUCTION CUP OR DISK

Connected as a directional element (Figure A.1).

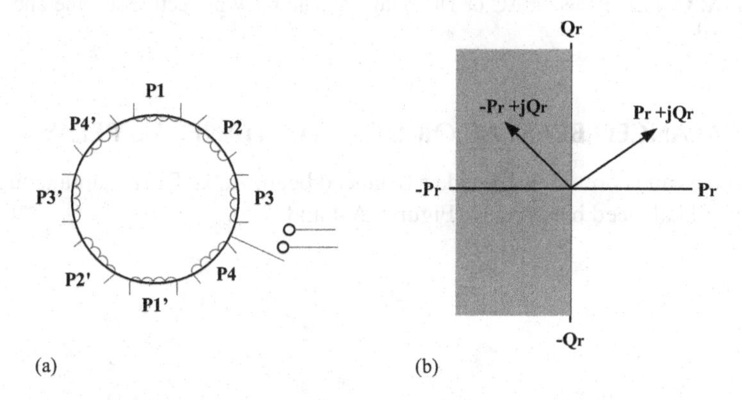

FIGURE A.1 Eight-pole induction cup relay. (a) The relay protective scheme and (b) the characteristics.

A.2 INSTANTANEOUS AC OR DC RELAY

The relay's current and voltage with a suitable number of turns of the windings are decided by Ohm's law (Figure A.2).

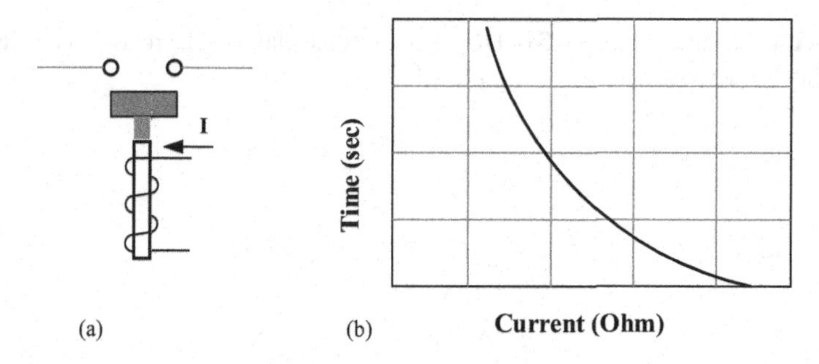

FIGURE A.2 Instantaneous AC or DC relay. (a) The relay protective scheme and (b) the characteristics.

A.3 INSTANTANEOUS AC OR DC RELAY

The relay's current and voltage with a suitable number of turns of the windings are decided by Ohm's law (Figure A.3).

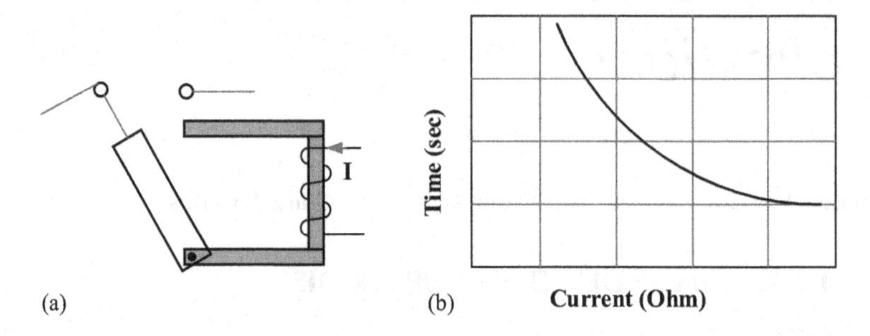

FIGURE A.3 Instantaneous AC or DC relay. (a) The relay protective scheme and (b) the characteristics.

A.4 BALANCED BEAM AC OR DC INSTANTANEOUS RELAY

The coil taps and core screw provide a balanced beam AC or DC instantaneous relay. High-speed balanced beam relay (Figures A.4 and A.5).

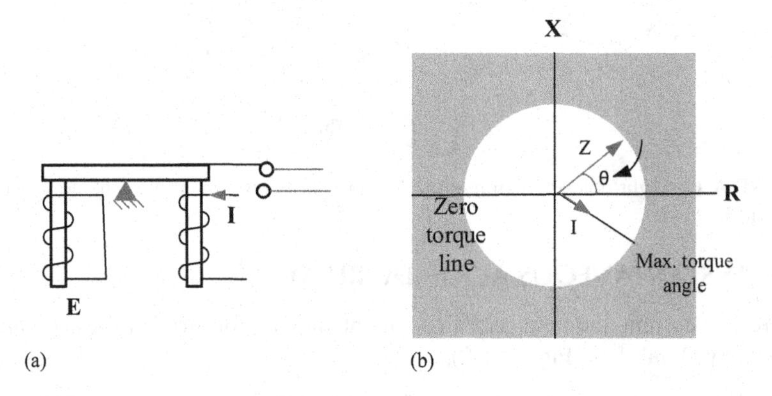

FIGURE A.4 Balanced beam AC or DC instantaneous relay. (a) The relay protective scheme and (b) the characteristics.

A.5 BALANCED BEAM AC OR DC INSTANTANEOUS RELAY

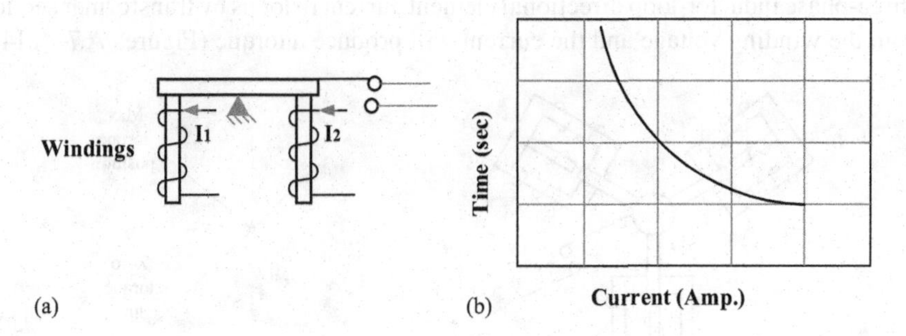

FIGURE A.5 Balanced beam AC or DC instantaneous relay. (a) The relay protective scheme and (b) the current versus time characteristics.

A.6 CURRENT PICK-UP DISK

The current pick-up disk runs with speed proportional to the spring force's building due to passing a current through it (Figure A.6).

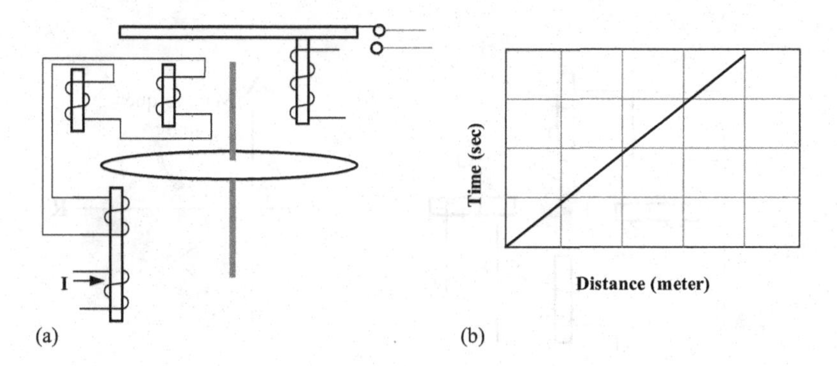

FIGURE A.6 Current pick-up disk. (a) The relay protective scheme and (b) the distance versus time characteristics.

A.7 INSTANTANEOUS THREE-PHASE RELAY

Three-phase inductor-loop directional element current in loops by transformer section from the winding voltage and the current will produce a torque (Figures A.7–A.14).

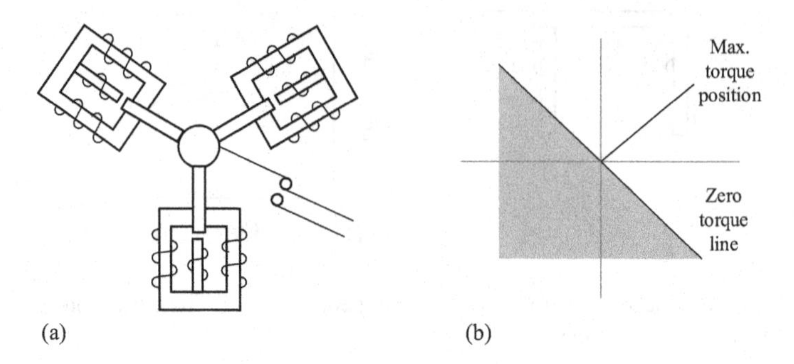

FIGURE A.7 Instantaneous three-phase relay. (a) The relay protective scheme and (b) the characteristics.

A.8 HIGH-SPEED FOUR-POLE INDUCTION CYLINDER OHM UNIT

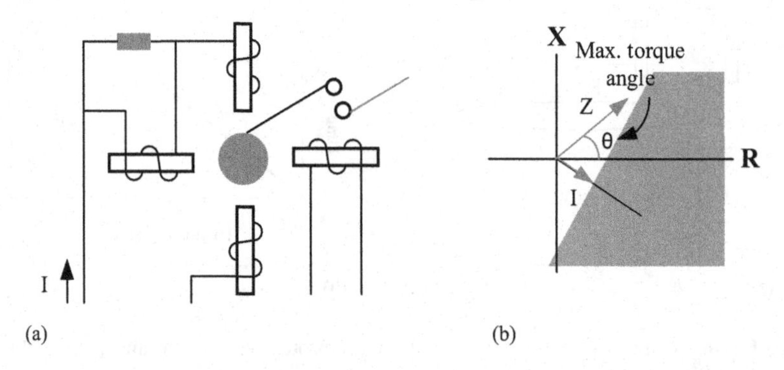

FIGURE A.8 High-speed four-pole induction cylinder Ohm unit. (a) The relay protective scheme and (b) the characteristics.

A.9 NEGATIVE-SEQUENCE DIRECTIONAL ELEMENT USING POTENTIAL

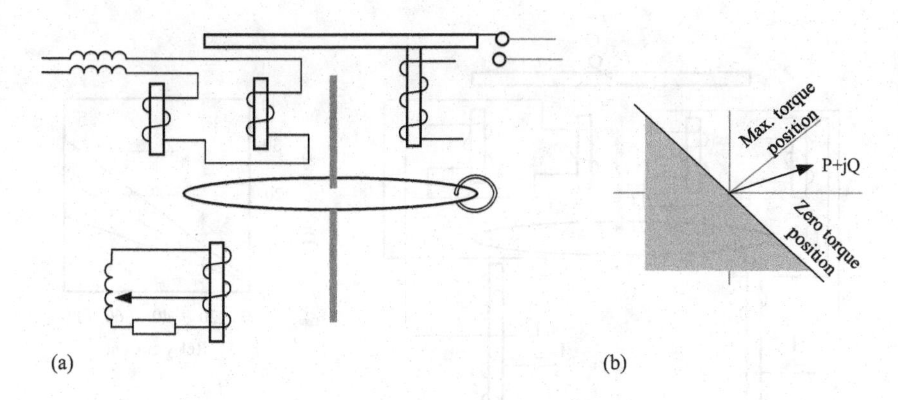

FIGURE A.9 Negative-sequence directional element using potential. (a) The relay protective scheme and (b) torque position characteristics.

A.10 INDUCTION IMPEDANCE ELEMENT

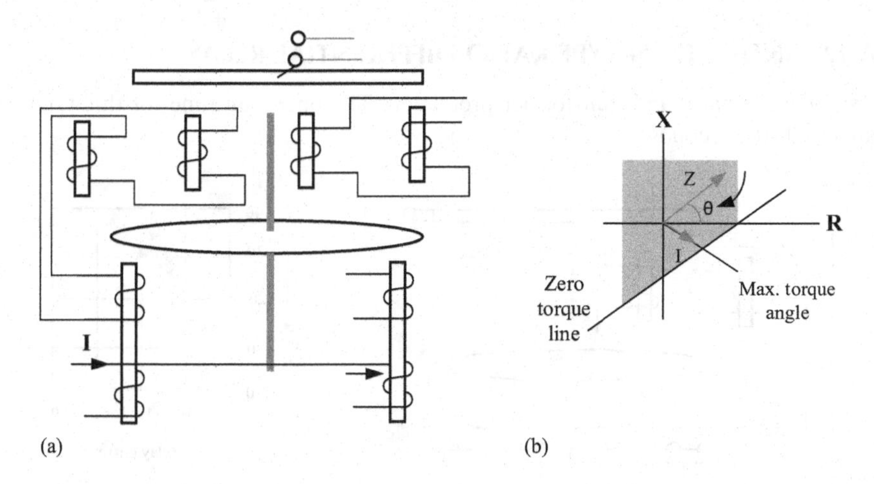

FIGURE A.10 Induction impedance element. (a) The relay protective scheme and (b) *R-X* characteristics.

A.11 THREE WINDINGS TRANSFORMER DIFFERENTIAL RELAY ADJUST BY TAPS ON OPERATING WINDINGS

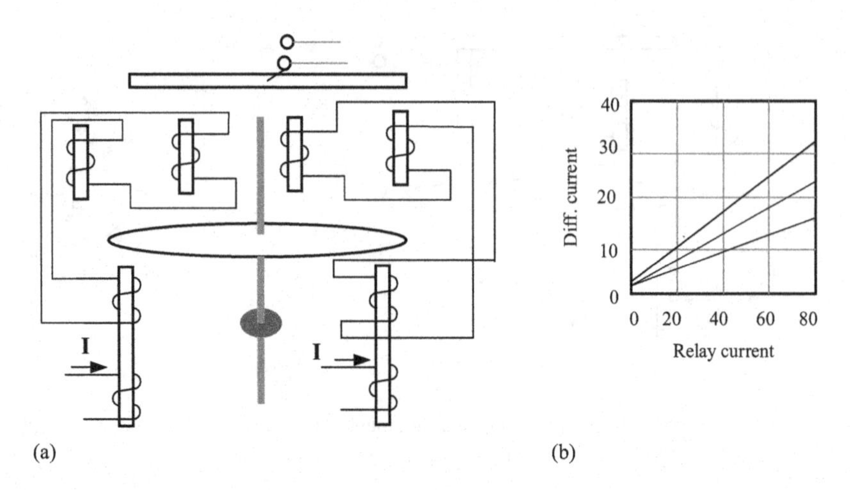

FIGURE A.11 Three windings differential transformer relay adjust by taps on operating windings. (a) The relay protective scheme and (b) the characteristics.

A.12 INDUCTION-TYPE RATIO DIFFERENTIAL RELAY

Used for generator and transformer protection. The operating time for this type is about 0.1–0.2 seconds.

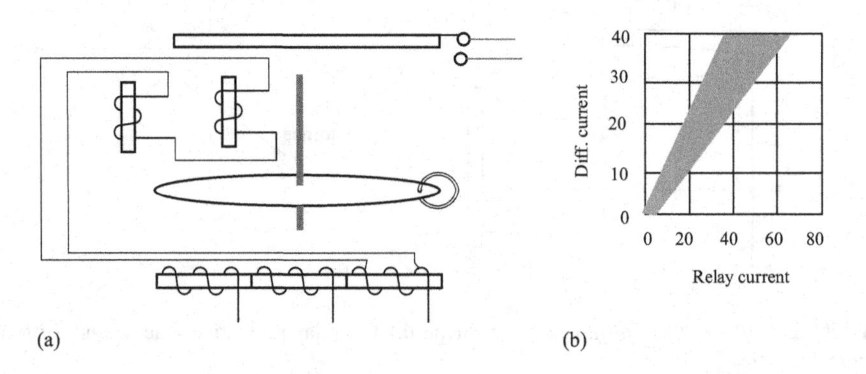

FIGURE A.12 Induction-type ratio differential relay. (a) The relay protective scheme and (b) the characteristics.

A.13 HIGH-SPEED FOUR-POLE INDUCTION CYLINDER-TYPE MHO RELAY

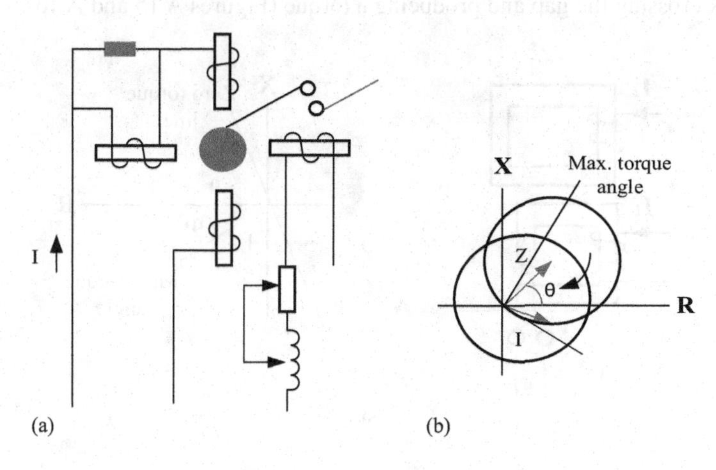

FIGURE A.13 High-speed four-pole induction cylinder-type MHO relay. (a) The relay protective scheme and (b) the characteristics.

A.14 POLARIZED ELEMENT DC OR AC RELAY

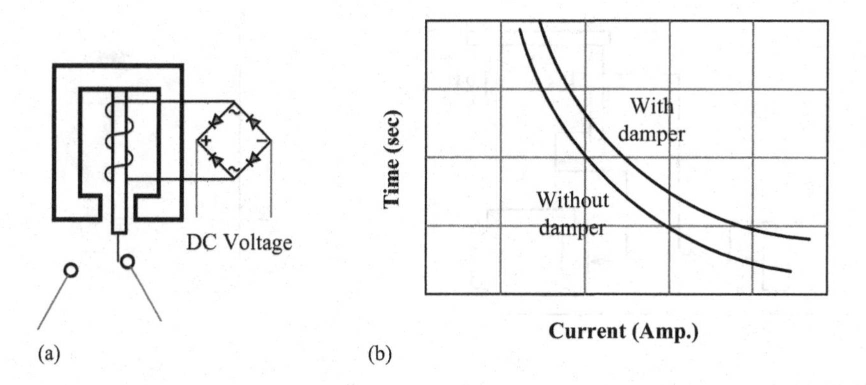

FIGURE A.14 Polarized element DC or AC relay. (a) The relay protective scheme and (b) the characteristics.

A.15 INDUCTOR-LOOP HIGH-SPEED DIRECTIONAL ELEMENT

The current in the loop by the transformer section from the voltage windings reacts with flux crossing the gap and producing a torque (Figures A.15 and A.16).

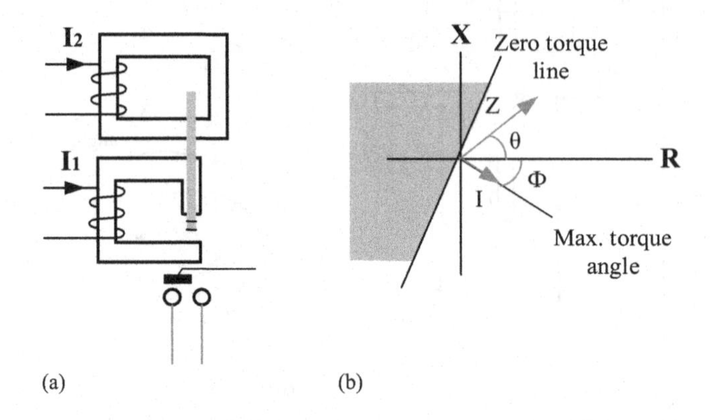

FIGURE A.15 Inductor-loop high-speed directional element. (a) The relay protective scheme and (b) the characteristics.

A.16 HIGH-SPEED BALANCED BEAM RELAY

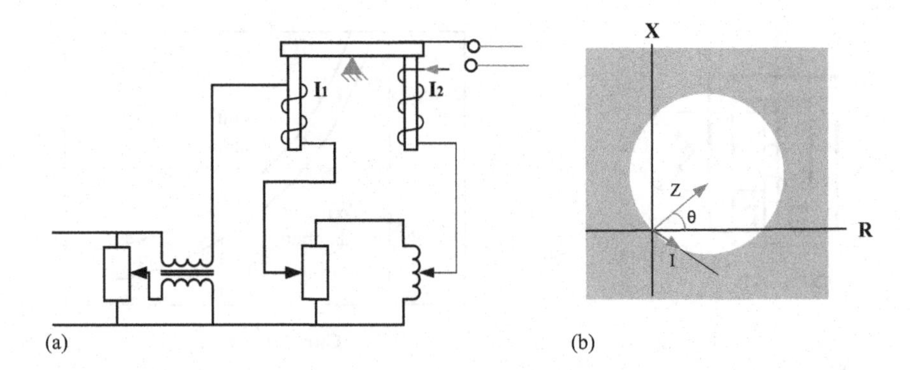

FIGURE A.16 High-speed balanced beam relay. (a) The relay protective scheme and (b) the characteristics.

A.17 INDUCTION DISK INVERSE TIME OVERCURRENT RELAY

The operating time for this type is adjustable at 0.1 or 0.05 seconds (Figure A.17).

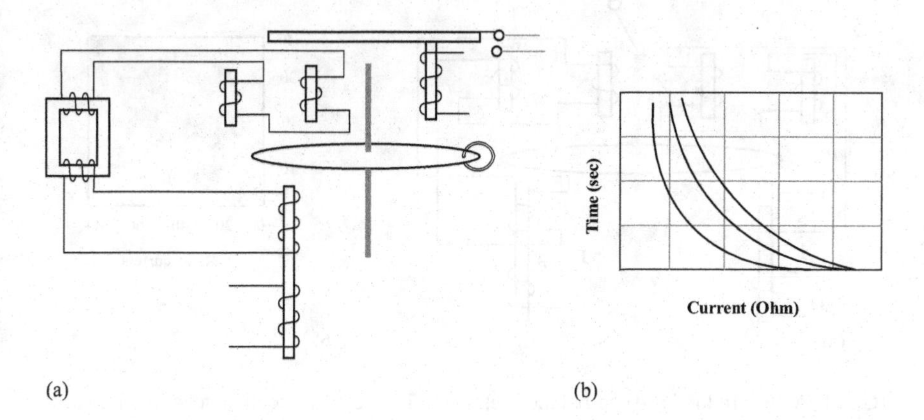

FIGURE A.17 Induction disk inverse time overcurrent relay. (a) The relay protective scheme and (b) the characteristics.

A.18 INDUCTION DISK OR CUP DIRECTIONAL RELAY

Activated by a phase shift coil (Figure A.18).

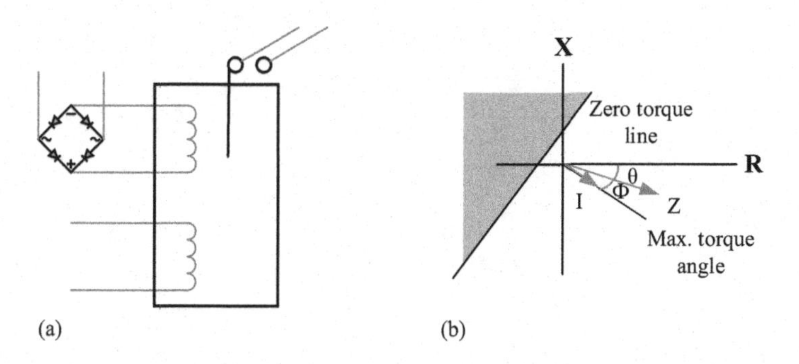

FIGURE A.18 Induction disk or cup directional relay. (a) The relay protective scheme and (b) the characteristics.

A.19 INDUCTION-TYPE BALANCE RELAY

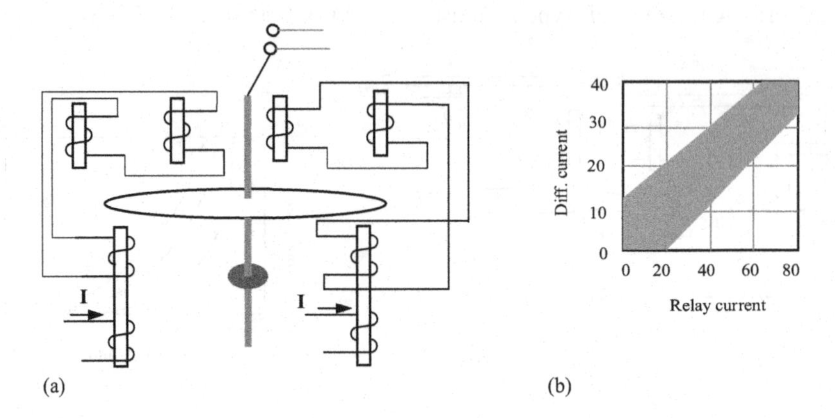

FIGURE A.19 Induction-type balance relay. (a) The relay protective scheme and (b) the characteristics.

Bibliography

1. Gilleßen, N. ILA course protective systems for high-voltage transmission lines ELP course number: SO2800-6L version 2.4.0.1. Lucas-Nülle GmbH Siemensstraße 2 · D-50170 Kerpen (Sindorf), www.lucas-nuelle.com.
2. Hileman, A. (1999). *Insulation Coordination for Power Systems.* Marcel Dekker, Inc., New York.
3. Pack, S., and Plesch, J. (2010). EMTP – RV course, Graz, Austria, 18–20 October, 2010.
4. Kisielewicz, T., Lo Piparo, G. B., Fiamingo, F., Mazzetti, C., Kuca, B., and Flisowski, Z. (2014). Factors affecting selection, installation, and coordination of surge protective devices for low voltage systems. *Electric Power Systems Research*, 113, 220–227.
5. IEEE Committee Report (1968). Proposed definitions of terms for reporting and analyzing outages of electrical transmission and distribution facilities and interruptions. *IEEE Transactions on Power Apparatus and Systems*, PAS-87(5), 1318–1323.
6. IEEE Committee Report (1978). Guidelines for developing a specific underground distribution system design standard. *IEEE Transactions on Power Apparatus and Systems*, PAS-97(3), 810–827.
7. IEEE (1973). IEEE standard definitions in power operations terminology. IEEE Standard 346-1973, November 2, 1973.
8. IEEE (1966). Proposed standard definitions of general electrical and electronics terms. IEEE Standard 270.
9. Pender, H., and Del Mar, W. A. (1962). *Electrical Engineers' Handbook—Electrical Power*, 4th edition. Wiley, New York.
10. National Electrical Safety Code (1977). ANSI C2, IEEE, New York, November 1977.
11. Fink, D. G., and Carroll, J. M. (eds.) (1969). *Standard Handbook for Electrical Engineers*, 10th edition. McGraw- Hill, New York.
12. IEEE (1972). IEEE standard dictionary of electrical and electronics terms, IEEE, New York.
13. US Department of Energy (1981). The national electric reliability study: technical study reports.
14. USDOE (1981). DOEIEP-0005, USDOE, Office of Emergency Operations, Washington, DC.
15. Clair, M. L. (1980). Annual statistical report. *Electro World,* 193(6), 49–80.
16. Energy Information Administration (1975–1978). Energy data reports—statistics of privately-owned electric utilities in the United States. US Department of Energy, Washington, DC.
17. Vennard, E. (1979). *Management of the Electric Energy Business.* McGraw-Hill, New York.
18. Institute of Electrical and Electronics Engineers Committee Report (1980). The significance of assumptions implied in long-range electric utility planning studies. *IEEE Transactions on Power Apparatus and Systems*, PAS-99, 1047–1056.
19. National Electric Reliability Council (1980). *Tenth Annual Review of Overall Reliability and Adequacy of the North American Bulk Power Systems.* NERC, Princeton, NJ.
20. Gonen, T. (1986). *Electrical Power Distribution System Engineering.* McGraw-Hill, New York.
21. Electric Power Research Institute (1979). *Transmission Line Reference Book: 345 kV and above.* EPRI, Palo Alto, CA.

22. Gonen, T. (2008). *Electrical Power Distribution System Engineering*, 2nd edition. CRC Press, Boca Raton, FL.
23. Gonen, T. (2009). *Electrical Power Transmission System Engineering: Analysis and Design*, 2nd edition. CRC Press, Boca Raton, FL.
24. Institute of Electrical and Electronics Engineers (1971). Graphic Symbols for Electrical and Electronics Diagrams, IEEE Standard 315-1971 [or American National Standards Institute (ANSI) Y32.2–1971]. IEEE, New York.
25. Anderson, P. M. (1973). *Analysis of Faulted Power Systems.* Iowa State University Press, Ames, IA.
26. Neuenswander, J. R. (1971). *Modern Power Systems.* International Textbook Company, Scranton, PA.
27. Fink, D. G., and Beaty, H. W. (1978). *Standard Handbook for Electrical Engineers*, 11th edition. McGraw-Hill, New York.
28. Wagner, C. F., and Evans, R. D. (1933). *Symmetrical Components.* McGraw-Hill, New York.
29. Weedy, B. M. (1972). *Electric Power Systems*, 2nd edition. Wiley, New York.
30. Concordia, C., and Rusteback, E. (1970). Self-excited oscillations in a transmission system using series capacitors. *IEEE Transactions on Power Apparatus and Systems,* PAS-89(no. 7), 1504–1512.
31. Elliott, L. C., Kilgore, L. A., and Taylor, E. R. (1971). The prediction and control of self-excited oscillations due to series capacitors in power systems. *IEEE Transactions on Power Apparatus and Systems,* PAS-90(no. 3), 1305–1311.
32. Kilgore, L., Taylor, E. R., Jr., Ramey, D. G., Farmer, R. G., and Schwalb, A. L. (1973). Solutions to the problems of subsynchronous resonance in power systems with series capacitors. *Proceedings of the American Power Conference,* 35, 1120–1128.
33. Bowler, C. E. J., Concordia, C., and Tice, J. B. (1973). Subsynchronous torques on generating units feeding series-capacitor compensated lines. *Proceedings of the American Power Conference,* 35, 1129–1136.
34. Schifreen, C. S., and Marble, W. C. (1956). Changing current limitations in the operation of high-voltage cable lines. *Transactions of the American Institute of Electrical Engineers,* 26, 803–817.
35. Wiseman, R. T. (1956). Discussions to charging current limitations in the operation of high-voltage cable lines. *Transactions of the American Institute of Electrical Engineers,* 26, 803–817.
36. Wright, A. (1990). Construction, behaviour and application of electric fuses. *Power Engineering Journal*, *4*(3), 141–148.
37. Electrical Technology. https://www.electricaltechnology.org/2014/11/fuse-types-of-fuses.html
38. Horowitz, S. H., and Phadke, A. G. (2008). *Power System Relaying,* 3rd edition. John Wiley & Sons Ltd, Boca Raton, FL.
39. Stevenson, W. D. (1989). *Elements of Power System Analysis*, 4th edition. Mc Graw - Hill, New York.
40. Guile, A. E., and Paterson, W. (1969). *Electric Power Systems,* vol. 1. Oliver & Boyd, Edinburgh.
41. Taylor, C. W. (1994). *Power System Voltage Stability.* McGraw Hill Book Company, New York.
42. Stagg, G. W., and El-Abiad, A. H. (1968). *Computer Methods in Power System Analysis.* McGraw-Hill Book Company, New York.
43. Venkataraman, P. (2001). *Applied Optimization with MATLAB Programming.* McGraw-Hill Book Company, New-York.
44. Boehle, O. B. (1988). *Switchgear Manual*, 8th edition, Trans. By David Stone. Asea Brown Boveri, Zürich, Switzerland.

45. Elgerd, O. I. (1982). *Electric Energy Systems Theory,* 2nd edition. McGraw-Hill, Inc., New York.
46. Saadat, H. (1999). *Power System Analysis.* McGraw-Hill, Inc., New York.
47. Zhang, Y., Wang, L., Xiang, Y., & Ten, C. W. (2016). Inclusion of SCADA cyber vulnerability in power system reliability assessment considering optimal resources allocation. *IEEE Transactions on Power Systems*, 31(6), 4379–4394.
48. Brown, H. E. (1985). *Solution of Large Networks by Matrix Methods,* 2nd edition. John Wiley & Sons, New York.
49. Lakervi, E., & Holmes, E. J. (1995). *Electricity distribution network design* (No. 212). IET.
50. Van Joolingen, W. R., and De Jon, T. (1992). Modeling domain knowledge for intelligent simulation learning environments. *Computer Education,* 18(1–3), 29–37.
51. Syed, N. A. (1996). *Electric Energy System.* Macmillan Publishing Company, New York.
52. Kron, G. (1939). *Tensor Analysis of Networks.* John Wiley & Sons, New York.
53. Jaysekara Menik, N. S. K. (2004). *Computer Simulation of Transient Stability Analysis of Power System.* University of Windsor, Canada.
54. Boldea, I., and Nasar, S. A. (1986). *Electric Machine Dynamics.* Macmillan, New York.
55. Kimbark, E. W. (1976). *Power System Stability Volume 1, Elements of Stability Calculations.* John Wiley & Sons, Inc., New York.
56. Vadhera, S. S. (1987). *Power System Analysis & Stability.* Khanna, Nai Sarak, Delhi.
57. Kothari, D. P. (2004). *Modern Power System Analysis.* Mc Graw-Hill, New Delhi.
58. Gupta, B. R. (1985). *Power System Analysis and Design.* S Chand & Company, New Delhi.
59. Stark, P. A. (1970). *Introduction to Numerical Methods.* Macmillan Publisher, New York.
60. Lewis, E. E. (1996). *Introduction to Reliability Engineering*, 2nd edition. John Wiley & Sons, Inc., New York.
61. Kalas, P. (1998). *Reliability for Technology, Engineering and Management.* Prentice-Hall Inc., Hoboken, NJ.
62. Milligan, M. M. (2002). *Modeling Utility – Scale Wind Power Plants Part 2: Capacity Credit.* National Renewable Energy Laboratory, Golden, CO.
63. O'Coonnor, P. D. T., Newton, D., and Bromley, R. (2002). *Practical Reliability Engineering*, 4th edition. John Wiley & Sons, Ltd., New York.
64. Endernyi, J. (1978). *Reliability Modeling in Electric Power System.* John Wiley & Sons, New York.
65. Shooman, M. L. (1990). *Probabilistic Reliability an Engineering Approach*, 2nd edition. McGraw-Hill, New York.
66. Weedy, B. M., and Cory, B. J. (1998). *Electric Power Systems*, 4th edition. John Wiley & Sons, New York.
67. Machowski, J., Lubosny, Z., Bialek, J. W., & Bumby, J. R. (2020). *Power System Dynamics: Stability and Control.* John Wiley & Sons, New-York.
68. Haupt, R. L., and Haupt, S. E. (2004). *Practical Genetic Algorithms,* 2nd edition. John Wiley & Sons, New York.
69. Goldberg, D. E. (1989). *Genetic Algorithm in Search, Optimization, and Machine Learning.* Addison-Wesley, Boston, MA.
70. Wood, J., and Wollenberg, B. F. (1996). *Power Generation Operation and Control.* John Wiley & Sons Book Company, New York.
71. Stott, B. (1974). Review of load flow calculation methods. *Proceedings of IEEE*, 62, 916–929.
72. Gross, C. A. (1979). *Power System Analysis.* Wiley, New York.
73. Miller, T. J. E. (1982). *Reactive Power Control in Electric Systems.* John Wiley & Sons, New York.

74. Adamson, C., and Hingorani, N. G. (1960). *High Voltage Direct Current Power Transmission.* Garaway Ltd, Roseau.
75. Cory, B. J. (1965). *High Voltage Direct Current Converters and Systems.* Macdonald, London.
76. Mahalanabis, A. K., Kothari, D. P., and Ahson, S. I. (1988). *Computer Aided Power System Anlysis and Control.* Tata McGraw-Hill, New Delhi.
77. Bergen, A. R., and Vittal, V. (2000). *Power System Analysis*, 2nd edition. Prentice-Hall Inc., Hoboken, NJ.
78. Billinton, R., and Allan, R. N. (1984). *Reliability Evaluation of Power System.* Plenum Press, New York.
79. Warwick, K., Kwue, A. E., and Aggarwar, R. (eds) (1997). *Artificial Intelligence Techniques in Power Systems.* Institution of Electrical Engineers, New York.
80. Yong-Hua, S. (ed) (1999). *Modern Optimization Techniquesin Power Systems.* Kluwer Academic Publishers, London.
81. Debs, A. S. (1988). *Modern Power Systems Control and Operation.* Kluwer Academic Publishers, New York.
82. Berrie, T. W. (1983). *Power System Economics.* IEEE, London.
83. Billinton, R. (1970). *Power System Reliability Evaluation.* Gordon and Breach, New York.
84. Billirrton, R., Ringlee, R. J., and Wood, A. J. (1973). *Power System Reliability Calculations.* The MIT Press, Boston, MA.
85. Kusic, G. L. (1986). *Computer Aided Power System Analysis.* Prentice-Hall, Hoboken, NJ.
86. Grunbaum, R., Noroozian, M., and Thorvaldsson, B. (1999). FACTS powerful systems for flexible power transmission. *ABB Review,* 5, 4–17.
87. Canadian Electrical Association (1984). *Static Compensators for Reactive Power Control.* Cantext Publications, Canada.
88. Thorborg, K. (1988). *Power Electronics.* Prentice-Hall International (UK) Ltd., London.
89. Rashid, M. H. (1992). *Power Electronics, Circuits Devices and Applications.* Prentice-Hall International Editions, London.
90. Mohan, N., Undeland, T. M., and Robbins, W. P. (1989). *Power Electronics: Converters, Applications, and Design.* John Wiley & Sons, New York.
91. Arrillaga, J., Bradley, D. A., and Bodger, P. S. (1989). *Power System Harmonics.* John Wiley & Sons, New York.
92. Murphy, J. M. D., and Turnbull, F. G. (1988). *Power Electronic Control of AC Motors.* Pergamon Press, Oxford.
93. Woodford, D. A. (1998). HVDC transmission, Professional Report from Manitoba HVDC Research Center, Winnipeg, Manitoba.
94. Meliopoulos, A. P. (1988). *Power System Grounding and Transient an Introduction.* Marcel Dekker, Inc., New York and Basel.
95. Grigsby, L. (2007). *Electric Power Generation, Transmission, and Distribution.* CRC Press, Boca Raton, FL.
96. Robert Eaton, J., and Cohen, E. (1991). *Electric Power Transmission Systems.* Prentice-Hall, Inc., Hoboken, NJ.
97. Van Cutsem, T., and Vournas, C. (1998). *Voltage Stability of Electrical Power Systems.* Kluwer Academic Publishers, Boston, MA.
98. Abdel-Azim, M., Salah, H. E. D., and Eissa, M. E. (2018). IDS against black-hole attack for MANET. *IJ Network Security,* 20(3), 585–592.
99. Amara Korba, A., Nafaa, M., and Ghanemi, S. (2016). An efficient intrusion detection and prevention framework for ad hoc networks. *Information & Computer Security,* 24(4), 298–325.

100. Anwar, S., Jasni, M. Z., Mohamad, F. Z., Inayat, Z., Khan, S., Anthony, B., and Chang, V. (2017). From intrusion detection to an intrusion response system: fundamentals, requirements, and future directions. *Algorithms*, 10(2), 39. doi: 10.3390/a10020039.
101. Assad, N., Elbhiri, B., Moulay, A. F., Ouadou, M., and Aboutajdine, D. (2015). Analysis of the deployment quality for intrusion detection in wireless sensor networks. *Journal of Computer Networks and Communications*, 5(2), 20. doi: 10.1155/2015/812613.
102. Baggili, I., and Rogers, M. (2009). Self-reported cyber-crime: an analysis of the effects of anonymity and pre-employment. *International Journal of Cyber Criminology*, 3, 550–565. Retrieved from http://www.cybercrimejournal.com.
103. Bahrami, M., and Bahrami, M. (2014). An overview of software architecture in the intrusion detection system. *Journal of Computer Networks and Communications*, 1(1), 1–8. doi: 10.7321/jscse.v1.n1.1.
104. Bailetti, T., Gad, M., and Shah, A. (2016). Intrusion learning: an overview of an emergent discipline. *Technology Innovation Management Review*, 6(2), 15–20. doi: 10.22215/timreview/964.
105. Barot, V., Sameer, S. C., and Patel, B. (2014). Feature selection for modeling intrusion detection. *International Journal of Computer Network and Information Security*, 6(7), 56–62. doi: 10.5815/ijcnis.2014.07.08.
106. Ben-Asher, N., and Gonzalez, C. (2015). Effects of cybersecurity knowledge on attack detection. *Computers in Human Behavior*, 48, 51–61. doi: 10.1016/j.chb.2015.01.039.
107. Bilal, M. B. (2014). A new classification scheme for intrusion detection systems. *International Journal of Computer Network and Information Security*, 6(8), 56–70. doi: 10.5815/ijcnis.2014.08.08.
108. Blanche, G. A. (2018). The cybersecurity workforce: profession or not? (order no. 13841062). Available from ProQuest Dissertations & Theses Global (2179183753).
109. Carter, N., Bryant-Lukosius, D., DiCenso, A., Blythe, J., and Neville, A. J. (2014). The use of triangulation in qualitative research. *Oncology Nursing Forum*, 41, 545–547. doi: 10.1188/14.onf.545-547.
110. Chebrolua, S., Abraham, B., and Thomas, J. (2005). Feature deduction and ensemble design of intrusion detection systems. *Computers and Security*, 24(4), 295–307. doi: 10.1016/j.cose.2004.09.008.
111. Chourasiya, R., Patel, V., and Shrivastava, A. (2018). Classification of cyber-attack using machine learning technique at Microsoft azure cloud. *International Journal of Advanced Research in Computer Science*, 6(1), 4–8.
112. Creswell, J. (2013). *Research Design: Qualitative, Quantitative, and Mixed Methods Approach*, 4th edition. Sage Publications, Inc., Thousand Oaks, CA.
113. Da Silva Cardoso, A. M., Lopes, R. F., Teles, A. S., and Magalhães, F. B. V. (2018, April). Real time DDoS detection based on complex event processing for IoT. *In 2018 IEEE/ACM Third International Conference on Internet-of-Things Design and Implementation (IoTDI)*, Orlando, FL, pp. 273–274. IEEE.
114. Denzin, N. K., and Lincoln, Y. S. (2005). *Introduction: The Discipline and Practice of Qualitative Research.* Sage Publications, Inc., Thousand Oaks, CA.
115. Desai, A., Prajapati, H., and Bhatti, D. (2011). Problems and challenges in wireless network intrusion detection. *National Journal of System and Information Technology*, 4(2), 175–189.
116. Duan, T. (2016). Analysis focusing on intrusion detection technology when an outside party breaks into computer database. *RISTI (Revista Iberica de Sistemas e Tecnologias de Informacao)*, 17B, 180–193.

Index

Note: **Bold** page numbers refer to tables, *italic* page numbers refer to figures.

Printed in the United States
by Baker & Taylor Publisher Services